Biosensors and Nanotechnology

Biosensors and Nanotechnology

Applications in Health Care Diagnostics

Edited by Zeynep Altintas

Technical University of Berlin, Berlin, Germany

WILEY

Registered Office
John Wiley & Sons, Inc, 111 River Street, Hoboken, NJ 07030, USA

Editorial Office
111 River Street, Hoboken, NJ 07030, USA

For details of our global editorial offices, customer services, and more information about Wiley products visit us at www.wiley.com.

Library of Congress Cataloging-in-Publication data applied for

ISBN: 9781119065012

Cover design by Wiley
Cover image: © petersimoncik/Gettyimages

Set in 10/12pt Warnock by SPi Global, Pondicherry, India

Printed in the United States of America

10 9 8 7 6 5 4 3 2 1

Contents

List of Contributors

Sinan Akgol
Department of Biochemistry, Faculty
of Science
Ege University
Izmir
Turkey

Deniz Aktas-Uygun
Department of Chemistry, Faculty of
Science and Arts
Adnan Menderes University
Aydin
Turkey

Zeynep Altintas
Technical University of Berlin
Berlin
Germany

Adina Arvinte
"Petru Poni" Institute of
Macromolecular Chemistry
Centre of Advanced Research in
Bionanoconjugates and Biopolymers
Iasi
Romania

Mohammad Asghari
Institute of Materials Science
and Nanotechnology, National
Nanotechnology Research Center
(UNAM)

Bilkent University
Ankara
Turkey

Jon Ashley
Department of Micro- and
Nanotechnology
Technical University of Denmark
Lyngby
Denmark

Eren Aydın
Department of Electrical and
Electronics Engineering
Middle East Technical University
Ankara
Turkey

Frank Davis
Department of Engineering and
Applied Design
University of Chichester
Chichester
UK

Ece Eksin
Department of Analytical Chemistry,
Faculty of Pharmacy
Ege University
Izmir
Turkey

Caglar Elbuken
Institute of Materials Science
and Nanotechnology, National
Nanotechnology Research Center
(UNAM)
Bilkent University
Ankara
Turkey

Arzum Erdem
Department of Analytical Chemistry,
Faculty of Pharmacy
Ege University
Izmir
Turkey

Wellington M. Fakanya
Atlas Genetics Ltd
Wiltshire
UK

Furkan Gökçe
Department of Electrical and
Electronics Engineering
Middle East Technical University
Ankara
Turkey

Mustafa Tahsin Guler
Department of Physics
Kirikkale University
Kirikkale
Turkey

Ziya Isiksacan
Institute of Materials Science
and Nanotechnology, National
Nanotechnology Research Center
(UNAM)
Bilkent University
Ankara
Turkey

Ali Kalantarifard
Institute of Materials Science
and Nanotechnology, National
Nanotechnology Research Center
(UNAM)
Bilkent University
Ankara
Turkey

Mustafa Kangül
Department of Electrical and
Electronics Engineering
Middle East Technical University
Ankara
Turkey

Ece Kesici
Department of Analytical Chemistry,
Faculty of Pharmacy
Ege University
Izmir
Turkey

Haluk Külah
Department of Electrical and
Electronics Engineering
Middle East Technical University
Ankara
Turkey

Giovanna Marrazza
Department of Chemistry "Ugo Schiff"
University of Florence
Florence
Italy

Noor Azlina Masdor
Cranfield University
Cranfield
UK
and
Malaysian Agricultural Research and
Development Institute (MARDI)

Kuala Lumpur
Malaysia

Ebru Özgür
Department of Electrical and
Electronics Engineering
Middle East Technical University
Ankara
Turkey

Andrea Ravalli
Department of Chemistry "Ugo Schiff"
University of Florence
Florence
Italy

Frieder W. Scheller
Institute of Biochemistry and Biology
University of Potsdam
Potsdam
Germany

Adama Marie Sesay
Unit of Measurement Technology,
Kajaani University Consortium
University of Oulu
Oulu
Finland

Flavio M. Shimizu
São Carlos Institute of Physics (IFSC)
University of São Paulo (USP)
São Carlos
Brazil

Yi Sun
Department of Micro- and
Nanotechnology
Technical University of Denmark
Lyngby
Denmark

Pirkko Tervo
Unit of Measurement Technology,
Kajaani University Consortium
University of Oulu
Oulu
Finland

Elisa Tikkanen
Unit of Measurement Technology,
Kajaani University Consortium
University of Oulu
Oulu
Finland

Murat Uygun
Department of Chemistry, Faculty of
Science and Arts
Adnan Menderes University
Aydin
Turkey

Özge Zorlu
Department of Electrical and
Electronics Engineering
Middle East Technical University
Ankara
Turkey

Preface

Multifactorial diseases such as cancer, cardiovascular disorders, and infectious diseases are the leading cause of death worldwide. This is mostly due to the lack of early diagnosis, which plays a key role in successful treatment and in elimination of huge costs required for the treatment. Today, common use of biosensor technology in the field of medical diagnostics and drug discovery has resulted in cost-effective, rapid, reliable, and easy-to-use sensing platforms. Biomedical sensors are analytical devices that utilize recognition elements such as antibodies, aptamers, peptides, and molecularly imprinted polymers for detection. They possess two main elements: (i) a biological recognition element (receptor) that supplies specific binding through a biochemical interaction of a target to a receptor and (ii) a signal transducer that converts this biochemical reaction into an easily measurable electrical signal. Other components of biosensors are the input/output systems to operate the sensing device and fluidics systems to handle reagents and samples necessary for the testing. In this book, we aim to describe a range of biosensor technologies for the detection of cancer, cardiac problems, and neurodegenerative and infectious diseases with the hope of helping the integration of biomedical sensors into common clinical usage.

Advancements taking place in nanotechnology, microelectronics, computational science, and biomedical engineering have led to new technologies and application-specific devices by bringing various disciplines together. However, there is still a gap between research and clinical applications. Taking this fact into account, the objective of this book is to provide a wide range of information from basic to the advanced applications in the biosensor area and impact of nanotechnology on the development of biosensors for healthcare. A significant up-to-date review of various sensor platforms, their use in cancer, cardiovascular system problems, neurodegenerative disorders, infectious diseases, and drug discovery with the implementations of smart nanomaterials is also given. This project is a comprehensive approach to the medical biosensors area presenting a thorough knowledge of the subject and an effective integration of these sensors on healthcare in order to appropriately convey the state-of-the-art fundamentals and applications of the most innovative technologies.

This book is comprised of 15 chapters written by 31 researchers who are actively working in Germany, the United Kingdom, Italy, Turkey, Denmark, Finland, Romania, Malaysia, and Brazil. The book covers four main sections: Section 1 describes general information on biosensors, recognition receptors, biomarkers, and disease diagnostics. Section 2 provides biosensor-based healthcare applications through various sensing systems, and it covers all main types of biosensors including surface plasmon resonance-, piezoelectric-, electrochemical-, microelectromechanical-, and lab-on-a-chip-based sensors in disease detection and diagnostics. Applications of nanomaterials in biosensors and diagnostics follow this part as the third main section, Section 3, and it talks about the application of quantum dots, carbon nanotubes, metal nanoparticles, molecularly imprinted nanostructures, magnetic nanomaterials, and graphene with the latest trends in the field. The last section, Section 4, is dedicated to organ-specific healthcare applications for disease cases using biosensors. In this part, optical biosensors and applications to drug discovery for cancer cases, and also DNA-based biosensors for anticancer drug detection, are covered.

The anticipated audience is researchers, scientists, regulators, consultants, and engineers. Furthermore, graduate students will find this book very useful since it provides a wide range of knowledge on biosensors for healthcare diagnostics. The contributors of the book were also asked to use a pedagogical tone to comply with the needs of novice researchers such as doctoral students and postdoctoral scholars as well as of senior researchers seeking new pathways. All related and significant subtopics are given in one book to provide a not only comprehensive but also easily understandable handbook in the area. Educational purposes were also considered while generating this book; hence it has a great potential to be used as a textbook in universities and research institutes. The complexity and flow of the book is suitable for all related and interested students in the area.

March 2017, Berlin

Zeynep Altintas
Department of Chemical Engineering
and
Department of Biomolecular Modelling
Technical University of Berlin, Germany

Acknowledgments

We are very thankful to all the authors for their participation and invaluable contributions in the making of this book. I also extend my thanks to Tom Scrace and Sumathi Elangovan of John Wiley who assisted me in all stages of preparing this book for the publication. Last, but not least, I dedicate this book to my parents, Ilyas and Eva, with sincere regards and my niece, Beren, who inspired me while working on the book.

Section 1

Introduction to Biosensors, Recognition Elements,
Biomarkers, and Nanomaterials

1

General Introduction to Biosensors and Recognition Receptors

Frank Davis[1] and Zeynep Altintas[2]

[1] Department of Engineering and Applied Design, University of Chichester, Chichester, UK
[2] Technical University of Berlin, Berlin, Germany

1.1 Introduction to Biosensors

There are laboratory tests and protocols for the detection of various biomarkers, which can be used to diagnose heart attack, stroke, cancer, multiple sclerosis, or any other conditions. However, these laboratory protocols often require costly equipment, and skilled technical staff, and hospital attendance and have time constraints. Much cheaper methods can provide cost-effective analysis at home, in a doctor's surgery, or in an ambulance. Rapid diagnosis will also aid in the treatment of many conditions. Biosensors generically offer simplified reagentless analyses for a range of biomedical [1–8] and industrial applications [9, 10]. Due to this, biosensor technology has continued to develop into an ever-expanding and multidisciplinary field during the last few decades.

The IUPAC definition of a biosensor is "a device that uses specific biochemical reactions mediated by isolated enzymes, immunosystems, tissues, organelles or whole cells to detect chemical compounds usually by electrical, thermal or optical signals." From this definition, we can gain an understanding of what a biosensor requires.

Most sensors consist of three principal components:

1) Firstly there must be a component, which will selectively recognize the analyte of interest. Usually this requires a binding event to occur between the recognition element and target.
2) Secondly some form of transducing element is needed, which converts the biochemical binding event into an easily measurable signal. This can be a generation of an electrochemically measurable species such as protons or

Biosensors and Nanotechnology: Applications in Health Care Diagnostics, First Edition.
Edited by Zeynep Altintas.
© 2018 John Wiley & Sons, Inc. Published 2018 by John Wiley & Sons, Inc.

H_2O_2, a change in conductivity, a change in mass, or a change in optical properties such as refractive index.

3) Thirdly there must be some method for detecting and quantifying the physical change such as measuring an electrical current or a mass or optical change and converting this into useful information.

There exist many methods for detecting binding events such as electrochemical methods including potentiometry, amperometry, and AC impedance; optical methods such as surface plasmon resonance; and piezoelectric methods that measure mass changes such as quartz crystal microbalance (QCM) and surface acoustic wave techniques. A detailed description of these would be outside the remit of this introduction, but they are described in many reviews and elsewhere in this book. Instead this chapter focuses on introducing the recognition receptors used in biosensors.

1.2 Enzyme-Based Biosensors

Leyland Clark coated an oxygen electrode with a film containing the enzyme glucose oxidase and a dialysis membrane to develop one of the earliest biosensors [11]. This could be used to measure levels of glucose in blood; the enzyme converted the glucose to gluconolactone and hydrogen peroxide with a concurrent consumption of oxygen. The drop in dissolved oxygen could be measured at the electrode and, with careful calibration, levels of blood glucose calculated. This led to the widespread use of enzymes in biosensors, mainly driven by the desire to provide detection of blood glucose. Diabetes is one of the major health issues in the world today and is predicted to affect an estimated 300 million people by 2045 [12]. The world market for biosensors was approximately $15–16 billion in 2016. In 2009 approximately half of the world biosensor market was for point-of-care applications and about 32% of the world commercial market for blood glucose monitoring [13].

Enzymes are excellent candidates for use in biosensors, for example, they have high selectivities; glucose oxidase will only interact with glucose and is unaffected by other sugars. Being highly catalytic, enzymes display rapid substrate turnovers, which is important since otherwise they could rapidly become saturated or fail to generate sufficient active species to be detected. However, they demonstrate some disadvantages: for instance, a suitable enzyme for the target of interest may simply not exist. Also enzymes can be difficult and expensive to extract in sufficient quantities and can also be unstable, rapidly denaturing, and becoming useless. They can also be subject to poisoning by a variety of species. Moreover, detection of enzyme turnover may be an issue, for instance, in the glucose oxidase reaction; it is possible to directly electrochemically detect either consumption of oxygen [11] or production of hydrogen peroxide. However in samples such as blood and saliva, there can be other electroactive

Figure 1.1 Schematic of a second-generation biosensor.

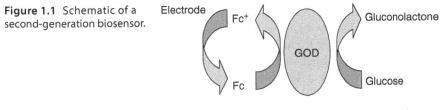

Fc = ferrocene derivative, God = glucose oxidase

substances such as ascorbate, which also undergo a redox reaction and lead to false readings. These types of biosensors are often called "first-generation biosensors." To address this issue of interference, a second generation of glucose biosensors was developed where a small redox-active mediating molecule such as a ferrocene derivative was used to shuttle electrons between the enzyme and an electrode [14]. The mediator readily reacts with the enzyme, thereby avoiding competition by ambient oxygen. This allowed much lower potentials to be used in the detection of glucose, thereby reducing the problem of oxidation of interferents and increasing signal accuracy and reliability. Figure 1.1 shows a schematic of a second-generation glucose biosensor.

Third-generation biosensors have also been developed where the enzyme is directly wired to the electrode, using such materials as osmium-containing redox polymers [15] or conductive polymers such as polyaniline [16]. More recently nanostructured materials such as metal nanoparticles, carbon nanotubes, and graphene have been used to facilitate direct electron transfer between the enzyme and the electrode as described in later chapters. As an alternative to glucose oxidase, sensors based on glucose dehydrogenase have also been developed.

The techniques for glucose sensing using glucose oxidase can be applied to almost any oxidase enzymes, allowing sensors to be developed based on cholesterol oxidase, lactate oxidase, peroxidase enzymes, and many others. Sensors have also been constructed using urease, which converts urea to ammonia, causing a change in local pH that can be detected potentiometrically or optically by combining the enzyme with a suitable optical dye. Enzyme cascades have also been developed; for example, cholesterol esters can be determined using electrodes containing cholesterol esterase and cholesterol oxidase. Applications of enzyme-containing biosensors have been widely reviewed [16–18].

1.3 DNA- and RNA-Based Biosensors

DNA is contained within all living cells as a blueprint for making proteins, and it can be thought of as a molecular information storage device. RNA also has a wide number of applications in living things, including acting as a messenger between DNA and the ribosomes that synthesize proteins and as a regulator of gene expression. Both DNA and RNA are polymeric species based on a sugar–phosphate backbone with nucleic bases as side chains, in DNA, namely, adenine, cytosine, guanine, and thymine. In RNA uracil is utilized instead of thymine. It is the specific binding between base pairs, that is, guanine to cytosine or adenine to thymine (uracil), that determine the structure of these polymers, in the case of DNA leading to a double helix structure (Figure 1.2) [19].

DNA sensors are usually of a format where one oligonucleotide chain is bound to a suitable transducer, that is, an electrode, surface plasmon resonance (SPR) chip, quartz crystal microbalance (QCM), and so on, and is exposed to a solution containing an oligonucleotide strand of interest [20]. The surface-bound oligonucleotide is selected to be complementary to the oligonucleotide of interest, and the bound and solution strands will undergo sequence-specific hybridization as the recognition event.

An in-depth review of DNA sensing is outside the scope of this introduction and has been reviewed elsewhere [20–24]; however, a few examples are given here. A method based on ruthenium-mediated guanine oxidation allowed selective electrochemical detection of messenger RNA from tumors at $500\,zmol\,L^{-1}$ levels [25]. A sandwich-type assay using magnetic beads and fluorescence analysis utilized a complementary nucleotide to dengue fever virus

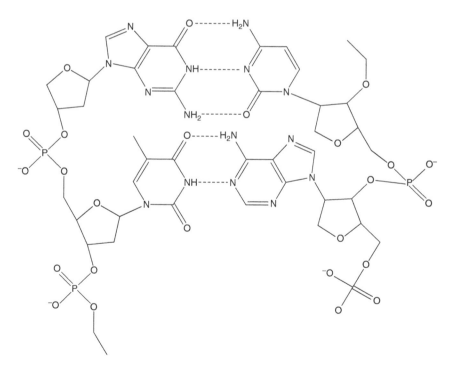

Figure 1.2 Schematic of interstrand binding in DNA.

RNA to allow detection at levels as low as $50\,pmol\,L^{-1}$ [26]. Five different probe DNAs could be immobilized onto an SPR-imaging chip and simultaneously used to determine binding of RNA sequences found in several pathogenic bacteria such as *Brucella abortus, Escherichia coli*, and *Staphylococcus aureus* [27] for use in food safety.

1.4 Antibody-Based Biosensors

Antibodies are natural Y-shaped proteins produced by living systems, usually as a defense mechanism against invading bacteria or viruses. They bind to specific species (antigens) with an extremely high degree of specificity by a mixture of hydrogen bonds and other non-covalent interactions, with the binding taking place in the cleft of the protein molecule [28]. One major advantage of antibodies is that they can be "raised" by inoculating laboratory animals with the target in question; the natural defense mechanisms of the animal are to develop antibodies to the antigen. These antibodies can then be harvested from animals. A range of animals are used including mice, rats, rabbits, and larger animals such as sheep or llamas. Therefore, it is possible to develop a selective antibody for almost any target. This high selectivity led to first the development of the Nobel prize-winning radioimmunoassay [29] and then later the enzyme-linked immunosorbent assay (ELISA) [30], which is commonly used today to quantify a wide range of targets in medical and environmental fields.

Once developed the antibody can be immobilized onto a transducer to develop a biosensor, shown schematically in Figure 1.3. One issue is that when antibodies bind to their antigens to form a complex, no easily measured by-products such as electrons or redox-active species are produced. There

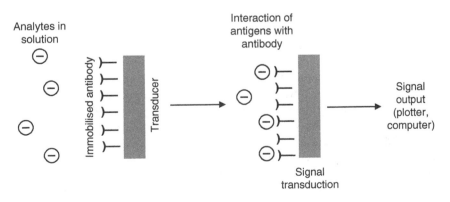

Figure 1.3 Schematic of an antibody-based immunosensor.

are several methods of addressing this drawback. For example, a sandwich immunoassay format can be used where an antibody is bound to the surface and an antigen bound to it from the solution to be analyzed. Development then occurs by exposing the sensor to a labeled secondary antibody, which binds to the antigen, and then the presence of the label is detected; this can be an enzyme or a fluorescent or electroactive species. Competitive assays where the sample is spiked with a labeled antigen and then the labeled and sample antigens compete to bind to the immobilized antibody are also used. However these require labeling of the antibody/antigen, which can be problematic, leading to loss of activity and requiring additional steps with their time and cost implications. Therefore, label-free detection methods have been widely studied that can simply detect the binding event directly without need for labeling. These include electrochemical techniques such as AC impedance, optical techniques such as SPR, and mass-sensitive techniques such as QCM [28].

Another issue is that the strong binding between antibody and antigen means that there is no turnover of substrate; the binding is essentially irreversible. In this case, the sensors are often prone to saturation and can only be used once. Although the antibody–antigen reaction can be reversed by extremes of pH or strongly ionic solutions, these can damage the antibody, leading to permanent loss of activity. However, if costs can be brought down far enough, the possibilities of simple single-shot tests for home use become possible. This led to the first commercially available immunoassay, the home pregnancy test, which detects the presence of human chorionic gonadotrophin (hCG). Initial tests simply detect its presence by showing a blue line, that is, pregnant or not pregnant; however later models incorporate an optical reader that measures the color intensity, thereby assessing the hCG level and giving an estimate of time since conception.

1.5 Aptasensors

Aptamers are a family of RNA/DNA-like oligonucleotides capable of binding a wide variety of targets [31] including proteins, drugs, peptides, and cells. When they bind their targets, the binding event is usually accompanied by conformational changes in the aptamer; for example, it may fold around a small molecule. These structural changes are often easy to detect, making aptamers ideal candidates for sensing purposes. Aptamers also display other advantages over other recognition elements such as enzymes and antibodies. They can be synthesized *in vitro*, requiring no animal hosts and usually with a high specificity and selectivity to just about any target from small molecules to peptides, proteins, and even whole cells [31]. The lack of an animal host means that aptamers can be synthesized to highly toxic compounds. Once a particular optimal aptamer for a certain target has been determined, it can be commercially synthesized in the pure state and often displays superior stability to other biological molecules, hence their nickname "chemical antibodies."

Aptamers can be sourced by firstly utilizing a library of random oligonucleotides. It is possible that within this library a number of the oligonucleotides will display an affinity to the target, whereas most of them will not. They are then subjected to a process called systematic evolution of ligands by exponential (SELEX) enrichment. In this process, the library is incubated with the target and then bound molecules, that is, oligonucleotide/target complexes separated and the unbound species discarded. The bound oligonucleotides are then released from the target and then subjected to polymerase chain reaction (PCR) amplification. This then forms a new library for the process to begin again. Over a number of cycles (6–12) [31], the oligonucleotides with the strongest affinity to the target are preferred in a manner similar to natural selection. After a number of cycles, these aptamers are cloned and expressed. Figure 1.4 shows a schematic of this process.

Aptamers bind to their targets with excellent selectivity and high affinity, dissociation constants often being nanomolar or picomolar [32]. Like antibodies, aptamers can be utilized in a variety of formats; for small molecules there is usually a simple 1:1 complex formed with the target encapsulated inside the aptamer. However with larger analytes the aptamer binds to the surface of the target, and different aptamers can be isolated, which bind to different areas [31].

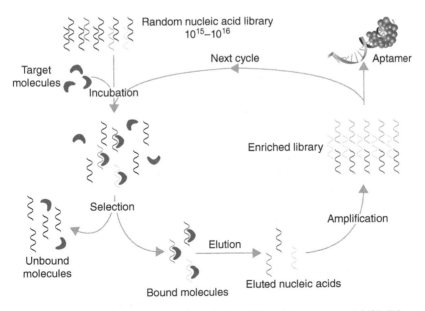

Figure 1.4 Scheme for the systematic evolution of ligands by exponential (SELEX) enrichment process. *Source:* Song et al. [31]. Reproduced with permission of Elsevier.

This allows for sandwich-type assays where two aptamers are used to enhance the biosensor response; there also exist mixed sandwich assays using an aptamer and an antibody.

One issue is that since aptamers simply form complexes with their counterparts, again there is no easily detectable product such as a redox-active species formed. However, the easy availability and stability of aptamers also allows their functionalization with labels such as enzymes, nanoparticles, fluorescent, or redox-active groups for use in labeled assays. Alternatively, label-free techniques such as AC impedance, SPR, and QCM can be used to detect binding events [31].

1.6 Peptide-Based Biosensors

Peptides are natural or synthetic polymers of amino acids and are built from the same building blocks as proteins. Since many proteins have the ability to bind targets with good selectivity and specificity, peptides of the correct amino acid sequence should be capable of doing the same [33]. Shorter peptides have a number of advantages over proteins; they will generally display better conformational and chemical stability than proteins and be much less susceptible to denaturing. Also they can be synthesized with specific sequences using well-known solid-phase synthesis protocols and can be easily substituted with labeling groups without affecting their activity. Especially popular is the labeling of one or both ends of the peptide with fluorescent groups [33].

These recognition receptors can be synthesized with a particular sequence or a library of peptides can be used to assess affinity to a particular target. For example, peptides can be made to specifically chelate certain metal ions even in the presence of other metal ions. Peptide-based sensors are especially effective systems for activity of certain enzymes such as proteases. Proteases can hydrolyze peptide bonds, and certain proteases are linked to many disease states. For example, matrix metallopeptidase-2 (MMP-2) and MMP-9 are thought to be important in a number of inflammatory and pathological processes as well as tumor metastasis [34–36]. Peptides can be used to assess proteinase activity. For example, quantum dots could be coated with peptides conjugated with a large number of dye molecules, fluorescence resonance energy transfer interactions occur between the dye molecules, and the dot, which quenches the dot fluorescence. When a proteinase is added, the peptide is hydrolyzed, the coating removed, and the dot fluorescence returned [37]. Activity of a variety of other materials such as kinases can also be assessed [33].

Libraries of short (<50 amino acids) peptides from random phage display can be screened against various targets as reviewed before [38]. Also *in silico* modeling of peptide strand interactions with targets of interest can be used to select possible receptor peptides, these can then be synthesized and assayed [38, 39].

One issue however is that immobilizing these onto a solid surface may lead to structural modifications, which remove its activity. Also peptide sequences that form the active sites of natural receptors can be synthesized and can retain the activity of the parent molecule.

1.7 MIP-Based Biosensors

Biosensors were initially made using biological molecules such as enzymes or antibodies; however, this led to issues such as cost, difficulty in purification and isolation, and stability. The use of semisynthetic materials such as aptamers and peptides that can be synthesized or selected has addressed this issue to some extent. However, another approach is to use totally synthetic materials that mimic the behavior of enzymes or antibodies. This has led to the development of molecularly imprinted polymers (MIPs), which although not biosensors *per se*, are a possible solution [40–42].

For manufacturing of MIPs, the analyte of interest (often biological in nature) is mixed with a variety of polymerizable monomers and some of these will interact with the analyte. Polymerization will then be initiated and a cross-linked polymer is formed containing entrapped analytes, which act as templates (Figure 1.5). Removal of the analyte will, if the polymer is sufficiently rigid, leave pores within the polymer, which not only match the template size and shape but also contain their internal surface groups, which will interact with the analyte [42–45]. Often this technique is combined with *in silico* modeling of the template interaction with a library of monomers, allowing selection of a monomer mixture that will interact strongly with the template [9, 10, 46]. MIPs display several advantages over biological materials; they have much higher stabilities and can be stored dry for months or years, synthesized in large quantities from readily available monomers, and used in nonaqueous solvents and over a range of temperatures [45].

A wide variety of protocols can be used. For example, inorganic polymers containing glucose were deposited onto a QCM by a sol–gel process, the glucose washed out, and the resultant system shown to act as a sensor, giving an

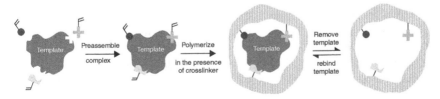
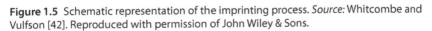

Figure 1.5 Schematic representation of the imprinting process. *Source:* Whitcombe and Vulfson [42]. Reproduced with permission of John Wiley & Sons.

increase in mass when exposed to aqueous glucose [47]. Polymers can also be deposited electrochemically onto electrode surfaces in the presence of a template. For example, poly(*o*-phenylenediamine) could be electrochemically deposited from template solutions onto a QCM chip to give sensors for atropine (with a linear range between 8×10^{-6} and 4×10^{-3} M) [48]. Much larger targets can also be used; for example, a number of enzymes can be incorporated into cross-linked polymers, then removed, and the resultant MIPs display strong binding affinities for those templates [49]. These types of system have even been successfully applied to the detection of viruses in tobacco plant sap using QCM chips [50].

Most of these MIPs have been utilized as solid films since the cross-linking reaction renders them completely insoluble. However, more recently methods of making nanoparticle MIPs, which are soluble, have come to the field [51, 52]. For example, nanosized MIPs toward a range of substrates could be synthesized and used in competitive ELISA assays, giving comparable or better performance than assays based on commercial antibodies with detection limits as low as 1 pM [50]. MIP-based biomimetic sensors have been successfully developed for viruses [51–53], toxins [9, 10, 54], and drugs [45, 46, 55] in recent years in the form of nanoparticles, which can be covalently immobilized on gold sensor chips. Moreover, regeneration of sensor surfaces using acidic and/ or basic solutions is also possible which allows use the same sensor multiple times and decreases the required cost and time substantially. A comprehensive research on adenoviruses has compared the sensing efficiency of antibodies and these MIPs by employing SPR biosensors [52], which indicates the promising future of these recognition receptors for many important analytes. The recent years have also witnessed the implementations of MIPs in biosensors for the detection of disease biomarkers, which are covered in Chapter 12 with detailed examples.

1.8 Conclusions

In this chapter, we have described the major groups of recognition elements used in biosensors. Initial studies used enzymes because of their specificity, high turnover, and the fact that they often produce an easily measured product such as hydrogen peroxide. Antibodies also show high specificity; although in their case measurement of the recognition event can be more complex. One major issue with these biological receptors is their fragility; since purification, immobilization, storage, and labeling may all abolish their activity. This drawback has led to the development of semisynthetic and synthetic analogues of these biological species, such as peptides, aptamers, and MIPs. These demonstrate much higher stabilities and can be produced in greater quantities for almost any target. However, in many cases the sensitivity and selectivity of these materials is still not as high as natural molecules. It can be concluded that the requirements of an assay may well determine the optimum recognition receptors to be used in any biosensor.

Acknowledgment

Z.A. gratefully acknowledges support from the European Commission, Marie Curie Actions and IPODI as the principle investigator.

References

1 Altintas, Z.; Kallempudi, S. S.; Sezerman, U.; Gurbuz, Y. *Sens. Actuators B Chem.* **2012**, 174, 187–194.

2 Altintas, Z.; Fakanya, W. M.; Tothill, I. E. *Talanta* **2014**, 128, 177–186.

3 Altintas, Z.; Tothill, I. E. *Nanobiosens. Dis. Diagn.* **2015**, 4, 1–10.

4 Altintas, Z.; Tothill, I. *Sens. Actuators B Chem.* **2013**, 188, 988–998.

5 Altintas, Z.; Uludag, Y.; Gurbuz, Y.; Tothill, I. E. *Talanta* **2011**, 86, 377–383.

6 Altintas, Z.; Tothill, I. E. *Sens. Actuators B Chem.* **2012**, 169, 188–194.

7 Kallempudi, S. S.; Altintas, Z.; Niazi, J. H.; Gurbuz, Y. *Sens. Actuators B Chem.* **2012**, 163, 194–201.

8 Altintas, Z.; Kallempudi, S. S.; Gurbuz, Y. *Talanta* **2014**, 118, 270–276.

9 Abdin, M. J.; Altintas, Z.; Tothill, I. E. *Biosens. Bioelectron.* **2015**, 67, 177–183.

10 Altintas, Z.; Abdin, M. J., Tothill, A. M.; Karim, K.; Tothill, I. E. *Anal. Chim. Acta* **2016**, 935, 239–248.

11 Clark L.; Lyons C. *Ann. N. Y. Acad. Sci.* **1962**, 102, 29–45.

12 Newman, J. D., Tigwell, L. J., Turner, A. P. F., Warner, P. J. *Biosensors: A Clearer View.* Proceedings of the 8th World Congress on Biosensors, Granada, Spain, May 24–26, **2004**, pp. 17–20.

13 Thusu, R. Strong Growth Predicted for Biosensors Market, **2010**, http://www.sensorsmag.com/specialty-markets/medical/strong-growth-predicted-biosensors-market-7640 (accessed on February 9, 2017).

14 Cass, A. E. G.; Davis, G.; Francis, G. D.; Hill, H. A.; Aston, W. J.; Higgins, I. J.; Plotkin, E. V.; Scott, L. D.; Turner, A. P. F. *Anal. Chem.* **1984**, 56, 667–671.

15 Degani, Y.; Heller, A. *J. Phys. Chem.* **1987**, 91, 1285–1289.

16 Davis, F.; Higson, S. P. J. *Biosens. Bioelectron.* **2005**, 21, 1–20.

17 Diaz-Gonzalez, M.; Gonzalez-Garcia, M. B.; Costa-Garci, A. *Electroanalysis* **2005**, 17, 1901–1918.

18 Rodriguez-Mozaz, S.; de Alda, M. J. L.; Barcelo, D. *Anal. Bioanal. Chem.* **2006**, 386, 1025–1041.

19 Watson, J. D.; Crick, F. H. C. *Nature* **1953**, 171, 737–738.

20 Davis, F.; Higson, S. P. J. DNA and RNA biosensors, in *Biosensors for Medical Applications*, Ed. Higson, S. P. J., Woodhead, Cambridge, **2012**, 161–190.
21 Gooding J. J. *Electroanalysis* **2002**, 14, 1149–1156.
22 Pividori, M. I.; Merkoci, A.; Alegret S. *Biosens. Bioelectron.* **2000**, 15, 291–303.
23 Wang, J.; Rivas, G.; Cai, X.; Palecek, E.; Nielsen, P.; Shiraishi, H.; Dontha, N.; Luo, D.; Parrado, C.; Chicharro, M.; Farias, P. A. M.; Valera, F. S.; Grant, D. H.; Ozsoz, M.; Flair, M. N. *Anal. Chim. Acta* **1997**, 347, 1–8.
24 Cagnin, S.; Caraballo, M.; Guiducci, C.; Martini, P.; Ross, M.; SantaAna, M.; Danley, D.; West, T.; Lanfraanchi, G. *Sensors* **2009**, 9, 3122–3148.
25 Armistead, P. M.; Thorp, H. H. *Bioconjug. Chem.* **2002**, 13, 172–176.
26 Zaytseva, N. V.; Montagna, R. A.; Baeumner, A. J. *Anal. Chem.* **2005**, 77, 7520–7527.
27 Piliarik, M.; Párová, L.; Homola, J. *Biosens. Bioelectron.* **1959**, 24, 1399–1404.
28 Holford, T. R. J.; Davis, F.; Higson, S. P. J. *Biosens. Bioelectron.* **2012**, 34, 12–24.
29 Yalow, R. S.; Berson, S. A. *Nature* **1959**, 184, 1648–1649.
30 Engvall, E.; Perlmann, P. *Immunochemistry* **1971**, 8, 871–874.
31 Song, S.; Wang, L.; Li, J.; Zhao, J.; Fan, C. *Trends Anal. Chem.* **2008**, 27, 108–117.
32 Jenison, R. D.; Gill, S. C.; Pardi, A.; Polisky, B. *Science* **1994**, 263, 1425–1429.
33 Liu, Q.; Wang, J.; Boyd B. J. *Talanta* **2015**, 136, 114–127.
34 Masson, V.; de la Ballina, L. R.; Munaut, C.; Wielockx, B.; Jost, M.; Maillard, C.; Blacher, S.; Bajou, K.; Itoh, T.; Itohara, S.; Werb, Z.; Libert, C.; Foidart, J. M.; Noel, A. *FASEB J.* **2004**, 18, 234–236.
35 Maatta, M.; Soini, Y.; Liakka, A.; Autio-Harmainen, H. *Clin. Cancer Res.* **2000**, 6, 2726–2734.
36 Ara, T.; Fukuzawa, M.; Kusafuka, T.; Komoto, Y.; Oue, T.; Inoue, M.; Okada, A. *J. Pediatr. Surg.* **1998**, 33, 1272–1278.
37 Medintz, I. L.; Clapp, A. R.; Brunel, F. M.; Tiefenbrunn, T.; Uyeda, H. T.; Chang, E. L.; Deschamps, J. R.; Dawson, P. E.; Mattoussi, H. *Nat. Mater.* **2006**, 5, 581–589.
38 Pavan, S.; Berti, F. *Anal. Bioanal. Chem.* **2012**, 402, 3055–3070.
39 Heurich, M.; Altintas, Z.; Tothill, I. E. *Toxins* **2013**, 5, 1202–1218.
40 Yan, M.; Ranstrom, O. *Molecularly Imprinted Materials: Science and Technology*, Taylor & Francis, New York, **2004**.
41 Alexander, C.; Andersson, H. S.; Andersson, L. I.; Ansell, R. J.; Kirsch, N.; Nicholls, I. A.; O'Mahony, J.; Whitcombe, M. J. *J. Mol. Recognit.* **2006**, 19, 106–180.
42 Whitcombe, M. J.; Vulfson, E. N. *Adv. Mater.* **2001**, 13, 467–478.
43 Altintas, Z. Molecular imprinting technology in advanced biosensors for diagnostics, in *Advances in Biosensor Research*, Ed. Everett, T. G., Nova Science Publishers Inc, New York, **2015**, 1–30.

44 Altintas, Z. Advanced imprinted materials for virus monitoring, in *Advanced Molecularly Imprinting Materials*, Eds. Tiwari, A., Uzun, L., Wiley-Scrivener Publishing LLC, Beverly, **2016**, 389–412.

45 Altintas, Z.; Guerreiro, A.; Piletsky, S. A.; Tothill, I. E. *Sens. Actuators B Chem.* **2015**, 213, 305–313.

46 Altintas, Z.; France, B.; Ortiz, J. O.; Tothill, I. E. *Sens. Actuators B Chem.* **2016**, 224, 726–737.

47 Lee, S. W.; Kunitake, T. *Mol. Cryst. Liq. Cryst.* **2001**, 37, 111–114.

48 Peng, H, Liang, C. D.; Zhou, A. H.; Zhang, Y. Y.; Xie, Q. J.; Yao, S. Z. *Anal. Chim. Acta* **2000**, 423, 221–228.

49 Hayden, O.; Bindeus, R.; Haderspock, C.; Mann, K. J.; Wirl, B.; Dickert, F. L. *Sens. Actuators B* **2003**, 91, 316–319.

50 Smolinska-Kempisty, K.; Guerreiro, A.; Canfarotta, F.; Cáceres, C.; Whitcombe, M. J.; Piletsky, S. F. *Nat. Sci. Rep.* **2016**, 6, 37638.

51 Altintas, Z.; Pocock, J.; Thompson, K.-A.; Tothill, I. E. *Biosens. Bioelectron.* **2015**, 74, 994–1004.

52 Altintas, Z.; Gittens, M.; Guerreiro, A.; Thompson, K.-A.; Walker, J.; Piletsky, S.; Tothill, I. E. *Anal. Chem.* **2015**, 87, 6801–6807.

53 Altintas, Z.; Guerreiro, A.; Piletsky, S. A.; Tothill, I. E. *Affinity molecular receptor for viruses capture and sensing*, **2014**, GB Patent, GB1413209.6.

54 Altintas, Z.; Abdin, M. J.; Tothill, I. E. *MIP-NPs for endotoxin filtration and monitoring*, **2014**, GB Patent, GB1413206.2.

55 Altintas, Z.; Tothill, I. E. *Molecularly imprinted polymer-based affinity nanomaterials for pharmaceuticals capture, filtration and detection*, **2014**, GB Patent, GB1413210.4.

2

Biomarkers in Health Care

Adama Marie Sesay, Pirkko Tervo, and Elisa Tikkanen

Unit of Measurement Technology, Kajaani University Consortium, University of Oulu, Oulu, Finland

2.1 Introduction

Biomonitoring for health-care purposes is predominantly aimed at appraising an individual's chronic or acute state of health, for example, bacterial or viral infection, nutritional deficiency, exposure to environmental agents (chemical or biological) capable of inducing adverse health effects, and eradication of a particular ailment or disease [1, 2]. In order for biomonitoring to transpire, biological samples (e.g., blood and sweat) are generally needed to be collected, or classical physical parameters measured (e.g., heart rate, temperature). Biomonitoring relies on measurable indicators or variations of chemical or biological states of the human body [3, 4]. These measurable indicators or parameters have been coined biomarkers [5].

The inherent nature of a biomarker is that its presence, concentration, or fluctuation is a result of a physiological and complex biological pathway that is related to a particular clinical diagnostic. Biomarkers play a vital role in disease detection and treatment follow-up. The detection of biomarkers in body fluids such as blood and urine is a powerful medical tool for early diagnosis and treatment of diseases [6]. The potential use of new biomarkers in health care is a growing area, which is still in the primary stages of discovery [2, 7].

Biomarkers are often present at very low concentrations within biological matrices and therefore may be difficult to identify or monitor due to interfering matrix effects. Often early detection of certain biomarker related to diseases or ill-health and the monitoring of these diseases near onset or at an early stage of appearance are generally easier to treat and to obtain a successful outcome [8]. Therefore, the early detection of biomarkers related to ill-health is very important especially in the case of cancer, cardiovascular disorders, and other

Biosensors and Nanotechnology: Applications in Health Care Diagnostics, First Edition.
Edited by Zeynep Altintas.
© 2018 John Wiley & Sons, Inc. Published 2018 by John Wiley & Sons, Inc.

pathological conditions. Early and timely detection of disease biomarkers can also prevent the spread of infectious diseases and drastically decrease the morbidity and mortality rate of people suffering from a variety of illnesses caused by viruses and other infectious agents [9].

In medical diagnostics, biomarkers are used as a reflection of a patient health and fitness status or an intervention outcome [10], for example, the presence and subsequent decrease of C-reactive protein (CRP) in blood, which is a good biomarker for monitoring inflammation in the body [11]. To date, blood is by far the most commonly used body fluid for the evaluation of systemic processes. However, other noninvasive biological sample matrices like urine, saliva, and sweat are also becoming popular [12–19].

2.2 Biomarkers

Biomarkers are in short biological markers and are important as they are able to indicate a biological or physiological medical diagnostic snapshot that can be measured and monitored thereafter. They are generally used as a physiological or molecular indicator related to a pathological state of health. Biomarkers are more than often used to identify the onset, progression, or endpoint of a pathological process or a response to a therapeutic or a pharmacological intervention [8, 20, 21]. The National Institutes of Health in 1988 set a definition stating that a biomarker is "A characteristic that is objectively measured and evaluated as an indicator of a normal biological processes, pathogenic processes, or pharmacologic responses to a therapeutic intervention." In the 1990s the World Health Organization (WHO) set a more general definition stating that a biomarker is "almost any measurement reflecting an interaction between a biological system and a potential hazard, which may be chemical, physical, or biological. The measured response may be functional and physiological, biochemical at the cellular level, or a molecular interaction" [22, 23].

2.2.1 Advantage and Utilization of Biomarkers

The use of biomarkers in health care has the potential to overcome certain problems related to symptom-based clinical assessment [4]. Biomarkers exist in a plethora of different forms that include antibodies, proteins, microbes, DNA, RNA, and so on [24–28] whereby any change in their occurrence, appearance/disappearance, concentration, structure, function, or action can be monitored and associated with the onset, progression, and even regression of a disease or disorder [15]. They are valuable indicators for screening, risk assessment, diagnosis/prognosis, and monitoring of a disease and can give a better understanding of an individual's personal biomarker signature that would be highly useful in determining the presence, location, risk, and treatment of a disease in a more personalized holistic manner [29–31].

Biomarkers that are used as surrogate endpoints are often cheaper and easier to measure than "true" endpoints [2]. As an example, it is easier to measure cholesterol in blood than to invasively investigate the amount of plaque buildup in the arteries of the heart when determining and assessing risk of heart disease. Other advantages of surrogate endpoint biomarkers are that they can be measured more frequently relatively quickly and results are available earlier. True clinical endpoints are less ideal to use as they are often associated with undesirable outcomes (e.g., death) and pose many ethical problematic situations. An example is in cases of paracetamol overdose by determining the concentration of paracetamol in plasma to decide whether or not to treat a patient would be better and more ethical than waiting for evidence of liver damage. Hence the use of the term "biomarker" in short may have different meanings depending on the field of science it is being used, but thankfully there is much overlap and the general definitions given would be relevant in most cases [2, 16].

2.2.2 Ideal Characteristics of Biomarkers

Biomarkers that are specific and respond to or are released only in one disease state or toxicological exposure event are highly useful and sought after. The time window a particular biomarker is present in a biological sample (e.g., blood, tissue, saliva) is important, because, when a biomarker response is too transient, it may be of limited value due to sampling regime and timing. Conversely, more persistent biomarkers with slower rates of clearance and recovery are highly demanded. In this case it is crucial to investigate the inherent variability of biomarker occurrence [1, 16].

An ideal biomarker can be classified under seven different criteria [12]:

- A major oxidative product modification that may be implicated directly in the development of a disease
- A stable product that is not easily lost, changed during storage, or susceptible to artifactual induction
- Representative of the balance between the generation of oxidative damage and clearance
- Determined by an analytical or bioanalytical assay that is specific, sensitive, reproducible, and robust
- Independent and free of confounding and interfering factors from dietary intake
- Associated and accessible in a targeted localized tissue or in a valid surrogate tissue or biofluids such as a leukocyte
- Detectable and measurable within the limits of detection of a reliable analytical procedure

Before a biomarker is accepted for medical diagnostic applications, it is important for it to be verified and validated. There are six essential requirements a new biomarker needs to go through before acceptance: (i) *in vitro* preclinical assay development, (ii) preliminary studies with real or spiked samples, (iii) feasibility studies on small preset group to determine discriminating parameters between healthy and diseased subjects, (iv) reference method validation for assay accuracy, (v) statistical analysis (determination for large populations), and (vi) final approval reporting and testing [12]. These steps would also be relevant when developing any new analytical assay, bioanalytical assay, and point-of-care (POC) devices used to monitor biomarkers for medical applications [31–33].

2.3   Biological Samples and Biomarkers

Several different biological matrices can be used to monitor biomarkers, for example, blood, sweat, saliva, and urine. Biological samples can be obtained actively (invasive) and passively (noninvasive) [16]. Biomarkers that are associated with the anatomical localization of a disease or ailment to be monitored hypothetically should be represented in the body fluid that resides in close proximity to it, for example, detecting bacteria in a urine sample to confirm a bladder infection or cancer cells in saliva to diagnose mouth and throat cancer [16, 34, 35].

The intrinsic nature of biological matrices is that they are complex fluids with a variety of compounds and molecules that can nonspecifically bind to the sensing surface of the monitoring device. Methods of collecting and handling samples are not a trivial pursuit [16, 34, 36, 37] with many old but classic methods still in use (e.g., venepuncture and swabs). However, new alternative sample collection and handling methods are still not well developed and lag behind the progress of innovative detection techniques, assay methodology, and nanotechnology. It is clear that research effort in this field needs to develop innovative biological fluid sampling techniques and devices that can be integrated to simplify the handling and downstream processing of diagnostic test procedures [16, 27, 36].

Biological fluids are highly complex sample matrices that contain a diverse and variable amount of proteins, cells, macromolecules, hormones, metabolites, and other small molecules [16]. There are up to 24 different biological fluids of which the common ones used for biological sampling are blood, saliva, sweat, and urine (Table 2.1). Sample collection methods vary widely across the different fluids and are dominated by bulk sampling [38].

Concentration profiles of proteins across the different fluids vary a lot with blood and plasma containing the most; however, other fluids like saliva and

Table 2.1 Main biological sample, sampling techniques, and key properties.

Body fluid	Sampling technique	pH	Total protein	Exclusive proteins
Blood	Needle/lancet	7.35–7.45	60–80 mg mL^{-1}	NA
Saliva	Mouth swab/ passive drool	6.2–7.4	0.2–5 mg mL^{-1}	~34
Urine	Passive collection/ catheter	4.5–8.0	<150 mg day^{-1} (excreted)/ <0.1 mg mL^{-1}	30
Sweat	Swab/direct	4.0–6.8	0.1–0.7 mg mL^{-1}	20
Tear	Swab/contact lenses	6.5–7.5	6–10 mg mL^{-1}	34
Breath	Bag/cold trap	7.5–7.65	1–4 mg mL^{-1}	—
Interstitial fluid (skin)	Microdialysis, microneedles/ iontophoresis	7.2–7.4	13–20 mg mL^{-1}	32

tears also have relatively high concentrations of proteins. It is also interesting to note that all body fluids possess a unique protein profile and proteome (20–40%) in relation to blood [15], therefore illustrating the need for correctly combining the monitoring of specific biomarkers with body fluid-specific sampling for medical applications. Nevertheless, there is also enough overlap for biomarkers to cross into other biological fluids that make it possible for biomarker monitoring to be conducted in more than one body fluid as long as there is enough crossover and concentration correlation [39].

The relatively low concentration of relevant biomarker in the respective biological samples can cause an enormous statistical sampling problem. As the concentration of the biomarker decreases, the probability that the collected sample will not contain the biomarker of interest increases. Hence, this could lead to an unpredictable distribution of false-negative results unrelated to the analytical device sensitivity or downstream sample processing and due solely to the fact that the sample just may not contain the biomarker [16]. This is a systematic problem with all endpoint sampling practices and the reason why real time *in situ* sampling would be beneficial. Early detection is paramount in the quest to diagnose and treat all diseases, and the analysis of biological fluid offers a window to detect disease at an early stage [17]. Currently there are many diagnostic tests for single biomarker screening (e.g., glucose, lactate, etc.). For future health-care needs, single biomarker detection will not be efficient enough for accurate clinical diagnosis, while simultaneous multi-analyte detection will be more and more sought after [19, 40–42].

2.4 Personalized Health and Point-of-Care Technology

The search, identification, and development of new biomarkers and novel easy-to-use analytical devices are essential components for predictive, preventive, and personalized medicine. Traditionally, analytical devices used for the determination and monitoring of biomarkers are performed in dedicated centralized laboratories, often on large, automated analytical instruments that require skilled personnel, which increase the costs and are intrinsically associated with extended waiting time to obtain results. Presently, there is a great push and need for smaller, faster, and cheaper devices that can be used at the very least as diagnostic screening tools. POC devices in health care are seen as technological disruptors and represent a paradigm shift that can completely change and replace these time-consuming central laboratory analyses for more rapid and small analytical tools. These POC devices can make available analytical diagnostics and results at the patient's bedside offering better and informed diagnostic decisions and treatment [31, 32, 40].

Point-of-care tests (POCT) are simple, rapid, and relatively inexpensive portable diagnostic assays and devices (Figure 2.1). The guidelines for developing POC given by the WHO are known as ASSURED, which is an acronym for *affordable, sensitive, user-friendly, rapid/robust, equipment-free or minimal, and delivered* to the greatest need [44]. They lend themselves easily to a means to cut health-care cost by reducing hospital stays, improving early diagnosis,

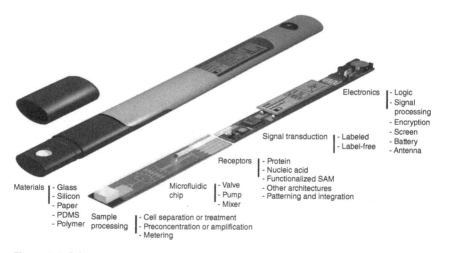

Figure 2.1 Schematic of an ideal POC diagnostic device. The ideal POC device would be able to quantitatively detect several analytes, within minutes, at ultrasensitive femtomolar sensitivity. Sample handling, preparation, detection, and electronics for displaying the diagnostic result would all be included. *Source:* Gervais et al. [43]. Reproduced with permission of John Wiley & Sons.

and enticing individuals to comply and be more involved in their health management, thereby improving adherence to treatment, early intervention, and avoidance of any complications that may arise. It is hoped that POC test would lead to a reduction in central hospital laboratory testing, sample transportation, and data analysis and reporting [29]. Some POCT are given in Table 2.2.

Table 2.2 Selected analyte tests available in POCT format that have been waived by Clinical Laboratory Improvement Amendments of 1988.

Analyte tests available in POCT format	
Alanine aminotransferase (ALT/SGPT)	Influenza A/B
Alcohol (saliva)	Ketone, blood, and urine
Amines	Luteinizing hormone (LH)
Amphetamines	Methamphetamines
Bladder tumor-associated antigen	Microalbumin
Cannabinoids (THC)	Nicotine and/or metabolites
Catalase, urine	Opiates
Cholesterol	Ovulation test (LH) by visual color comparison
Cocaine metabolites	pH
Creatinine	Phencyclidine (PCP)
Erythrocyte sedimentation rate (unautomated)	Prothrombin time (PT)
Estrone 3-glucuronide	Spun microhematocrit
Ethanol (alcohol)	Streptococcus, group A
Fecal occult blood	Triglyceride
Follicle-stimulating hormone (FSH)	Urine dipstick or tablet analytes, unautomated
Fructosamine	*Helicobacter pylori* antibodies
Gastric occult blood	*H. pylori*
Glucose	Hematocrit
Glucose monitoring device (FDA cleared/ home use)	Hemoglobin by copper sulfate, unautomated
Glucose, fluid (approved by the FDA for prescription home use)	Hemoglobin, single-analyte instrument with self-control
Glycosylated hemoglobin (Hgb A1c)	Infectious mononucleosis antibodies (mono)
Human chorionic gonadotropin (HCG, urine)	Vaginal pH
HDL cholesterol	

Source: Nichols [45]. Reproduced with permission of Elsevier.

Generally, the two main types of available POC formats are small-scale benchtop analyzers and handheld devices [29, 32]. Small-scale benchtop analyses may be more portable and more user-friendly but are basically scaled-down analytical instruments found in central hospital laboratories. Handheld devices are often highly innovative integrated analytical platforms, which perform several unit operations, for example, sample cleanup, separation, analysis, and data reporting [29, 43, 46–49].

Biosensors and biosensing techniques are probably some of the most promising technology platforms fitting to solve many of the challenges concerning the current and future needs to obtain highly sensitive, fast, and cost-effective methods for biomarker analysis in medical diagnostics. Novel highly sensitive biosensors and biosensing technologies could allow biomarkers related to health care to be tested reliably in a decentralized setting (i.e., home, doctor's office, care home) [50, 51]. However, there are still many obstacles and challenges that need to be improved, and many limitations remain, such as design, sample choice and handling, selection of relevant biomarker for a particular disease, identification, quantification, and interpretation of translational clinical results. Great research effort is ongoing and new devices are forecasted to make their way into clinical laboratories and doctors' offices [32, 33, 40, 52].

2.5 Use of Biomarkers in Biosensing Technology

The use of biomarkers for diagnostic and prognostic monitoring of health ultimately requires reliable, specific, sensitive, and cost-effective tools. The most effective analytical platform that can fulfill these criteria is analytical biosensors and biosensing technologies. The definition of a chemical sensor/biosensor by the International Union of Pure and Applied Chemistry is "a device that transforms chemical information, ranging from the concentration of a specific sample component to total composition analysis, into an analytically useful signal" where too a chemical sensor is based on the integration of two units (a receptor and a physiochemical transducer) [7, 34]. A biosensor, therefore, can be described in the same way but has selective biological components (e.g., antibody, enzyme, DNA, etc.) as its receptor [53–55]. The interactive binding event of the receptor transforms the analyte/biomarker concentration into a chemical or physical signal that is proportional to the amount of analyte/biomarker in the sample. It is therefore highly important that the receptor is highly specific and selective toward the analyte of interest especially in the presence of interfering molecules in order to avoid false-positive results. The transducer is also important as it converts the biomarker/receptor binding event to an often digitalized readable value. Biosensors can be divided into two subcategories: (i) enzymatic/catalytic sensors and (ii) affinity-based sensors. In the same manner they can

Figure 2.2 Schematic of a typical biosensor.

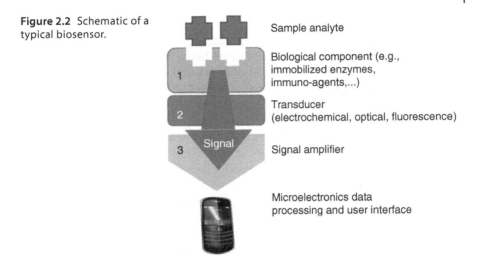

Sample analyte

Biological component (e.g., immobilized enzymes, immuno-agents,...)

Transducer (electrochemical, optical, fluorescence)

Signal amplifier

Microelectronics data processing and user interface

be classed according to their transducer (e.g., electrochemical, optical, piezoelectric, mass, or colorimetric sensors) (Figure 2.2).

An ideal biosensor for medical applications should measure and monitor biomarker(s) in real time directly and continuously in the body fluid where the relevant biomarker of interest can be found [33, 56]. This way, frequency in sample collection and problems associated with "when and where" are eliminated. Although there are optical biosensing systems that are considered to provide real-time sensing (e.g., real-time binding kinetics—surface plasmon resonance (SPR) sensing), true real-time monitoring has to be connected with real-time sampling in the future to realize the ultimate medical and clinical application [16, 33, 57].

The effectiveness of using biosensors for POC monitoring relies on three integrated unit operations: (i) sample collection/handling, (ii) assay chemistry and methodology, and (iii) detection and recording of a quantifiable signal (biological recognition coupled with signal transduction) [16]. The most successful biosensor platform to date that has been translated from central clinical laboratories to POC use for personal diagnostics is the electrochemical enzyme biosensor (blood glucose and lactate). To detect larger molecules, specific affinity receptors (e.g., antibodies, protein receptors) are often used to selectively bind to the biomarker of interest. Affinity receptor techniques based on immunoassays have often been translated into POC devices such as lateral flow devices (e.g., pregnancy test) [31, 40].

Enzyme-linked immunosorbent assays (ELISA) is a common technique used in clinical settings for biomarker detection and monitoring. The most common ELISA format used is the sandwich assay in which the biomarker of interest is captured between two antibodies in a sandwich-like complex, which has a quantifiable label (e.g., enzyme) that can generate a signal from converting a substrate into a detectable form (color or fluorescent signal) that is proportionally equal to the concentration of the analyte being detected [6].

Currently, many POC-based biosensors face challenges from nonspecific adsorption of the body fluid components onto the bioreceptor sensing surface to the limited availability of receptors for many known biomarkers of medical interest. This has meant that a restricted number of POC biosensors have been fully developed and are currently available for commercial use. However, progress in supporting technologies, for example, Systematic evolution of ligands by exponential enrichment (SELEX) production for recombinant antibodies, aptamers, synthetic antibodies otherwise known as molecularly imprinted polymers (MIPs), and ultrasensitive affinity-based sensors are paving their way into new exciting medical diagnostic applications [53, 54, 58, 59]. The rapid development and integration of antibody-based biosensors and nanotechnology is greatly predicted to lead a new era of POC devices and personalized diagnostics [6, 29, 60].

2.6 Biomarkers in Disease Diagnosis

Considering the enormous amount of antigens and endogenous biomarkers that can be used for medical diagnostics, it would be unrealistic to try and list them all. Table 2.3 lists major clinically important biomarkers and associated disease state and the biological sample fluids within which they can be monitored [20].

Biomarkers can be used to screen cancer and chronic diseases [61, 64, 74], predict risk, and aid to develop targeted therapies [62]. Biomarkers for cancers are used to test and follow bodily processes to predict how an individual may be at risk of developing a certain disease and the likelihood on how they would respond to a drug or therapeutic treatment [8]. Inflammatory cytokines are one of the most important groups as they are responsible in triggering or suppressing a biochemical inflammation cascading response that can be acute (e.g., peanut allergy) or chronic (autoimmune) [25, 54, 74].

Cytokine release changes the behavior of cells around them and cytokines are involved in autocrine cell signaling, paracrine cell-to-cell signaling, and endocrine hormone signaling for immunomodulating and regulatory pathways. Cytokines include chemokines, interferons, interleukins, lymphokines, and tumor necrosis factors (but not including hormones or growth factors) [12]. The presence of certain antibodies in blood can indicate whether an individual has been exposed to a particular disease-causing antigen, for example, bacteria/microbe, virus (HIV), or nut protein (peanut allergy). Patients with suspected HIV infection are screened for HIV-1 and HIV-2 by affinity-based immunoassays [15]. The increased concentration of salivary IgA can be used as an indicator of respiratory infection and stress. Conversely an individual can

Table 2.3 Clinically important biomarkers and associated disease states.

Biomarkers	Disease type and state	Body fluid
Cancer [28, 61–63]		
Alpha-fetoprotein		Blood
Angiogenin		Blood
Cancer antigen 125	Ovarian	Blood
Carbohydrate antigen 15-3	Epithelial ovarian cancer	Blood
Carcinoembryonic antigen (CEA)	Colon cancer	Blood
Epidermal growth factor receptor	Cancer, breast	Blood
Human ferritin		Blood
Human phosphatase of regenerating liver 3	Gastric	Blood
Insulin-like growth factor I		Blood
Insulin-like growth factor II	Epithelial ovarian cancer	Blood
MicroRNA		Blood, saliva, urine
Mucin 1		Blood
Murine double minute 2	Brain	Blood
Neuron-specific enolase		Blood
Osteopontin		Blood
p53 antibody	Prognostic cancer	Blood
p53 protein		Blood
Prostrate-specific cancer (PSA)	Prostate	Blood
TP53 mutation	Lung	Blood
Inflammation, immunity, and stress [37, 57, 64–68]		
Autoimmune antibodies (ANA, β-2m, anti-tTG, anti-dsDNA)	Lupus erythematosus, Sjögren's syndrome, SLE	Blood, saliva, tissue
Cardiac troponin-I	Cardiac injury	Blood
Cardiac troponin-T	Acute myocardial infarction, cardiac injury	Blood
Cortisol	Stress	Blood, saliva
C-reactive protein	Cardiovascular diseases and cardiac inflammation	Blood, saliva

(Continued)

Table 2.3 (Continued)

Biomarkers	Disease type and state	Body fluid
Cytokines		Blood, tissue, saliva
Proinflammatory IL-1α, IL-1β, IL-2, IL-6, IL-8, IL-12, TNF-α, and interferon-γ (IFNγ)	Inflammation promoting	
Anti-inflammatory (IL-4, IL-5, IL-10, TGF) abilities	Inflammation suppressive	
HIV-1 and HIV-2	HIV infection	Saliva, blood
IFNγ	Autoinflammatory, autoimmune disease	Blood
Low-density lipoprotein	Coronary heart diseases	Blood
Lysozyme	Atherosclerotic cardiovascular disease, rheumatoid arthritis	Blood
Matrix metallopeptidase 9 (MMP-9)	Wound infection	Blood
Myoglobin	Acute myocardial infarction, cardiovascular disease	Blood
Nucleocapsid protein	Severe acute respiratory syndrome (SARS)	Blood
Salivary IgA	Stress, upper respiratory infection	Blood, saliva
Salivary α-amylase	Stress	Saliva
	Metabolic disorders [69–71]	
Alanine aminotransferase	Hepatotoxicity	Blood
Adiponectin	Metabolic syndrome	Blood
Hemoglobin A1c	Diabetes	Blood
Human serum albumin	Liver function	Blood
Insulin	Diabetes	Blood, saliva
Nicotinamide phosphoribosyl	Obesity-related metabolic diseases	Blood
Transferrin	Protein–calorie malnutrition	Blood
	Neurological disorders [20, 72, 73]	
Alpha-synuclein autoantibody	Parkinson's disease	Blood
Dopamine		
Amyloid β peptides	Alzheimer's disease	Blood

also produce antibodies against itself, and the presence of these can indicate an autoimmune response or disease. Sjögren's syndrome is a systemic autoimmune disease that affects the moisture-producing glands of the body and is often displayed as dry mouth and eye moisture disorders. Therefore, the presence of an autoimmune antibody (β_2m) is a diagnostic indicator for the disease [74].

Hormones are important regulatory small molecules that are carried by the circulatory system to organs around the body to regulate physiology and behavior. They have diverse chemical structures that can be steroids (cortisol, estrogen) or amino acid derivatives (amines, peptides, and proteins). They also primarily assist the communication between organs and tissues for physiological regulation and behavioral attributes, such as digestion, metabolism, respiration, tissue function, sensory perception, sleep, excretion, lactation, stress, growth and development, movement, and reproduction [70]. There are two major stress biomarkers that have been identified: salivary α-amylase and cortisol. The most used for evaluating stress is cortisol, which is analytically monitored using immunoassay technique for detection [37, 70].

A metabolic disorder occurs when an abnormal chemical reaction in the body alters the normal metabolic process. Diabetes, insulin resistance, and metabolic syndrome [71] can be classed as metabolic disorders. It is important for diabetics to control and manage their blood glucose levels. Therefore, in this case blood glucose concentration is used as a regulatory biomarker. However, hemoglobin A1c is also used as an indicator for the disease [20].

2.7 Conclusions

Biomarkers based on proteins, antibody fragments, DNA and RNA molecules have been combined with nanoparticles and used as targets in medical diagnostic applications [24, 28, 59]. It is highly anticipated that in the near future, we might be able to detect cancer at a very early stage, providing a much higher chance of treatment [75] due to the unique optical, magnetic, mechanical, chemical, and physical properties of nanomaterials. Rapid technological advancement in the area of materials science and nanotechnology has made it possible to synthesize and fabricate nanoparticles and surfaces with highly desirable properties that are not evident in the bulk material [76]. The integration and application of nanoparticles and nanotechnology to biomarker detection and health-care applications offers great potential in improving medical diagnosis and therapies. These technologies are predicted to provide and enable personalized and yet more affordable health care while simultaneously providing improved quality of life [21].

A vast number of studies and developments in the nanotechnology area have been demonstrated and many nanomaterials have been utilized to detect a variety of biomarkers for health application. Nanoparticles like quantum dots, gold nanoparticles, magnetic nanoparticles, carbon nanotubes, gold nanowires, and a plethora of other materials have been developed in conjunction with the discovery of a wide range of biomarkers that can lower the detection limit of biomarkers and improve diagnosis of diseases [6, 76–81]. It is with great anticipation that future technological advances will be born from the combination of discovered novel biomarkers, innovative POC devices, and nanotechnology.

References

1 Chambers, J. E., Boone, J. S., Carr, R. L. *Human and Ecological Risk Assessment*, **2002**, 8, 165–176.
2 Aronson, J. K. *British Journal of Clinical Pharmacology*, **2005**, 59, 491–494.
3 Strimbu, K., Tavel, J. A. *Currant Opinions on HIV AIDS*, **2010**, 5, 463–466.
4 Lopresti, A. L., Maker, G.-L., Hood, S. D. *Progress in Neuro-Psychopharmacology and Biological Psychiatry*, **2004**, 48, 102–111.
5 Hill, A. B. *Proceedings of the Royal Society of Medicine*, **1965**, 58, 295–300.
6 Andreasson, U., Blennow, K., Zetterberg, H. *Alzheimer's and Dementia Diagnosis*, **2016**, 3, 98–102.
7 Nimse, S. B., Sonawane, M. D., Song, K. *Analyst*, **2016**, 141, 740–755.
8 Karley, D., Gupta, D., Tiwari, A. *Journal of Molecular Biomarkers and Diagnosis*, **2011**, 2, 118.
9 Dijkstra, S., Mulders, P. F. A., Mena-Bravo, A. *Journal of Pharmaceutical and Biomedical Analysis*, **2014**, 90, 139–147.
10 Williamson, S., Munro, C., Pickle, R. *Nursing Research and Practice*, **2012**, 2012, 4.
11 Azar, R., Richard, A. *Journal of Inflammation*, **2011**, 8, 37.
12 Liu, J., Duan, Y. *Oral Oncology*, **2012**, 48, 569–577.
13 Miller, C. S., et al. *JDR Clinical Research Supplement*, **2014**, 93, 72S–79S.
14 Farnaud, S. J.C., Kosti, O., Getting, S. J. *The Scientific World Journal*, **2010**, 10, 434–456.
15 Yoshizawa, J. M., Schafer, C. A., Schafer, J. J. *Clinical Microbiology Reviews*, **2013**, 26, 781–791.
16 Corrie, S. R., Coffey, J. W., Islam, J. *Analyst*, **2015**, 140, 4350–4364.
17 Spielmann, N., Wong, D. T. *Oral Diseases*, **2011**, 17, 345–354.
18 Jadoon, S., Karim, R., Akram, A. K. *International Journal of Analytical Chemistry*, **2015**, 2012, 7.
19 Heikenfeld, J. *Nature*, **2016**, 529, 475–476.
20 Luo, X., Davis, J. J. *Chemical Society Reviews*, **2013**, 42, 5944–5962.
21 Reinke, J. *Biomarkers in drug development*, John Wiley & Sons, Inc, Hoboken, **2010**, pp. 709–730.
22 WHO. **1993**, Biomarkers and Risk Assessment: Concepts and Principles. http://www.inchem.org/documents/ehc/ehc/ehc155.htm (accessed on June 30, 2017).

23 WHO. **2001**, Biomarkers in Risk Assessment: Validity and Validation. http://www.inchem.org/documents/ehc/ehc/ehc155.htm (accessed on June 30, 2017).

24 Hayes, J., Peruzzi, P. P., Lawler, S. *Trends in Molecular Medicine*, **2014**, 20, 460–469.

25 Gramlich, O. W., Bell, K., von Thun und Hohensterin-Blaul, N. *Current Opinion in Pharmacology*, **2013**, 13, 90–97.

26 Cardoso, A. R., Moreira, F. T., Fernandes, R. *Biosensors and Bioelectronics*, **2016**, 80, 621–630.

27 Obayashi, K. *Clinica Chimica Acta*, **2013**, 425, 196–201.

28 Altintas, Z., Tothill, I. E. *Sensors and Actuators B: Chemical*, **2012**, 169, 188–194.

29 McDonnell, B., Hearty, S., Leonard, P. *Clinical Biochemistry*, **2009**, 42, 549–561.

30 Lee, S., et al. **2016**, *Lab on a Chip*, 16, 2408–2417.

31 Gubala, V., Harris, L., Ricco, A. *Analytical Chemistry*, **2011**, 84, 487–515.

32 Syedmoradi, L., Deandshpour, M., Alvandipour, M. *Biosensors and Bioelectronics*, **2017**, 87, 373–387.

33 Vaddiraju, S., Tomazox, I., Burgess, D. J. *Biosensors and Bioelectronics*, **2010**, 25, 1553–1565.

34 Kaushik, A., Vasudev, S. K., Arya, S. K. *Biosensors and Bioelectronics*, **2014**, 53, 499–512.

35 Rathnayake, N., et al. *PLoS One*, **2013**, 8(4), e61356, 1–5.

36 Sesay, A. M., Kruhne, U., Sonny, S. *14th International Conference on Miniaturized Systems for Chemistry and Life Sciences*, Curran Associates, Inc, Red Hook, **2011**, 1520–1522.

37 Sesay, A. M., Micheli, L., Tervo, P. *Analytical Biochemistry*, **2013**, 434, 308–314.

38 Bandodkar, A. J., Wang, J. *Trends in Biotechnology*, **2014**, 32, 363–371.

39 Yeh, C., Christodoulides, N. L., Floriano, P. N. *Texas Dental Journal*, **2010**, 127, 651–661.

40 Warsinke, A. *Analytical and Bioanalytical Chemistry*, **2009**, 393, 1393–1405.

41 Gao, W., Emaminejad, S., Nyein, H. Y. Y. *Nature*, **2016**, 529, 509–514.

42 Altintas, Z., Kallempudi, S. S., Gurbuz, Y. *Talanta*, **2014**, 118, 270–276.

43 Gervais, L., de Rooij, N., Delamarche, E. *Advanced Materials*, **2011**, 123, H151–H176.

44 Gormez, F. A. *Bioanalysis*, **2013**, 5, 1–3.

45 Nichols, J. H. *Clinics in Laboratory Medicine*, **2007**, 24, 893–908.

46 Nie, C., Frijns, M., Zevenbergen, M. *Sensors and Actuators B: Chemical*, **2015**, 227, 427–437.

47 Munje, R., Muthukumar, S., Selvam, A. P. *Scientific Reports*, **2015**, 5, 14586.

48 Vashist, S. K., Schneider, E. M., Luong, J. H. *Diagnostics*, **2014**, 4, 104–128.

49 Roda, A., Michelini, E., Cevenini, L. *Analytical Chemistry*, **2014**, 86, 7299–7304.

50 Bahadir, E. B., Sezgintürk, M. K. *Biosensors and Bioelectronics*, **2015**, 68, 62–71.

51 Oncescu, V., O'Dell, D., Erickson, D. *Lab on a Chip*, **2013**, 13, 3232–3238.

52 Arduini, F., Micheli, L., Moscone, D. *Trends in Analytical Chemistry*, **2016**, 79, 114–126.

53 Chao, J., et al. *Biosensors and Bioelectronics*, **2016**, 76, 68–79.

54 Weng, X., Neethirajan, S. *Biosensors and Bioelectronics*, **2016**, 85, 649–656.

55 Cieplak, M., Kutner, W. *Trends in Biotechnology*, **2016**, 34, 922–940.

56 Wang, S. Q., Chinnasamy, T., Lifson, M. A., Inci, F., Demirci, U. *Trends in Biotechnology*, **2016**, 34, 909–921.

57 Das, C., Wang, Q., Ledden, B. *Sensors and Actuators B: Chemical*, **2017**, 238, 633–640.

58 Jimenez, A. M., Rodrigo, V., Milosavljevic, V. *Sensors and Actuators B: Chemical*, **2017**, 240, 503–510.

59 Wang, X., Liu, L., Wang, Z. *Journal of Electroanalytical Chemistry*, **2016**, 781, 351–355.

60 Altintas, Z., Guerreiro, A., Piletsky, S., Tothill, I. E. *Sensors and Actuators B: Chemical*, **2015**, 213, 305–313.

61 Altintas, Z., Uludag, Y., Gurbuz, Y., Tothill, I. E. *Talanta*, **2011**, 86, 377–383.

62 Rusling, J. F., Kumar, C. V., Gutkind, J. S. *Analyst*, **2010**, 135, 2496–2511.

63 Altintas, Z., Uludag, Y., Gurbuz, Y., Tothill, I. E. *Analytica Chimica Acta*, **2012**, 712, 138–144.

64 Altintas, Z., Fakanya, W. M., Tothill, I. E. *Talanta*, **2014**, 128, 177–186.

65 Yoon, A. J., Cheng, E., Philipone, R. *Journal of Clinical Periodontology*, **2012**, 39, 434–440.

66 Whiteley, W., Tseng, M., Sandercock, P. *Stroke*, **2008**, 10, 2902–2909.

67 Vanmassenhove, J., Vanholder, R., Nagler, E. *Nephrology, Dialysis, Transplantation*, **2013**, 28, 254–273.

68 Pawula, M., Altintas, Z., Tothill, I. E. *Talanta*, **2016**, 146, 823–830.

69 Arvinte, A., Westerman, C., Sesay, A. M. *Sensors and Actuators B: Chemical*, **2010**, 150, 756–763.

70 Mahosenaho, M., et al. *Microchimica Acta*, **2010**, 170, 243–249.

71 Klünder-Klünder, M., Flores-Huerta, S., Garcia-Macedo, R. *BMC Public Health*, **2013**, 88, 13.

72 Diaz-Diestra, D., Tahpa, B., Beltran-Huarac, B. R. *Biosensors and Bioelectronics*, **2017**, 87, 693–700.

73 Scarano, S., Lisi, S., Ravelet, C. *Analytica Chimica Acta*, **2016**, 940, 21–37.

74 Hu, S., Gao, K., Pollard, M. *Arthritis Care and Research*, **2010**, 62, 1633–1638.

75 Choi, Y., Kwak, J., Park, J. *Sensor (Basel)*, **2010**, 10, 428–455.

76 Kumar, S., Ahlawat, W., Kumar, R. *Biosensors and Bioelectronics*, **2015**, 70, 498–503.

77 Altintas, Z., Gittens, M., Pocock, J., Tothill, I. E. *Biochimie*, **2015**, 115, 144–154.

78 Yang, N., Chen, T., Ren, P. *Sensors and Actuators B: Chemical*, **2015**, 207, 690–715.

79 Wang, J. *Biosensors and Bioelectronics*, **2016**, 76, 234–242.

80 Deng, H., Liu, Q., Wang, R. *Biosensors and Bioelectronics*, **2017**, 87, 931–940.

81 Tuteja, S. K., Chen, R., Kukkar, M. *Biosensors and Bioelectronics*, **2016**, 86, 548–556.

3

The Use of Nanomaterials and Microfluidics in Medical Diagnostics

Jon Ashley and Yi Sun

Department of Micro- and Nanotechnology, Technical University of Denmark, Lyngby, Denmark

3.1 Introduction

There has been an ever-increasing demand for more sensitive, cheaper, and faster diagnostic tests in health care that would allow health-care professionals to make rapid and informed decisions. It is crucial that diseases and conditions can be diagnosed at early stage as this would increase the prognosis of a patient and allow for quick and effective treatments. Currently diagnostics is limited by several factors, which include the ability to detect a disease biomarker at a low concentration, the time taken to complete a test, the stability of components within the test (antibody, protein, etc.), the cost of analysis, and the availability of analytical instrumentation.

Advances in nanotechnology have led to the development of cheaper, faster, more sensitive, and accurate assays and devices that have found applications in a number of areas of medical diagnostics. At present, we are starting to see diagnostic tests utilizing nanotechnology that are capable of detecting cells and biomolecules down to the single molecule sensitivity. In proteomics, nanotechnology has proven to be able to identify disease biomarkers even in low abundance, which provides a very useful tool for screening diseases at early stage. It also has the potential to improve upon microscopy techniques that can be used in hospitals to routinely diagnose patients and as follow-up procedures after diagnosis and/or treatment to monitor patient prognosis. The use of nanotechnology in medical devices such as biosensors, microarrays, and microfluidic devices could allow for simple and rapid tests that are especially important for the diagnosis of diseases at point-of-care (POC) settings such as in developing countries that lack adequate medical facilities.

Biosensors and Nanotechnology: Applications in Health Care Diagnostics, First Edition.
Edited by Zeynep Altintas.
© 2018 John Wiley & Sons, Inc. Published 2018 by John Wiley & Sons, Inc.

There are generally two types of approaches that scientists use to engineer nanotechnology in medical diagnostics. The bottom-up approach is where nanomaterials are made atom by atom to achieve controlled particle size and size distribution. The top-down approach refers to the manipulation of bulk material to fabricate nano- and microfluidic devices. The two approaches have shown to be effective in the medical diagnostic applications, and the field is rapidly evolving as new nanomaterials are discovered and nanodevices are engineered.

3.2 Nanomaterials in Medical Diagnostics (Bottom-Up Approach)

Nanoparticles are classified as having at least one dimension that is less than 100 nm; however, the definition can be quite vague, meaning that particles up to 1 μm can also be considered nanoparticles, which display unique properties as we go from bulk materials down to the nanosize. For instance, gold in the bulk form is quite chemically inert, but as we go down to the nanoscale, the properties of gold change and suddenly the chemical properties of gold become more interesting. This in part is due to an increase in surface area being observed. These changes in properties are not just restricted to gold. A large number of elements display interesting optical, conductive, and magnetic properties all dependent on size. These effects are not just confined to single elements; composite nanomaterials also display unique properties that scientists would like to exploit. In addition the shape can also affect the properties of nanoparticles. For example, gold nanospheres can display remarkably different properties compared with gold nanorods.

There are a large number of different types and forms of nanomaterials that have been reported in the scientific community. This makes the classification of these nanomaterials difficult due to complexity and number of different forms of nanoparticles. Nevertheless, we can arrange nanomaterials into three main groups: carbon nanomaterials, metal-based nanoparticles, and polymer-based nanoparticles as shown in Figure 3.1.

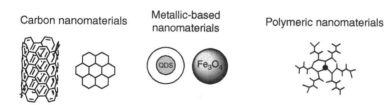

Carbon nanomaterials   Metallic-based nanomaterials   Polymeric nanomaterials

Figure 3.1 Classification of nanomaterials for use in medical diagnostics.

3.2.1   Carbon Nanomaterials

Ever since the discovery of fullerene back in 1985, there has been intense research into the use of carbon nanomaterials and their application in diagnostics [1]. These supramolecular C_{60} molecules display unique properties that have been exploited for diagnostic applications. Fullerenes are highly symmetrical in nature and can be easily made from small aromatic molecules via chemical synthesis or evaporation of graphite using either laser or heat. These C_{60} nanoparticles are insoluble in water, but functionalization with hydrophilic groups can increase their solubility and biocompatibility. These nanoparticles can be easily functionalized, which allows for the conjugation of biomolecules such as antibodies. These in turn are useful in targeting biomarkers.

Carbon nanotubes (CNTs) (those same sheets rolled up into tubes) have greatly expanded the applications of carbon nanomaterials in medical diagnostics. They can exist as either a single layer known as single-walled carbon nanotubes (SWCNTs) or multiple coaxial layers known as multiwalled carbon nanotubes (MWCNTs). SWCNTs can be synthesized using a number of different top-down approaches including arc discharge, laser ablation, and chemical vapor deposition [2]. CNTs display high tensile stress, which have allowed them to be used in a number of applications in materials science. Other interesting properties of CNTs relevant to medical diagnostics include optical, thermal, and electrical conductivity. Due to the bandgap in SWCNTs, unique optical properties have been demonstrated. For instance, SWCNT can absorb and emit photons in the visible and near-infrared range, giving rise to autofluorescence properties.

Graphenes (two-dimensional (2D) sheets of carbon atoms) are a relatively new type of carbon nanomaterial. They are effectively thin films, consisting of sp^2-hybridized carbon atoms in a hexagonal configuration. They were discovered in 2003 by researchers at Manchester University when they used scotch tape to exfoliate layers of graphene from graphite [3]. More sophisticated and effective methods for making graphene have been developed, which include chemical and mechanical exfoliations, unzipping of CNTs, epitaxial growth, and chemical synthesis [4]. Due to the ability of electrons within graphene to act as Dirac fermions, a number of unusual physical properties have been observed in graphene including high thermal conductivity, ambipolar electric field effects, and room temperature quantum hall effects. In addition, functionalization of graphene is possible to produce graphene derivatives such as graphene oxide displaying interesting optical properties such as fluorescence.

Carbon-based quantum dots are the newest addition to the carbon nanomaterial class and can be split into two types, namely, graphene quantum dots (QGDs) and carbon quantum dots (CQDs), although the structure of carbon-based QDs is similar to that of graphene. These less than 10 nm-sized nanoparticles were discovered in 2004 when they were purified from SWCNTs.

Subsequent synthesis methods were developed including chemical and laser ablation, electrochemical carbonization, microwave irradiation, hydrothermal/solvothermal treatment, and pyrolysis [5]. Carbon-based QDs display interesting optical properties such as high luminescence, optical adsorption in the UV region, and high solubility in water as well as low toxicity and good biocompatibility, which could eventually replace metal QDs. Their unique luminescence properties come about due to quantum confinement, which is discussed in more detail later in the chapter.

Nanodiamonds are another emerging class of new carbonaceous nanomaterials that also display interesting optical, mechanical, and chemical properties and have sizes between 2 and 10 nm. They were first discovered in the residue soot left by Soviet nuclear detonation tests in the 1960s [6]. Unlike other classes of carbon nanomaterials, they consist of sp^3-hybridized carbon atoms. They are highly photostable fluorophores, allowing them to be used as labels in bioimaging. Currently the only method used to make nanodiamonds is to extract them from the controlled detonation of explosives.

Carbon nanomaterials are perhaps the most studied type of nanomaterial in medical diagnostics. The fluorescent properties of fullerenes, graphene, carbon dots, CNTs, and nanodiamonds have been of interest in bioimaging due to them displaying photoluminescence in the visible and near-infrared regions of the electromagnetic spectrum [7]. The properties of carbon nanomaterials have resulted in the development of new exciting applications of carbon nanomaterials.

Fullerene cages have been shown to be nontoxic and resistant to body metabolism, which potentially allow their use in medical diagnostics as contrast agents in optical microscopy [8]. Highly soluble fluorescent fullerenes were recently investigated as contrast agents in cancer cell bioimaging, and it was found that by functionalizing the surface with polyhydroxylated fullerene, their high surface charge allowed them to penetrate into breast cancer cells [9]. SWCNTs demonstrate unique optical properties that can be exploited for use as fluorophore labels as contrast agents for deep tissue fluorescence imaging [10, 11]. The use of CQDs as contrast agents in bioimaging was first demonstrated in 2009 where images obtained using fluorescence microscopy demonstrated their binding to *E. coli* ATCC 25922 [12, 13]. The same group went on to further demonstrate their use as contrast agents by injecting aqueous solutions of CQDs into mouse specimens.

Magnetic resonance imaging (MRI) is a diagnostic technique that has allowed doctors to diagnose a large number of different neurological, musculoskeletal, cardiovascular, and gastrointestinal diseases through the visualization of the body at the molecular and cellular levels [14]. The technique uses large magnetic fields and radio frequencies to produce high-resolution images of soft tissues. Carbon nanomaterials that are capable of being functionalized or can encapsulate metal ion species such as fullerenes and MWCNTs have been investigated as potential contrast agents in MRI scanning, X-ray imaging, and radiopharmaceuticals [15].

The use of electron microscopy has also been a growing area of interest in diagnostics, which allows human tissue samples to be visualized. Graphene was recently shown to protect cells that could lead to higher-resolution images [16].

Researchers have extensively studied the use of carbon nanomaterials in a large number of biosensing applications for medical diagnostics, taking advantage of their optical and electrical properties to develop more sensitive and selective POC devices. For example, the electrical conductivity properties of graphene and graphene derivative-based materials have resulted in their use in a large number of different biosensing platforms including electrochemical and optical immunoassays [17]. In particular a number of electrochemical based biosensors based on the use of graphene as conductive surfaces for electrochemical reactions and electrodes have been developed for the detection of pathogens and biomarkers in blood samples [18]. Overall carbon nanomaterials have proven to be one of the most interesting and remarkable nanomaterial in medical diagnostics.

3.2.2 Metallic Nanoparticles

There are many examples of three-dimensional (3D) transition metals and noble metals displaying unique properties down at the nanometer scale. The most commonly investigated types of metal-based nanoparticles include QDs, metal oxides such as iron oxide, and gold and silver nanoparticles. They all display very unique features that can be exploited for medical diagnostic applications. These metal-based nanomaterials can be combined with other materials to produce nanocomposites and conjugated with labels that have led to new multifunctional nanoparticles with more interesting properties and potential uses.

3.2.2.1 Quantum Dots

QDs are semiconductor nanocrystals. These 2–10 nm-sized nanoparticles typically consist of an alloy core such as CdS, coated by a shell layer such as ZnS. The unique optical properties occur when the diameter of the semiconductor nanocrystal is below the electron–hole Bohr radius, which leads to the quantum confinement effect that becomes more dominant as the size of the dots decrease. The photoluminescence comes about as a result of an excited electron relaxing to the ground state followed by binding to the hole and releasing electromagnetic energy with a narrow frequency range in the UV or near-infrared range. This gives rise to large Stokes shifts that are far superior to those observed for organic dyes. In addition QDs have higher quantum yields, better photostabilities, and high molar extinction coefficients when compared with traditional fluorophores. They also display unique electronic properties due to their charge carriers occupying at discrete energy states, analogous to the electrons in a single atom. If these electrons are excited with enough energy, then they will be excited from the valence band to the conduction band. The electron subsequently loses energy and relaxes down to the energy levels near the bottom of the conduction band and the top of their valence bands. As the electron falls back across the bandgap to recombine with the hole, energy is released. QDs are unique in the fact that they can create multiple electron–hole pairs known as multiple exciton generation. They are sensitive to the presence of additional charges on the surface or in the surrounding environment, which alters the QD's adsorption and photoluminescence properties. These additional charges can lead to a quenching effect.

The first reported synthesis of QDs came in 1983 when Rossetti et al. developed a method for making CdS nanoparticles [19]. It was over a decade later that the first repeatable synthesis route for colloidal QDs was reported [20]. Subsequent methods have been developed that include both bottom-up and top-down approaches [21]. The top-down approaches where bulk semiconductor material is reduced in size to 30 nm include molecular beam epitaxy, ion implantation, e-beam lithography, and X-ray lithography [5]. Bottom-up approaches involve either wet-chemical methods, which include microemulsion, sol–gel, competitive reaction chemistry, hot-solution decomposition, microwave-assisted, and electrochemistry, or vapor-phase methods, which include self-assembly-based molecular beam epitaxy, spluttering, liquid metal ion-based methods, and aggregation of gaseous monomers. Current research is looking at developing less toxic cadmium-free QDs such as carbon-based QDs, which is discussed in Section 3.2.1.

QDs have been exploited in immunoassays due to their unique optical properties and induced quenching effects. In 1998, they were first demonstrated as fluorescent probes in biological diagnostics when they were used to stain mouse fibroblasts [22]. Warren et al. further demonstrated the use of QDs in biological detection by conjugating biomolecules that selectively bind to a biomarker of interest acting as a fluorescent probe [23]. Due to the narrow excitation and emission wavelengths, multicolor assays were developed to differentiate different antigens within the same sample, making multiplexed detection using QDs a reality. In addition, a number of modes for detection including Förster resonance energy transfer (FRET), bioluminescence resonance energy transfer (BRET), chemiluminescence resonance energy transfer (CRET), nanosurface energy transfer (NSET), and dipole to metal particle energy transfer (DMPET) have allowed for more interesting biosensor platforms to be developed. QDs can undergo conjugation with bioreceptors such as antibodies, aptamers, enzymes, or proteins to produce sensitive probes that quench the fluorescence signal upon interaction with an analyte of interest, making QDs attractive biosensors.

3.2.2.2 Magnetic Nanoparticles (Fe_2O_3, FeO, and Fe_3O_4)

Magnetic-based nanoparticles such as iron (III) oxide (Fe_2O_3) and magnetite (Fe_3O_4) display superparamagnetic properties and large surface areas that are useful in medical diagnostics. Sizes can vary from 10 nm up to 1000 nm, with size distributions dependent on the synthesis method. There are several methods described for the synthesis of magnetic nanoparticles in the literature. These include synthesis by microemulsions, sol–gel synthesis, sonochemical and hydrothermal reactions, and coprecipitation [24]. Magnetic nanoparticles tend to be unstable in colloidal solutions, resulting in nanoparticle aggregation. Therefore stabilizers need to be added, which minimize the effect and allow additional functionalization of the surface to allow the conjugation of biomolecules. They include binding monomer-based stabilizers such as carboxylates and citrates, coatings such as silica and gold, or polymer-based stabilizers such as polyethylene glycol (PEG), polyvinyl alcohol (PVA), and chitosan.

Research into the use of magnetic nanoparticles in MRI diagnosis has shown great potential due to their high superparamagnetic properties. The use of these nanoparticles in MRI could lead to noninvasive diagnostic tests with higher resolution. These nanoparticles can also be used to label the site of interest through the attachment of bioreceptors such as antibodies, aptamers, enzymes, or proteins, which can then selectively bind to the biomarker or site of interest [25]. In 2008, magnetofluorescent nanoparticles were employed to selectively target plectin-1, a biomarker of pancreatic ductal adenocarcinoma (PDAC). Scientists were able to distinguish PDAC cells from normal cells using both confocal microscopy and MRI [26]. More recently, magnetic nanoscale metal–organic frameworks have been proposed as possible imaging and contrast agents in cancer diagnostics [27].

Magnetic nanoparticles also have huge potential in sample preparation for medical diagnostics as recently demonstrated in nucleic acid analysis, pathogen detection, and solid-phase extraction [28, 29]. Another major application of magnetic nanoparticles is as a means to amplify the signal of electrochemical, optical, piezoelectric, and magnetic-based biosensors [30].

3.2.2.3 Gold Nanoparticles

Gold nanoparticles have also been commonly utilized in diagnostics due to their high surface area, ease of attaching biomolecules, and unique optical properties. For nanoparticles of less than 100 nm, an intense red color is observed; the color turns to yellow as the size increases. These optical properties are due to their interaction with light. Gold nanoparticles oscillate with respect to their metal lattice. This results in the confinement of a surface plasmon in the nanoparticle in a process known as localized surface plasmon resonance (LSPR). After absorption, the surface plasmons decay, resulting in light scattering and heat. This effect is also affected by the shape of gold nanoparticles with gold nanorods showing more pronounced color changes than spheres. The optical properties of gold nanoparticles can change depending on the degree of aggregation of these nanoparticles. This property can be exploited in biosensor development and colorimetric immunoassays. Monodispersed gold nanoparticles of about 10–20 nm size ranges are most commonly synthesized by chemical reduction of hydrogen tetrachloroaurate ($HAuCl_4$) using citrate as the reducing agent [31].

Multifunctional gold nanoparticles are currently being developed in a number of diagnostic applications such as nucleic acid detection and protein detection through the use of various biosensing and imaging platforms [32]. In 1997, researchers used gold nanoparticles to detect polynucleotides using a colorimetric-based assay that displayed a lower limit of 10 fmol for oligonucleotides [33]. In 2005, El-Sayed et al. demonstrated the unique LSPR properties of gold nanoparticles to develop a novel imaging biosensor for the detection of oral epithelial living cancer cells in both *in vitro* and *in vivo* samples [34].

3.2.2.4 Silver Nanoparticles

Silver nanoparticles also show unique properties that are of interest to researchers. Like gold nanoparticles, they display absorption at the LSPR frequency, giving rise to different colored solutions that are dependent on the size of the nanoparticles. As with gold nanoparticles, the optical properties of silver nanoparticles can change depending on the size, shape, and degree of aggregation and shape of the nanoparticle. In addition, silver nanoparticles are unique in the fact that they display antimicrobial properties, which have found use in a range of antibacterial products. They also display interesting electrical properties. A number of different methods have been reported for synthesizing silver nanoparticles, which include top-down approaches such as evaporation–condensation and laser ablation. These approaches give narrow size distributions. Bottom-up approaches such as chemical reduction, microemulsion, UV-initiated photoreduction, and microwave-assisted synthesis have been used with chemical reduction being the most common method [35]. Silver nanoparticles have been used as substrates in surface-enhanced Raman scattering (SERS) sensors in a number of diagnostic applications [36].

3.2.2.5 Nanoshells

The unique optical properties of gold and silver have been exploited in composite-based nanoparticles. Researchers from Rice University discovered nanoshells in 2004 [37]. These spherical particles consist of a dielectric core, usually silica with a layer of either gold or silver surrounding them. The dimensions of these nanoparticles can be precisely altered to produce optical resonance over a broad range of wavelengths. These nanoparticles can be easily synthesized by firstly using the Strober process to form the SiO_2 core of the nanoparticle after which a layer of gold is then formed around the silica [38]. Different forms of nanoshells do exist such as hollow shells where the core has been removed. Nanoshells were first investigated in cancer therapeutics due to their ability to undergo the enhanced permeability and retention (EPR) effect, which means that these nanoshells tend to aggregate in cancer lesion sites. Upon supplying electromagnetic radiation, the nanoshells heat up, causing the tumor cells to be selectively killed off.

Nanoshells have also found use in biosensing applications due to the optical properties and ability of these nanoparticles to aggregate depending on their conditions. For example, in 2008, a label-free biosensor based on a nanoshell self-assembly monolayer allowed for the monitoring of biomolecular interactions in whole blood through near-infrared detection [39]. Other examples of using nanoshells in biosensors include their use in pH sensing and amplifying fluorescence signals [40, 41]. In SERS-based sensors, nanoshells act as substrates for biosensor development. In addition, the use of these nanoshells with SERS could prove to be an effective platform for *in vivo* detection [42, 43].

3.2.2.6 Nanocages

Nanocages are a class of novel nanostructure with noble metal hollow interiors and porous walls, ranging in size from 10 to 150 nm. The optical properties of these cube-shaped nanoparticles differ from those displayed for spherical gold nanoparticles in the fact that they absorb light at the near-infrared region. The degree of LSPR can depend on the amount of silver precursor added to the reaction. Nanocages are synthesized from the noble metal nanoparticles reacting with chloroauric acid in water [44]. Gold nanocages have found various applications in biosensors due to their ability to absorb and scatter near-infrared light. For example, these properties were utilized by Wang et al. in combination with radioluminescence for use *in vivo* imaging [45].

3.2.2.7 Nanowires

Nanowires are typically one-dimensional structures that have lengths exceeding the thickness by more than 1000 times. Their properties largely depend on the type of nanowire prepared. Nanowires consisting of semiconducting materials such as InP, Si, and GaN and dielectric-based materials such as SiO_2 and TiO_2 display unique thermal and electrical conductance properties. Nanowires synthesized from silver and gold display LSPR-like properties that can be tailored depending on the thickness of the nanowire [46]. When aligned in arrays, they are capable of penetrating lipid bilayer of cells acting very much like nanoneedles [47]. This means that they have the potential to be used in cell biosensing and other medical diagnostic applications. There are a number of reported methods for making nanowires based on both top-down and bottom-up approaches, which include metal nanoparticle-mediated methods, direct deposition, and template-directed and template-oriented attachment [48]. Silicon oxide nanowires that are often incorporated into field-effect transistor (FET)-based biosensors due to their conductive properties have shown to be a promising biosensor platform for medical diagnostics [49]. Silicon-based nanowires are used as surface substrate for the bioconjugation of receptors in FET devices. As analytes bind to the receptor on the nanowire, a change in the electron density and hence a signal transduction occur, which allows for both qualitative and quantitative analysis. These silicon nanowire-based FET biosensors have shown promise in the area of medical diagnostics due to their rapid analysis times, sensitive detection, and potential for miniaturization. A silicon nanowire-based sensor for the rapid diagnosis of flu was proposed by Shen et al. in combination with magnetic nanoparticles that took at least two orders of magnitude less time compared with conventional RT-PCR analysis [50]. Researchers recently used a silicon nanowire-based FET biosensor for the detection of dengue virus-specific DNA with femtomolar detection limits [51].

3.2.3 Polymer-Based Nanoparticles

Polymer-based nanoparticles have been of growing interest in areas of drug delivery and medical diagnostics. This in part is due to their inert nature, flexibility in design, and biocompatibility. In addition they are relatively cheap to produce and thermally stable. Molecularly imprinted polymer-based nanoparticles (MIP-NPs) that are cross-linked polymers with the ability to selectively bind to biomolecules have shown potential to be biomimetic receptors in medical diagnostics as an alternative to antibodies. A number of different methods have been demonstrated to manufacture MIP-NPs including precipitation, emulsion polymerization, and living polymerization. These MIP-NPs have been demonstrated as effective receptors in a number of biosensing applications as synthetic antibodies [52].

Dendrimers have also been of interest to researchers in the application of clinical diagnostics to control the properties of clinically relevant biomolecules [53]. These highly branched star-shaped macromolecules were first synthesized in the 1980s [54]. Typically consisting of a central core, an interior consisting of the dendritic structure, and an exterior surface, changes to any one of these three groups can give rise to different properties such as shape, size, and functionality. These nanoparticles are synthesized in stepwise chemical reactions. The core can consist of either a metal nanoparticle, polymer, or the dendrimer itself. The dendritic structure can be added stepwise, which gives rise to distinct layers termed generations (G0, G1, G2, etc.). The well-controlled synthesis of the dendrimers allows for uniform monodisperse particles with low molecular weight ranges to be obtained. Dendrimers can be functionalized with an array of different biomolecules or fluorophores, which have allowed them to be used in a number of biomedical applications such as biosensoring and drug delivery or as contrast agents in MRI [55]. A recent publication demonstrated the use of dendrimers with gold nanodiscs in a nanoplasmonic sensor for diagnosis of allergies to amoxicillin in clinical samples [56].

3.3 Application of Microfluidic Devices in Clinical Diagnostics (Top-Down Approach)

In the top-down approach, microfabrication techniques are used to produce biosensing devices on which the analysis of fluids can occur on a micro- or nanoscale level. These devices have critical operational lengths in the sub-µm to a few hundred µm range and are generally referred to as microfluidic devices. Microfluidic chips are commonly fabricated in silicon, glass, and polymer materials [57]. The first generation of microfluidic devices was made on silicon and glass by cleanroom methods such as photolithography and etching, which originated from the microelectronic industries. Then polymers such as cyclic olefin copolymer, polystyrene, and polydimethylsiloxane (PDMS) quickly emerged as alternative materials due to simple fabrication process (e.g., molding, embossing, and printing), wide range of surface properties, and low cost. New materials such as paper and cotton thread have recently been introduced to produce simple and low-cost microfluidic platforms. Patterning paper or cotton thread with hydrophobic and hydrophilic areas by wax printing or oxygen plasma provides a convenient way for fluid manipulation.

3.3.1 Unique Features of Microfluidic Devices

Microfluidic biosensing devices are extremely attractive not only because they can scale down macroscopic processes to microscale but also because the scaling effects lead to new effects and phenomena that permit entirely new applications not accessible to classical liquid handling platforms [58]. The unique properties of microfluidic devices are listed as follows:

- **Well-controlled laminar flow.** In ultralow dimensions of micrometers, fluid flow is completely laminar and can be precisely controlled by adjusting the flow rate. This distinct property gives rise to efficient and accurate mass delivery in controlled time and space. For instance, diffusion between adjacent liquid streams is highly predictable and can be exploited to create well-defined concentration gradients [59].
- **Dominance of surface force.** Another characteristic of microfluidic devices is the high surface area-to-volume ratio (S/V). At the micro-/nanoscale, surface tension becomes dominant, while inertial and body forces are greatly reduced [60]. This feature can be used for a variety of tasks in microfluidic architecture, such as using capillary forces to passively actuate liquid in a narrow channel or porous material or generating monodisperse droplets in multiphase fluid streams.
- **Miniaturization and parallelization of reactions.** Microfluidic devices enable the miniaturization of reactions by compartmentalizing reactions in nano-, pico-, or even femtoliter volumes [61]. Many biological and chemical

applications are boosted by miniaturization, which both minimizes reagent costs and opens up entirely new experimental approaches such as single-cell analysis. Furthermore, miniaturization enables highly parallelized experiments, thus drastically increasing the throughput.

- **Integration of functional units.** With the efficient development of fabrication and interfacing technology, a set of functional units can be integrated in microfluidic platforms to form the so called lab-on-a-chip (LOC) devices or micro-total analysis systems (μTAS; [62]). The incorporation of interconnected fluidic microchannel networks, reaction chambers, and micropumps/valves enables a number of operations such as fluidic transport, mixing, valving, separation, amplification, and so on to be performed on a single chip. The functional integration minimizes human intervention and paves a way for automation of biochemical processes [63].

3.3.2 Applications of Microfluidic Devices in Medical Diagnostics

The microfluidic devices provide numerous advantages over conventional analytical instruments, such as reduced sample volumes, low reagent consumption, decreased processing time, low-cost analysis, and high portability [64]. In recent years, microfluidic technology has found an important niche in *in vitro* diagnostics (IVD), especially in POC diagnostics [65]. Microfluidic systems can be designed to obtain measurements from small volumes of complex fluids with efficiency and speed and without the need for an expert operator. This unique set of capabilities is precisely what is needed to create portable POC medical diagnostic systems [66]. The microfluidic POC devices are most advantageous in the following settings:

- At acute care settings such as in the emergency department, where clinical information is urgently needed since time is critical
- In clinical settings such as a physician's office or retail clinics, where prompt diagnosis is necessary for doctors to prescribe the appropriate treatment very quickly
- At resource-limited settings such as at home, remote locations, or some developing countries, where conventional laboratory facilities are unavailable or impractical

Numerous microfluidic devices have been developed and commercialized for infectious diseases, cardiac markers, diabetes, lipids, coagulation, and hematology tests [67]. Advancements have also been made to make testing simple enough to be correctly performed by moderately trained or non-trained staff. Many tests that previously required a laboratory for testing can now be accurately performed at the point of care. Microfluidics is transforming the health-care landscape by decentralizing laboratory testing and making health care more patient centered [68].

3.3.2.1 Types of Microfluidic POC Devices

Although all the microfluidic POC devices are focusing on providing simple and rapid testing, requirements of these systems may deviate significantly in terms of cost, power consumption, and portability. Therefore appropriate technologies must be developed for different applications, end users, and settings [69]. For instance, reliable electricity available in developed countries may not be present in developing countries, so the devices for use in these areas should exploit passive or low-power approaches rather than active methods for fluid manipulation. Additionally, the microfluidic devices used for glucose monitoring need to provide quantitative data, while a "yes/no" answer is sufficient when used to screen bacterial infection. Currently, microfluidic POC devices can be broadly classified into three categories: benchtop instruments with sophisticated built-in fluidics that are often tailored to hospitals and other clinical testing sites; small, lightweight devices with small cartridges or strips; and simple un-instrumented paper-based devices. The latter two are best for testing in remote and resource-limited environments.

3.3.2.2 Benchtop Microfluidic Instruments

Many clinical tests are complex and involve multiple analytical steps. Though the trend of laboratory automation has led to several automated or semiautomated instruments, these stand-alone analyzers typically can only perform specific tasks. Manual efforts are still needed to prepare the samples and transfer the samples after each step. The benchtop microfluidic instruments are designed to incorporate all the processes associated with analysis, from the pretreatment to the analysis itself. A popular format of such an integrated system comprises a disposable microfluidic cassette with reaction chambers and interconnecting channels as well as required reagents and a benchtop analyzer for fluid control, heating, detection, and data processing [70]. Operators simply load patient samples into the cassettes and insert the cassettes into the analyzers, and results are printed or sent to doctors' computers. Fluid is automatically moved through various stages with minimal concerns over sample loss and cross-contamination, and no user intervention is required during the whole process. These microfluidic instruments eliminate the need for multiple pieces of existing equipment, helping to make the testing process quicker, more efficient, and less likely to result in human error. But considering the cost and size of the analyzer and cartridges, these systems are not really portable. They may be best suited for high- or moderate-infrastructure settings (e.g., core laboratories, satellite hospital laboratories, or physician office laboratories). The following are some examples of benchtop microfluidic POC instruments developed for immunoassays and nucleic acid-based assays:

Immunoassay: Immunoassays are frequently used clinical tests for detection and quantification of analytes in biological liquids such as serum or urine.

However, conventional immunoassays such as ELISA are extremely labor intensive, time consuming, and prone to human error. To improve the efficacy and accuracy of immunoassay in clinical settings, robust and standardized platforms are needed [71]. A couple of microfluidic devices have been demonstrated that possess the ability to run the multistep protocols in an automatic way. For example, ProteinSimple has developed a fully automated, hands-free immunoassay platform [72]. The system includes a self-contained disposable microfluidic cartridge integrated with a simple desktop analyzer. Multiple reagents are stored in the assay cartridge and delivered in a controlled manner to run multistep protocols; hence it eliminates any manual washes or tedious reagent additions. A key feature of the cartridge is that each sample is analyzed within a unique microfluidic circuit consisting of four channels. Since each channel containing an immunoassay for a specific analyte, zero cross-reactivity is ensured. In addition, fluorescence detection in a glass nanoreactor gives high sensitivity of picogram per milliliter and a 4–5 log dynamic range. A single cartridge enables the simultaneous quantification of four analytes from 16 individual samples in an hour. More than 20 cytokines and angiogenic markers designed for oncology and ophthalmology have been validated for clinical use. This excellent example shows that the microfluidic approach not only offers unparalleled ease of use but also contributes to enhanced performance. These microfluidic platforms will make immunoassay a more powerful technique in clinical diagnostics.

Nucleic acid assay: With the rapid evolution of molecular biology, there is increasing clinical demand for detection of DNA and RNA signatures for diagnosis and monitoring of patients. Nucleic acid test is one of the most complicated clinical assays due to multiple steps required for sample pretreatment (e.g., cell sorting, isolation, and lysis, as well as nucleic acid extraction), nucleic acid amplification (e.g., polymerase chain reaction (PCR)), and target detection (e.g., electrophoresis). Using conventional methods, the whole process takes 1–2 days. The scenario is likely to change by developing microfluidic-based nucleic acid assays [64]. An early pioneer of integrated molecular diagnostics is Cepheid. The company launched GeneXpert, which simplifies molecular testing by fully integrating and automating the three processes (sample preparation, amplification, and detection) in one platform [73]. The system uses a cartridge containing all elements necessary for the reaction, including lyophilized reagents, liquid buffers, and wash solutions. Once sample mixture is loaded, fluids are automatically driven by pneumatic actuation, and a rotary valve controls the fluid movement among multiple reagent chambers and PCR tube. The analyzer performs ultrasonic lysis of filter-captured organisms and then mixes DNA molecules with onboard PCR reagents. Target detection and characterization is performed in real time using a six-color laser detection device. The integrated benchtop analyzer provides results from unprocessed sputum samples in less than 2 hours and has been certified for detection of MRSA, *C. difficile*, influenza, and tuberculosis. Since such microfluidics-based analyzers allow for rapid accurate diagnosis and reduced cross-contamination risk, they will greatly help to promote routine and widespread use of nucleic acid-based assays.

3.3.2.3 Small, Lightweight Microfluidic Devices

Compared with the benchtop instruments, the lightweight microfluidic devices have much reduced size and complexity. Instead of complicated micropumps/valves, they often exploit plungers or low-power pumping techniques to reduce power consumption [74]. With the advances in microfabrication technology, the disposables are usually in the form of plastic cartridges fabricated by thin-film lamination or injection molding or strips fabricated by screening printing [70]. These systems are in the size of a shoe box and can be operated with a battery, which makes them truly portable. Many of them have obtained Clinical Laboratory Improvement Amendment (CLIA) waiver, which allows them to be used by personnel without laboratory training. One drawback of these small microfluidic devices is that they are generally less analytically sensitive than their laboratory counterparts due to the miniaturized hardware and lower computer processing power. However, this is not likely to be a big concern when considering the clinical application and setting where the test is to be deployed. For example, POC glucose tests are not as sensitive as laboratory-based instrumentation, but they are still suitable for home management of stable diabetic patients. Hundreds of tests, such as blood chemistries, immunoassays, nucleic acid tests, and flow cytometry, which were considered too complex to be done outside the laboratory, are now routinely performed on portable microfluidic devices. Their widespread applications range from critical care (e.g., test of cardiac markers, blood cell count in emergency ward) to primary care (e.g., detection of infectious diseases, tumor markers at the doctor's office) to home care (e.g., monitoring of glucose in patient's home) [75]. Two examples of small microfluidic POC devices for blood chemistry and hematology are highlighted:

Blood chemistry: Analysis of blood chemistry can provide important information about the function of the kidneys and other organs. The conventional laboratory test usually draws a few milliliters of blood from the vein and takes at least 2 h to get results. The Abaxis Piccolo Xpress is a portable clinical chemistry system designed for on-site patient testing [76]. The system uses a disc-based approach to analyze blood chemistry. The injection-molded plastic discs contain an aqueous diluent in the center and dry reagents beads in cuvettes around the disc periphery. It requires only a few drops of blood from fingerstick to perform the test. Plasma separation, mixing, and volumetric measurements are driven by centrifugal and capillary forces on the disc. Care providers simply pipette the sample into the disc and insert the disc into the analyzer, and results come out in 12 min. The Piccolo® features 31 blood chemistry tests that range from liver, kidney, and metabolic functions to lipids, electrolytes, and other specialty analytes. These 31 tests are conveniently configured into 16 completely self-contained reagent discs, 11 of which are CLIA waived. Using such portable blood analyzers, care providers can diagnose, monitor, and treat their patients more efficiently than through traditional reference laboratory means.

Hematology: Blood test, particularly complete blood count, ranks among the most ordered tests in medical diagnostics. Currently, flow cytometer has dominated the arena of hematology analyzers. Besides being expensive and cumbersome, the high blood volume requirements and costly reagents necessary for sample pretreatment confer several more disadvantages onto these machines. The miniaturization of the traditional flow cytometer to the microfluidic level presents exciting future possibilities for chip-size hematology analyzers [77]. Among the leaders toward disposable microflow hematology is the Chempaq series of blood analyzers. This device operates via small disposable microfluidic cartridges containing several chambers for dilution, separation, and enumeration of blood components. Only one drop of blood (from fingerstick) is needed for analysis. The counting is based on the Coulter principle that sizes and counts suspended particles by measuring their electrical resistance. After the blood-loaded cartridge is inserted into the analyzer, the results of the complete blood count are displayed within 3 min. The Chempaq line of blood analyzers is measured just over a square foot in footprint, whereas the precision and reliability are comparable with their laboratory-confined precursors. Such microfluidic blood analyzers have been used for emergency prehospital care and patient monitoring in hospitals and nursing facilities. They may soon replace traditional flow cytometers as the next "gold standard" in clinical hematology.

3.3.2.4 Simple Un-instrumented Microfluidic Systems

During the past decade, near-patient testing using POC devices has become well established in developed countries. The interest in moving to a more patient-centered approach is also rising in the developing world. However, the cost of the microfluidic devices must be kept extremely low if they are to be applicable to the developing countries. Considering the relatively high cost related to instrument and disposability, most of the aforementioned compact microfluidic devices are not likely to be successful in the impoverished settings [78]. To fulfill the requirements of POC diagnostics in these areas, simple un-instrumented microfluidic systems have been developed. They often appear as simple kits with no fixed instrument, or sometimes small readers are equipped to quantify the signal [79]. The fluid is manipulated passively, typically by capillary force, and no external processor is needed. The devices rely on inexpensive materials and manufacturing processes, as well as affordable off-the-shelf components and reagents. They are meant for use by non-trained personnel in settings where electricity, refrigeration, and other resources might not be readily available. These devices can be formatted for detection of antigens, antibodies, or nucleic acids in a wide range of specimens. Many rapid and cost-effective tests have been developed to address the major challenges of global public health, such as HIV, tuberculosis, malaria, diarrheal diseases, and lower respiratory infections.

Lateral Flow Strips (LFA) LFA-based POC devices are among the most rapidly growing strategies for qualitative and semiquantitative analysis [80]. There are a number of variations of the technology that have been developed into commercial products, but they all operate using the same basic concept. LFA is basically performed over a strip that consists of sample pad (inlet and filtering), conjugate pad (reactive agents and detection molecules), nitrocellulose membrane (analytes detection), and adsorption pad (liquid actuation). The liquids are driven by the capillary force, and the movements are controlled by wettability and feature size of the porous material (mainly nitrocellulose). Pre-immobilized reagents at different parts of the strip become active upon flow of the liquid sample. The sample is placed into a sample well and migrates across the zone where the antigen or antibody is immobilized. The results are read after a certain amount of time has passed, and the readouts are quite often implemented as a color change in the detection area that can be seen by the naked eye. Lateral flow tests are the simplest type of POC devices, requiring only very minimal familiarity with the test and no equipment to perform, since all of the reactants and detectors are included in the test strip.

One good application of lateral flow tests is the development of malaria rapid diagnostic tests (RDTs) that assist in the diagnosis of malaria by detecting evidence of malaria parasites (antigens) in the human blood [81]. Malaria RDTs detect specific antigens (proteins) produced by malaria parasites in the blood of infected individuals. Some RDTs can detect only one species (*Plasmodium falciparum*), while others detect multiple species (*P. vivax*, *P. malariae*, and *P. ovale*). They permit a reliable detection of malaria infections within 20 min, which is particularly suitable for use in remote areas with limited access to good-quality microscopy services.

Paper-Based Analytical Devices (PADs) Lateral flow strips have received enormous attention for POC applications since they are low cost, rapid, simple to use, equipment-free, and easily mass-produced. However, conventional lateral flow tests are still limited in several important aspects, including low sensitivity, inability to manipulate fluid flow, and inability to detect multiple targets per strip. Recently, paper-based microfluidics that can enable fluid handling and quantitative analysis has emerged as a multiplexed POC platform [82].

Though it is still in its early development stages, paper-based microfluidics might transcend the capabilities of existing assays in resource-poor settings.

Analogous to traditional microfluidic devices that are fabricated by etching or molding channels into glass, silicone, or plastics, microfluidic channels can also be created on paper by patterning sheets of paper into hydrophilic channels bounded by hydrophobic barriers [83]. The patterning process defines the width and length of the channels, while the thickness of the paper defines the height. The hydrophilic cellulose fibers of paper enable aqueous fluids to wick along the channels. The fluidic control can be automated in porous network devices with proper network topology, channel geometry for changing flow rates, on/off switches for flow, or delays for flow. The invention of 2D PADs made it possible to carry out separations and simultaneously detect multiple analytes using a single sample reservoir. The subsequent development of 3D PADs led to even more sophisticated operations. The 3D fluidic network developed by layered construction of wax-patterned papers provides different fluidic paths for the sequential delivery of multiple fluids without the need for peripheral equipment. The ability to incorporate multistep processes such as rinsing and signal amplification steps makes the PADs much more sensitive than conventional lateral flow tests.

A pioneer in paper-based microfluidic device is the nonprofit company Diagnostics for All. They use patterned papers to develop instrument-free tests for targeting diseases of greatest concern in the developing world. Their initial prototypes have been designed to assess liver damage from HIV medication (e.g., albumin, transaminases, and lactate dehydrogenase) [84]. The low-cost paper network platforms have shown great potential for high-quality diagnostics in low-resource settings in the developed and developing worlds.

3.4 Integration of Microfluidics with Nanomaterials

The use of microfluidic devices and nanomaterials in medical diagnostics has seen huge benefits, but their use is not mutually exclusive from one another. The combination of the two types of technology has seen improvements in medical diagnostic devices for labeled and non-labeled diagnostic tests. Microfluidic-based chips still suffer from significant problems such as issues in sensitivity, selectivity, and low signal-to-noise (SN) ratios. Nanomaterials can overcome these problems by utilizing their unique properties to enhance the S/N ratios, provide sample pretreatment, and act as substrates for detection [85]. They have shown to improve signal enhancement in a number of paper-based biosensors [86]. For instance, QDs were used in a paper-based lateral flow device to detect nitrated ceruloplasmin, a significant biomarker for cardiovascular disease [87]. Other types of microfluidic device have also benefited from the use of nanomaterials. Recently Chin et al. demonstrated the use of an

ELISA-based microfluidic device coupled with antibody-labeled gold nano-particles that could detect HIV and syphilis using just 1 µL of blood [88]. Researchers at the Center for Systems Biology in Massachusetts demonstrated the combined use of magnetic nanoparticles with a miniaturized NMR device to allow for the simultaneous detection of a number of bacterial species to the single bacterial detection limit [89].

3.5 Future Perspectives of Nanomaterial and Microfluidic-Based Diagnostics

Nanomaterials have proven to be a versatile tool in the development of medical diagnostics. Researchers have continued to exploit their unique properties to produce more sensitive selective, robust, and cheaper diagnostic tests. However, key challenges need to be addressed in order to see their widespread use in medical diagnostics. These challenges include reducing their toxicity and developing synthetic methods that can result in uniform particle sizes and narrow particle size distributions. The nanomaterials market is currently worth $5.5 billion, and the increase in research into their use within the area of medical diagnostics will only serve to increase their current market value [90]. The current scope for nanomaterials is huge in terms of exploiting their unique properties. As nanomaterials are still a relatively new science, there are likely to be more interesting nanomaterials with desirable properties for medical diag-nostic applications, yet to be discovered.

Microfluidic diagnostics has had explosive growth in the last 20 years. According to Lux Research, the overall health-care market for microfluidics will swell to nearly $4 billion by 2020, growing at a compound annual growth rate (CAGR) of 13% [91]. Microfluidic POC instrumentation has been applied to several of the commonly centralized laboratory techniques, such as blood chemistries, immunoassays, molecular biology testing, and cell analysis [92]. The most attractive feature of microfluidics is the ability to rapidly and accu-rately deliver relevant clinical information in a POC situation [93]. Driven by the need to deliver patient-centered health care, the menus of microfluidic POC devices, both un-instrumented and instrumented, will continue to evolve and expand.

From the technological point of view, more efforts are needed in the minia-turization of the peripheral components, integration of sample preparation, and preprocessing steps, as well as developing new technologies for improved analytical performance. Overall the combination of nanomaterials and micro-fluidic devices has proven to be effective in making the ideal POC diagnostic devices a reality. In addition to the current applications such as glucose and metabolic monitoring, infectious disease testing, and cardiac and tumor marker detection, microfluidic systems in the future will also play a big role in the prevention and early detection of diseases (e.g., by detection of mutations, screening microRNA patterns, or analyzing circulating tumor cells), as well as management of multiple chronic conditions (e.g., monitoring chronic respiratory conditions at home). We strongly believe that POC diagnostic systems will eventually revolutionize the practice of medical diagnosis and dramatically reduce health-care costs.

References

1 Kroto, H. W.; Heath, J. R.; O'Brien, S. C.; Curl, R. F.; Smalley, R. E. *Nature* **1985**, 318 (6042), 162.

2 Prasek, J.; Drbohlavova, J.; Chomoucka, J.; Hubalek, J.; Jasek, O.; Adam, V.; Kizek, R. *J. Mater. Chem.* **2011**, 21 (40), 15872.

3 Geim, A. K.; Novoselov, K. S. *Nat. Mater.* **2007**, 6 (3), 183–191.

4 Avouris, P.; Dimitrakopoulos, C. *Mater. Today* **2012**, 15, 86–97.

5 Wang, Y.; Hu, A. *J. Mater. Chem. C* **2014**, 2, 6921–6939.

6 Mochalin, V. N.; Shenderova, O.; Ho, D.; Gogotsi, Y. *Nat. Nanotechnol.* **2012**, 7 (1), 11–23.

7 Hong, G.; Diao, S.; Antaris, A. L.; Dai, H. *Chem. Rev.* **2015**, 115 (19), 10816–10906.

8 Chen, H. H.; Yu, C.; Ueng, T. H.; Chen, S.; Chen, B. J.; Huang, K. J.; Chiang, L. Y. *Toxicol. Pathol.* **1998**, 26 (1), 143–151.

9 Xie, R.; Wang, Z.; Yu, H.; Fan, Z.; Yuan, F.; Li, Y.; Li, X.; Fan, L.; Fan, H. *Electrochim. Acta* **2016**, 201, 220–227.

10 Hong, G.; Lee, J. C.; Robinson, J. T.; Raaz, U.; Xie, L.; Huang, N. F.; Cooke, J. P.; Dai, H. *Nat. Med.* **2012**, 18 (12), 1841–1846.

11 Welsher, K.; Sherlock, S. P.; Dai, H. *Proc. Natl. Acad. Sci. U. S. A.* **2011**, 108 (22), 8943–8948.

12 Yang, S. T.; Cao, L.; Luo, P. G.; Lu, F.; Wang, X.; Wang, H.; Meziani, M. J.; Liu, Y.; Qi, G.; Sun, Y. P. *J. Am. Chem. Soc.* **2009**, 131 (32), 11308–11309.

13 Yang, S. T.; Wang, X.; Wang, H.; Lu, F.; Luo, P. G.; Cao, L.; Meziani, M. J.; Liu, J. H.; Liu, Y.; Chen, M.; Huang, Y.; Sun, Y. P. *J. Phys. Chem. C* **2009**, 113 (42), 18110–18114.

14 Estelrich, J.; Sánchez-Martín, M. J.; Busquets, M. A. *Int. J. Nanomed.* **2015**, 10, 1727–1741.

15 Bakry, R.; Vallant, R. M.; Najam-ul-Haq, M.; Rainer, M.; Szabo, Z.; Huck, C. W.; Bonn, G. K. *Int. J. Nanomed.* **2007**, 2 (4), 639–649.

16 Wojcik, M.; Hauser, M.; Li, W.; Moon, S.; Xu, K. *Nat. Commun.* **2015**, 6, 7384.

17 Yang, Y.; Asiri, A. M.; Tang, Z.; Du, D.; Lin, Y. *Mater. Today* **2013**, 16 (10), 365–373.

18 Shao, Y.; Wang, J.; Wu, H.; Liu, J.; Aksay, I. A.; Lin, Y. *Electroanalysis* **2010**, 22 (10), 1027–1036.

19 Rossetti, R.; Nakahara, S.; Brus, L. *J. Chem. Phys.* **1983**, 79 (May 2016), 1086–1088.

20 Murray, C. B.; Norris, D.; Bawendi, M. G. *J. Am. Chem. Soc.* **1993**, 115 (4), 8706–8715.

21 Pisanic, T. R.; Zhang, Y.; Wang, T. H. *Analyst* **2014**, 139 (12), 2968–2981.

22 Bruchez Jr., M.; Moronne, M.; Gin, P.; Weiss, S.; Alivisatos, A. P. *Science* **1998**, 281 (5385), 2013–2016.

23 Chan, W. C. W.; Nie, S. *Science* **1998**, 281 (5385), 2016–2018.

24 Laurent, S.; Forge, D.; Port, M.; Roch, A.; Robic, C.; Vander Elst, L.; Muller, R. N. *Chem. Rev.* **2008**, 108 (6), 2064–2110.

25 Na, H. B.; Song, I. C.; Hyeon, T. *Adv. Mater.* **2009**, 21 (21), 2133–2148.

26 Kelly, K. A.; Bardeesy, N.; Anbazhagan, R.; Gurumurthy, S.; Berger, J.; Alencar, H.; DePinho, R. A.; Mahmood, U.; Weisslcder, R. *PLoS Med.* **2008**, 5 (4), 0657–0668.

27 Ray Chowdhuri, A.; Bhattacharya, D.; Sahu, S. K. *Dalton Trans.* **2016**, 45 (7), 2963–2973.

28 Sun, H.; Zeng, X.; Liu, M.; Elingarami, S.; Li, G.; Shen, B.; He, N. *J. Nanosci. Nanotechnol.* **2012**, 12 (1), 267–273.

29 Shinde, S. B.; Fernandes, C. B.; Patravale, V. B. *J. Control. Release* **2012**, 164–180.

30 Rocha-Santos, T. A. P. *TrAC Trends Anal. Chem.* **2014**, 62, 28–36.

31 Turkevich, J.; Cooper, P. H. J. *Discuss. Faraday Soc.* **1951**, 55 (c), 55–75.

32 Mieszawska, A. J.; Mulder, W. J. M.; Fayad, Z. A.; Cormode, D. P. *Mol. Pharm.* **2013**, 10 (3), 831–847.

33 Elghanian, R.; Storhoff, J. J.; Mucic, R. C.; Letsinger, R. L.; Mirkin, C. A.; Razin, S.; Hacia, J. G.; Mansfield, E. S.; Wang, J.; Kreibig, U.; Genzel, L.; Dusemund, B.; Brust, M.; Grabar, K. C.; Yang, W.-H.; Schatz, G. C.; Duyne, R. P. Van; Grabar, K. C.; Mirkin, C. A.; Weisbecker, C. S.; Gryaznov, S. M.; Letsinger, R. L.; Urdea, M. S. *Science* **1997**, 277 (5329), 1078–1081.

34 El-Sayed, I. H.; Huang, X.; El-Sayed, M. A. *Nano Lett.* **2005**, 5 (5), 829–834.

35 Iravani, S.; Korbekandi, H.; Mirmohammadi, S. V.; Zolfaghari, B. *Res. Pharm. Sci.* **2014**, 9 (6), 385–406.

36 Vo-Dinh, T.; Yan, F.; Wabuyele, M. B. *J. Raman Spectrosc.* **2005**, 36 (6–7), 640–647.

37 Loo, C.; Lin, A.; Hirsch, L.; Lee, M.-H.; Barton, J.; Halas, N.; West, J.; Drezek, R. *Technol. Cancer Res. Treat.* **2004**, 3 (1), 33–40.

38 Duff, D. G.; Baiker, A.; Edwards, P. P. *Langmuir* **1993**, 9 (96), 2301–2309.

39 Wang, Y.; Qian, W.; Tan, Y.; Ding, S. *Biosens. Bioelectron.* **2008**, 23 (7), 1166–1170.

40 Wei, H.; Willner, M. R.; Marr, L. C.; Vikesland, P. J. *Analyst* **2016**, 141 (17), 5159–5169.

41 Ayala-Orozco, C.; Liu, J. G.; Knight, M. W.; Wang, Y.; Day, J. K.; Nordlander, P.; Halas, N. J. *Nano Lett.* **2014**, 14 (5), 2926–2933.

42 Ochsenkühn, M. A.; Campbell, C. J. In *Raman Spectroscopy for Nanomaterials Characterization*; Springer: Berlin/Heidelberg, 2012; pp. 51–74.

43 Henry, A.-I.; Sharma, B.; Cardinal, M. F.; Kurouski, D.; Van Duyne, R. P. *Anal. Chem.* 2016, 88 (13), 6638–6647.

44 Skrabalak, S. E.; Chen, J.; Sun, Y.; Lu, X.; Au, L.; Cobley, C. M.; Xia, Y. *Acc. Chem. Res.* 2008, 41 (12), 1587–1595.

45 Wang, Y.; Liu, Y.; Luehmann, H.; Xia, X.; Wan, D.; Cutler, C.; Xia, Y. *Nano Lett.* 2013, 13 (2), 581–585.

46 Hua, J.; Wu, F.; Fan, F.; Wang, W.; Xu, Z.; Li, F. *J. Phys. Condens. Matter* 2016, 28 (25), 254005.

47 Berthing, T.; Sørensen, C. B.; Nygård, J.; Martinez, K. L. *J. Nanoneurosci.* 2009, 1 (1), 3–9.

48 Dasgupta, N. P.; Sun, J.; Liu, C.; Brittman, S.; Andrews, S. C.; Lim, J.; Gao, H.; Yan, R.; Yang, P. *Adv. Mater.* 2014, 26 (14), 2137–2183.

49 Zhang, G. J.; Ning, Y. *Anal. Chim. Acta* 2012, 749, 1–15.

50 Shen, F.; Wang, J.; Xu, Z.; Wu, Y.; Chen, Q.; Li, X.; Jie, X.; Li, L.; Yao, M.; Guo, X.; Zhu, T. *Nano Lett.* 2012, 12 (7), 3722–3730.

51 Nuzaihan, M. M. N.; Hashim, U.; Md Arshad, M. K.; Kasjoo, S. R.; Rahman, S. F. A.; Ruslinda, A. R.; Fathil, M. F. M.; Adzhri, R.; Shahimin, M. M. *Biosens. Bioelectron.* 2016, 83, 106–114.

52 Bedwell, T. S.; Whitcombe, M. J. *Anal. Bioanal. Chem.* 2016, 408, 1735–1751.

53 Singh, P. *Biotechnol. Appl. Biochem.* 2007, 48 (Pt 1), 1–9.

54 Tomalia, D. A.; Naylor, A. M.; Goddard, W. A. *Angew. Chem. Int. Ed. Engl.* 1990, 29 (2), 138–175.

55 Abbasi, E.; Aval, S. F.; Akbarzadeh, A.; Milani, M.; Nasrabadi, H. T.; Joo, S. W.; Hanifehpour, Y.; Nejati-Koshki, K.; Pashaei-Asl, R. *Nanoscale Res. Lett.* 2014, 9 (1), 247.

56 Soler, M.; Mesa-Antunez, P.; Estevez, M.-C.; Ruiz-Sanchez, A. J.; Otte, M. A.; Sepulveda, B.; Collado, D.; Mayorga, C.; Torres, M. J.; Perez-Inestrosa, E.; Lechuga, L. M. *Biosens. Bioelectron.* 2015, 66, 115–123.

57 Prakash, S.; Pinti, M.; Bhushan, B. *Philos. Trans. R. Soc. A Math. Phys. Eng. Sci.* 2012, 370 (1967), 2269–2303.

58 Sackmann, E. K.; Fulton, A. L.; Beebe, D. J. *Nature* 2014, 507 (7491), 181–189.

59 Toh, A. G. G.; Wang, Z. P.; Yang, C.; Nguyen, N. T. *Microfluid. Nanofluid.* 2014, 16 (1–2), 1–18.

60 Pennathur, S.; Meinhart, C. D.; Soh, H. T. *Lab Chip* 2008, 8 (1), 20–22.

61 Vyawahare, S.; Griffiths, A. D.; Merten, C. A. *Chem. Biol.* 2010, 17 (10), 1052–1065.

62 Sin, M. L.; Gao, J.; Liao, J. C.; Wong, P. K. *J. Biol. Eng.* 2011, 5 (1), 6.

63 MacPherson, M.; Ravichandrian, M. *Univ. West. Ont. Med. J.* 2011, 80 (1), 24–26.

64 Kulinsky, L.; Noroozi, Z.; Madou, M. *Methods Mol. Biol.* 2013, 949 (1), 3–23.

65 Kumar, S.; Kumar, S.; Ali, M. A.; Anand, P.; Agrawal, V. V.; John, R.; Maji, S.; Malhotra, B. D. *Biotechnol. J.* 2013, 8 (11), 1267–1279.

66 Yager, P.; Edwards, T.; Fu, E.; Helton, K.; Nelson, K.; Tam, M. R.; Weigl, B. H. *Nature* **2006**, 442 (7101), 412–418.

67 Chin, C. D.; Linder, V.; Sia, S. K. *Lab Chip* **2012**, 12 (12), 2118–2134.

68 Wilding, P.; Verpoorte, S.; Allen Northrup, M.; Yager, P.; Quake, S.; Landers, J. *Clin. Chem.* **2010**, 56 (4), 508–514.

69 Mark, D.; Haeberle, S.; Roth, G.; von Stetten, F.; Zengerle, R. *Chem. Soc. Rev.* **2010**, 39 (3), 1153–1182.

70 St John, A.; Price, C. P. *Clin. Biochem. Rev.* **2014**, 35 (3), 155–167.

71 Han, K. N.; Li, C. A.; Seong, G. H. *Annu. Rev. Anal. Chem. (Palo Alto. Calif.)* **2013**, 6, 119–141.

72 ProteinSimple, "Ella, Simple Plex," **2016**. [Online]. Available at http://www.proteinsimple.com/ (accessed on June 24, 2017).

73 Jenkins, G.; Mansfield, C.D. In *Microfluidic Diagnostics: Methods and Protocols*; Springer Science-Business Media, LLC: New York, **2013**.

74 Spencer, D. H.; Sellenriek, P.; Burnham, C. A. D. *Am. J. Clin. Pathol.* **2011**, 136 (5), 690–694.

75 Abhari, F.; Jaafar, H.; Md Yunus, N. A. *Int. J. Electrochem. Sci.* **2012**, 7, 9765–9780.

76 Jung, W.; Han, J.; Choi, J. W.; Ahn, C. H. *Microelectron. Eng.* **2014**, 132, 46–57.

77 Abaxis, "Piccolo Xpress," **2016**. [Online]. Available at http://www.piccoloxpress.com/ (accessed on June 24, 2017).

78 Heikali, D.; Di Carlo, D. *J. Assoc. Lab. Autom.* **2010**, 15 (4), 319–328.

79 Mohammed, M. I.; Haswell, S.; Gibson, I. *Procedia Technol.* **2015**, 20 (July), 54–59.

80 Rozand, C. *Eur. J. Clin. Microbiol. Infect. Dis.* **2014**, 33 (2), 147–156.

81 Sajid, M.; Kawde, A. N.; Daud, M. *J. Saudi Chem. Soc.* **2015**, 19 (6), 689–705.

82 Maltha, J.; Gillet, P.; Jacobs, J. *Clin. Microbiol. Infect.* **2013**, 19 (5), 399–407.

83 Martinez, A. W.; Phillips, S. T.; Whitesides, G. M.; Carrilho, E. *Anal. Chem.* **2010**, 82 (1), 3–10.

84 Cunningham, J. C.; Degregory, P. R.; Crooks, R. M. *Annu. Rec. Anal. Chem.* **2016**, 9 (1), 183–202.

85 Pollock, N. R.; Rolland, J. P.; Kumar, S.; Beattie, P. D.; Jain, S.; Noubary, F.; Wong, V. L.; Pohlmann, R. A.; Ryan, U. S.; Whitesides, G. M. *Sci. Transl. Med.* **2012**, 4 (152), 152ra129.

86 Liang, W.; Lin, H.; Chen, J.; Chen, C. *Microsyst. Technol.* **2016**, 22 (10), 2363–2370.

87 Ge, X.; Asiri, A. M.; Du, D.; Wen, W.; Wang, S.; Lin, Y. *TrAC Trends Anal. Chem.* **2014**, 58, 31–39.

88 Li, Z.; Wang, Y.; Wang, J.; Tang, Z.; Pounds, J. G.; Lin, Y. *Anal. Chem.* **2010**, 82 (16), 7008–7014.

89 Chin, C. D.; Laksanasopin, T.; Cheung, Y. K.; Steinmiller, D.; Linder, V.; Parsa, H.; Wang, J.; Moore, H.; Rouse, R.; Umviligihozo, G.; Karita, E.; Mwambarangwe, L.; Braunstein, S. L.; van de Wijgert, J.; Sahabo, R.; Justman, J. E.; El-Sadr, W.; Sia, S. K. *Nat. Med.* **2011**, 17 (8), 1015–1019.

90 Chung, H. J.; Castro, C. M.; Im, H.; Lee, H.; Weissleder, R. *Nat. Nanotechnol.* **2013**, 8 (5), 369–375.

91 Freedonia. World Nanomaterials to 2016—Industry Market Research, Market Share, Market Size, Sales, Demand Forecast, Market Leaders, Company Profiles, Industry Trends and Companies including Arkema, BASF and Bayer. Available at http://www.freedoniagroup.com/World-Nanomaterials.html (accessed on June 24, 2017).

92 Lux Research, Inc, Health Care Microfluidics Market to Grow to Nearly $4 Billion in 2020, **2016**. [Online]. Available at http://www.luxresearchinc.com/ news-and-events/press-releases/read/health-care-microfluidics-market-grow-nearly-4-billion-2020 (accessed on June 24, 2017).

93 Chin, C. D.; Chin, S. Y.; Laksanasopin, T.; Sia, S. K. Low-cost microdevices for point-of-care testing. In Issadore, D., Westervelt, R. (eds), *Point-of-Care Diagnostics on a Chip. Biological and Medical Physics, Biomedical Engineering*; Springer: Berlin/Heidelberg, **2013**.

Section 2

Biosensor Platforms for Disease Detection and Diagnostics

4

SPR-Based Biosensor Technologies in Disease Detection and Diagnostics

Zeynep Altintas[1] and Wellington M. Fakanya[2]

[1] *Technical University of Berlin, Berlin, Germany*
[2] *Atlas Genetics Ltd, Wiltshire, UK*

4.1 Introduction

Ever since it was first presented in 1983 by Liedberg et al., surface plasmon resonance (SPR) has been rapidly adopted for use in the design of label-free optical biosensing applications [1]. The introduction of the first commercial SPR platform by Biacore in 1990 signaled a start in the rapid adoption of the technology for use in biosensors, lab on a chip, and other applications aimed at point-of-care diagnostics [2]. SPR-based biosensors are useful in providing information on non-covalent interactions of bio-molecules and are applicable to investigations that involve protein-to-protein interactions or protein-to-small molecule interactions such as enzyme–substrate, antibody–antigen, protein–nucleic acids, and protein–polysaccharides [3]. The diagram given in Figure 4.1 provides a summary of the diverse array of applications and biological targets for reported SPR biosensors.

One of the main advantages of SPR biosensors is their ability to noninvasively detect and monitor real-time binding events [4]. SPR detection is also capable of making direct determination of affinity and kinetic constants of bimolecular interactions without the use of fluorescence or radioactive labels that are known to interfere with the binding processes [5]. The advantages and disadvantages of SPR biosensors are summarized in Table 4.1.

The performances of the SPR sensors in terms of instrumentation, data processing, and analysis are continuously improved, which allow a better understanding of the binding kinetics, assay sensitivity, and high-throughput systems [6], through integrating robust sensing surfaces with precision microfluidics. This has contributed to the move toward compact instrumentation

Biosensors and Nanotechnology: Applications in Health Care Diagnostics, First Edition.
Edited by Zeynep Altintas.
© 2018 John Wiley & Sons, Inc. Published 2018 by John Wiley & Sons, Inc.

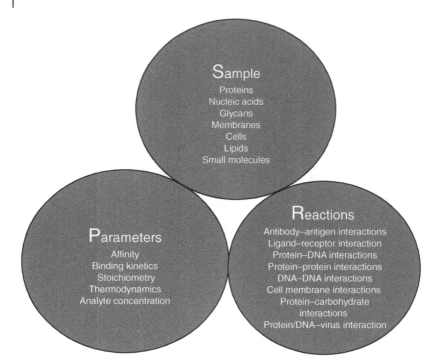

Figure 4.1 Applications of SPR including sample targets, measurable parameters, and reactions that facilitate measurement. (*See insert for color representation of the figure.*)

Table 4.1 Advantages and disadvantages of SPR biosensors.

Advantages of SPR biosensors	Disadvantages of SPR biosensors
Regeneration of sensor chips	High cost of sensor devices and chips
Real-time assay and continuous measurement	Nonspecific binding of target to non-sensor surfaces
Rapid and suitable for label-free detection	Quality of immobilization affects sensor performance
High sensitivity	Steric hindrance related to binding events
Multiplex assay capability	Complex data analysis
Small quantity of sample	Limited evanescent wave penetration depth

and miniaturization as well as multi-SPR biosensor platforms [7, 8]. Plasmonic-based sensor platforms have also managed to gain ground due to the spectacular progress currently being made in micro- and nano-fabrication technology [9, 10]. These developments have collectively helped to generate interest in the medical diagnostics and other areas where there is a need to detect biological and chemical presence, for example, in environmental monitoring, drug discovery, food security, polymer engineering, veterinary, and military applications [11, 12].

In this chapter the technology enabling the development of SPR biosensors is briefly described and the main approaches are highlighted. The focus of this chapter will be limited to the impact of SPR technology in disease detection and diagnostics; emphasis will be given to some of the developments that facilitate high-throughput and increased sensitivity.

4.2 Basic Theoretical Principles

SPR is a physiochemical phenomenon that occurs when plane-polarized light hits a thin metal film under total internal reflection conditions [13]. It utilizes surface plasmons or polaritons, which are particles that pose wave–particle duality and primarily exist on a surface either as photons or phonons. These surface plasmons are primarily found on the surface of substances containing abundant free electrons including high conducting metals like gold [14]. The interactions of photons from a plane-polarized light hitting the metal surface with the surface plasmons will excite them and produce SPR [13]. The building blocks of SPR biosensors consist of a sampling system connected to a microfluidic network with reservoirs enclosed within a chip (microflow cell), an optical sensor system, data capturing, and analysis software [15]. Bio-recognition is facilitated by the attachment of a ligand that is specific to the target of interest onto a polymer matrix using a suitable surface coupling technique; the sample with the target is then passed through a flow cell over the immobilized ligand matrix. The interaction between the ligand and the target-rich sample will lead to changes in mass, resulting in changes in the angle of the incident light needed to generate the SPR state at the gold polymer interface. This is measured as an energy or reflectance dip as a function of pixels, which translates to response units (RUs) over time [16]. While there are several transduction mechanism associated with SPR, which will be discussed later, the diagram in Figure 4.2 depicts a basic structure of an SPR biosensor.

The change in RU is directly proportional to the changes in molecular mass, enabling for both the binding kinetics and stoichiometry measurements to be recorded in real time. The optical sensor system for SPR biosensors consists of the light source, the photo detector, and the optical coupling mechanism that may be based on a prism, grating, waveguide, or optical fiber system [12]. The basic concept of SPR measurements is engraved in the ability of a surface plasma wave to switch to an exited state at the interface between a metal film and a dielectric medium. When surface plasma wave changes from an optically induced state into an excited state, it is referred to as SPR, resulting in changes

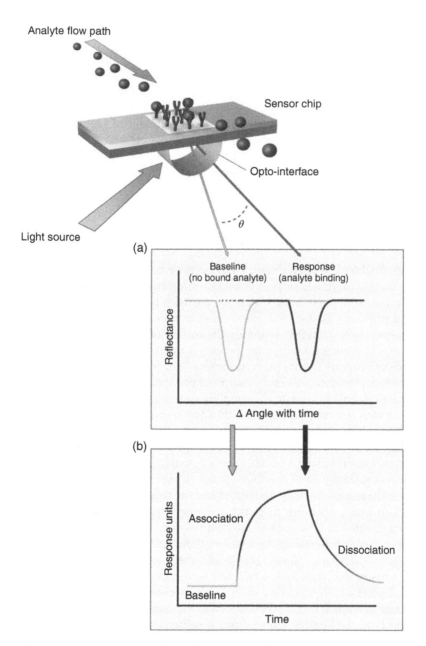

Analyte flow path

Sensor chip

Light source

Opto-interface

θ

(a)

Baseline
(no bound analyte)

Response
(analyte binding)

Reflectance

Δ Angle with time

(b)

Response units

Association

Dissociation

Baseline

Time

Figure 4.2 Generic principle of SPR. The receptor of interest is immobilized to a polymer matrix using a well-established surface chemistry. The analyte is then passed through a flow cell over the receptor-derivatized matrix. Any change in mass following the interaction between the receptor and the analyte is detected as a change in the angle of the incident light needed to generate the surface plasmon resonance phenomenon at the gold polymer interface (a). This is measured as an energy or reflectance dip as a function of pixels, which translates to response units (RU) over time. The RU change is directly proportional to molecular mass change, and so binding kinetics and stoichiometry can be measured in real time without any label (b). *Source:* Helmerhorst et al. [16]. Reproduced with permission from The Australian Association of Clinical Biochemists.

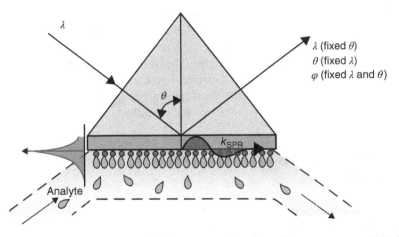

Figure 4.3 Interrogation modes for commercial surface plasmon resonance (SPR)-based instruments. *Source:* Copyright permission granted from Kabashin et al. [17].

in the refractive index, which can be measured. This change is also responsible for the shift in the propagation constant of the coupling conditions between the surface plasmon wave and a light wave. This shift can be measured through observing changes on the characteristics of an optical wave interacting with the surface plasmon wave. SPR biosensors can be classified according to which characteristic of the light wave interacting with the surface plasmon is measured. These sensors may fall under any one of the four subgroups—angular, wavelength, phase, and amplitude [17] modulation—as shown in Figure 4.3.

Amplitude modulation for SPR is performed at a fixed incidence angle and wavelength, with the refractive index variation being detected because of the changes in the resonance intensity. SPR transduction according to the phase detections is based on the principle that, under SPR, the phase of light can cause a dip in the angular dependence of the phase on the p-polarized light. In some instances this dip can be sharp and detected through a probing beam and a reference beam. In this setup the reference beam is compared with the s-polarized portion of the main beam. The phase shifts $\Delta\varphi$ due to interference can be observed through spatial displacement of the light beam. The phase shift under SPR conditions $\Delta\varphi_{max}$ produces a change in the refractive index n of the medium, so that the phase derivative $\Delta\varphi/\Delta n$ can be measured [17].

SPR biosensors with angular modulation are the most common type of modulation and have been developed into commercial products. They are based on the identification of the best angle at which the strongest SPR occurs. Their setup consists of a beam of monochromatic p-polarized light that is used to excite the surface plasmon wave, whose propagation constant is altered and changes are determined by measuring the intensity of reflected light at various angles of incidence to determine which yields the strongest coupling. This is achieved by using a scanning source or a rotating prism or light source but on a specific angle. In the case of biosensors with wavelength modulation, a beam of polychromatic light is fixed at a specific angle of incidence and by modulating the wavelength of the reflected light. The resonant condition can be achieved in a prism configuration through attenuated total reflection, resulting in a reflected intensity dip, which in turn is measured against any changes in the refractive index over a range of incident wavelengths.

4.3   SPR Applications in Disease Detection and Diagnostics

Despite of a lot of attention given to the development of biosensors based on SPR technology, most of the efforts have still remained in the research arena with limited breakthrough in the detection of analytes of relevance to laboratory medicine [16]. Although progress is being made and a great number of publications on the technology are being reported, the practical application of SPR for disease detection and diagnostics, which utilizes classical samples such as whole blood, serum, plasma, urine, or saliva, is still somewhat restricted [18]. Most of the SPR biosensor devices for the disease detection and diagnostics are still on the prototype or proof-of-concept stage where the conditions of use are yet to achieve the standards that will make them suitable for use in a clinical environment, let alone pass the stringent performance requirements necessary for regulatory approvals. This should not be construed to mean that the technology is not progressing toward the fabrication of devices that are of clinical significance in disease detection and diagnostics. SPR biosensors for the detection of a wide array of clinical conditions including heart disease, cancer, infectious diseases, and other clinical states have been reported. This section will summarize some of the achievements of SPR in disease detection and diagnostics.

4.3.1   SPR Biosensors in Cancer Detection

SPR biosensors are actively being developed to enable for the detection and diagnosis of several tumor markers for cancer diagnosis. Cancer biomarkers in serum samples have been detected using this sensing technology [19]. SPR biosensors for carcinoembryonic antigen, which is associated with colorectal and lung cancers, have been reported [20]. Another recently reported example, SPR-based biosensor for ultrasensitive detection of prostate-specific antigen (PSA) from clinical samples using microcontact imprinting-based SPR biosensor managed to achieve high sensitivity and specificity as well as acceptable levels of sensor stability and regeneration [21]. SPR biosensors have been used to measure two known tumor markers, human chorionic gonadotropin and activated leukocyte cell adhesion molecules, in diluted blood plasma [4]. The detection of breast [22] and oral cancers [23] has also been reported using this technology. Law et al. developed a nanoparticle-enhanced SPR biosensor by integrating both the nanoparticles and immunoassay sensing into a phase interrogation SPR system for detecting tumor necrosis factor alpha (TNF-α) antigen at femtomolar range. It was reported that the plasmonic field extension generated from the gold film to gold nanorod (GNR) provided a significant sensitivity enhancement. Antibody-functionalized biosensing film, together with antibody-conjugated GNRs, was employed as a plasmonic coupling partner that can be used as a powerful ultrasensitive sandwich immunoassay for cancer-related disease detection. The developed sensor demonstrated 40-fold sensitivity compared with the conventional SPR technique. This research also indicates that SPR technology can be successfully improved with the use of nanomaterials and integration of novel nanotechnology approaches. The obtained detection limit (0.03 pM) can be proficient in monitoring small variations of TNF-α to understand the cancer biology, and the GNR method can be employed for the monitoring and detection of other cancer biomarkers using an SPR-based device [24]. Moreover, protein biomarkers of cancer have been detected using RNA aptamer microarrays and enzymatically amplified SPR imaging that is able to quantify picomolar concentrations. This research relies on the adsorption of proteins onto the RNA microarray to be detected by the formation of a surface aptamer–protein–antibody complex. The SPR imaging signal was then amplified with a localized precipitation reaction catalyzed by the enzyme horseradish peroxidase, which was conjugated to the antibody. This method was first tested with human thrombin at a concentration of 500 fM and then transferred to the detection of vascular endothelial growth factor (VEGF), which is a serum biomarker for lung cancer, breast cancer, colorectal cancer, and rheumatoid arthritis and is also related to age-related macular degeneration. The SPR imaging techniques was capable of detecting the biomarker in its biologically relevant concentration of 1 pM [25].

SPR biosensors can also be constructed based on optical fiber approach. Jang et al. reported on an optical fiber SPR sensor for the detection of PSA by employing sandwich assay using surface and detector antibodies. The optical sensitivity of the developed sensor was determined as 2.5×10^{-6} RIU, and a detection range of $0.1–10 \, \mu g \, mL^{-1}$ was successfully investigated by measuring SPR wavelength shifts of the sensor [26].

Another setup of the SPR technology using SPR scattering approach was reported for oral cancer detection. In this research SPR scattering images and SPR absorption spectra were measured both from colloidal gold nanoparticles and from gold nanoparticles conjugated to monoclonal anti-epidermal growth factor receptor (anti-EGFR) antibodies after incubation in cell cultures with a nonmalignant epithelial cell line (HaCaT) and two malignant oral epithelial cell lines (HOC 313 clone 8 and HSC 3). Colloidal gold nanoparticles were observed in dispersed and aggregated forms within the cell cytoplasm and supplied anatomic labeling information (Figure 4.4). However, their uptake is normally nonspecific for cancer cells. Using the anti-EGFR antibody-conjugated nanoparticles, the specific binding was achieved to the cancer cell surface with 600% higher affinity than to the normal cells. This specific binding assay produced a relatively sharper SPR absorption band with a redshifted maximum compared with that observed when added to the normal cells. Therefore, SPR scattering imaging or SPR absorption spectroscopy generated from antibody-conjugated gold nanoparticles can be efficiently employed for the diagnosis of oral epithelial living cancer cells *in vivo* and *in vitro* [27].

In case of labeling superparamagnetic particles with biomarker-specific antibody and using those by SPR, a very sufficient and reliable platform can be developed. Such method was employed for the detection of PSA and led to the ultrasensitive detection with an LOD of $10 \, fg \, mL^{-1}$ (ca. $300 \, aM$) due to the enhanced mass and refractive index from aggregates of $1 \, \mu m$ magnetic particles on the SPR chip [28]. SPR technology also offers a promising future for the detection of single-nucleotide polymorphism that occurs on central genes in cancer cases. By employing convenient surface chemistry and optimal assay conditions, codon mutations on p53 gene that are responsible for lung cancer can be detected. Moreover, since these mutations are quite specific for disease subtypes, even the discrimination of small cell lung cancer and non-small cell lung cancer is possible [29]. SPR-based biosensors have also been reported in the detection of pituitary hormones in serum and urine samples. Some of the hormones detected and measured using SPR include thyroid-stimulating hormone (TSH), growth hormone, follicle-stimulating hormone, and luteinizing hormone [30].

4.3.2 SPR Sensors in Cardiac Disease Detection

Measurements of several cardiac markers in serum including C-reactive protein [31], myoglobin, cardiac troponin T (cTnT) [32], cardiac troponin I [33], and B type natriuretic peptide [34] using SPR biosensors have been reported. Recently, SPR detection of cTnT was reported in serum samples using direct and sandwich assay methods. The immunosensor demonstrated good reproducibility for cTnT detection in the ranges of $25–1000 \, ng \, mL^{-1}$ and $5–400 \, ng \, mL^{-1}$ for the direct and sandwich assays in buffer, respectively. When the methodology was transferred to serum sample detection, a very high nonspecific binding was initially observed. However, the researchers investigated various compositions of additives to reduce the nonspecific interaction and detected cTnT in the concentration range of $10–200 \, ng \, mL^{-1}$ using sandwich assay in the presence of surface and detector antibodies, each of which is specific for cTnT. The optimized nonspecific buffer added into the serum includes $0.5 \, M$ NaCl and $200 \, \mu g \, mL^{-1}$ BSA in PBS-T buffer. To further decrease the

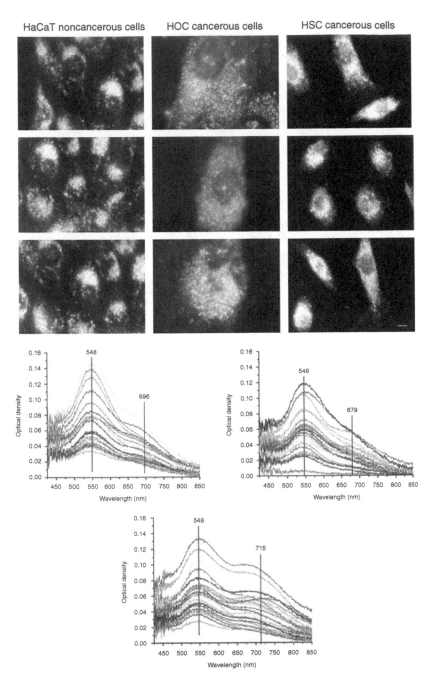

Figure 4.4 Light-scattering images and microabsorption spectra of HaCaT noncancerous cells, HOC cancerous cells, and HSC cancerous cells after incubation with unconjugated colloidal gold nanoparticles. The images display that the particles are inside the cells in the cytoplasm region but do not seem to adsorb strongly on the nuclei of the cells. The absorption spectra were measured for 25 different single cells of each kind. They show that nanoparticles have an SPR absorption maximum around 548 nm, independent of the cell type. The broad long wavelength tails in the absorption spectra suggest the presence of aggregates. It also shows that no specific difference is observed in either the scattering images or the absorption spectra of the gold nanoparticles in the cancerous and the noncancerous cells. *Source:* El-Sayed et al. [27]. Reproduced with permission of American Chemical Society.

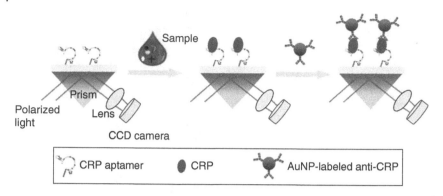

Figure 4.5 Schematic illustration of the AuNP-enhanced SPR biosensor with an aptamer–antibody sandwich assay. *Source:* Wu et al. [35]. Reproduced with permission of Royal Society of Chemistry. (*See insert for color representation of the figure.*)

detection limit of target biomarker, gold nanoparticle (AuNP)-modified sandwich assay was performed by conjugating detector antibody with AuNPs that allowed to quantify $0.5\,ng\,mL^{-1}$ cTnT in serum samples. The developed methodology can be used as a reference point to establish a point-of-care test for the early diagnosis of acute myocardial infarction (AMI) [32]. Aptamer–antibody sandwich assays have also been developed for cardiac disease diagnosis (Figure 4.5). Wu et al. investigated CRP detection using nanoparticle-enhanced SPR with this approach. High affinity DNA aptamers against CRP were selected using microfluidic chip and were immobilized on the surface as the receptor to capture the target molecule. To further increase the signal and decrease the detection limit, antibody-modified AuNPs were employed as signal amplification agent to conduct the sandwich assay that allowed to detect CRP in the concentration range of 10 pM to 100 nM in diluted human serum [35].

Novel interdisciplinary approaches result in continuous improvements in the field. One good example is the application of plasma-treated parylene-N film on SPR biosensor chip for the sensitive detection of CRP due to the high protein immobilization property of parylene-N film after plasma treatment [36]. CRP detection was also investigated by employing a plastic optical fiber biosensor based on SPR. The sensor was integrated into a thermostabilized microfluidic system to stabilize the conditions during the experiments and to avoid any thermal and/or mechanical fluctuation. The method allowed to quantify CRP in serum in the range of $6\,ng\,mL^{-1}$ to $70\,\mu g\,mL^{-1}$ with a detection limit of $9\,ng\,mL^{-1}$ [37].

Site-specific antibody immobilization plays a key role in direct detection of biomarkers although majority of bioassays still lack this. When applied in direct immunoassays, their analytical features depend strongly on the antibody immobilization strategy. A strategy to immobilize the antibodies correctly was developed using ProLinker™, and the method was optimized in SPR sensor in terms of stability, sensitivity, selectivity, and reproducibility. Special care was given to avoid manipulation of antibody, to prevent nonspecific adsorption, and to attain a robust surface with regeneration capability. ProLinker-based approach could fulfill the important requirements with the aid of PEG-derivative compounds, and it demonstrated an efficient performance for direct detection of biomarkers in biological fluids. ProLinker system was employed for the detection of several disease markers such as CRP and focal adhesion kinase (FAK). The novel antibody immobilization strategy allowed to detect CRP and FAK with LOD of 23 ng mL^{-1} ($R^2 = 0.9718$) and 86 ng mL^{-1} ($R^2 = 0.9799$), respectively. Moreover, this method demonstrated superior results when compared with protein G-based antibody immobilization at the same concentration of antibody (LOD$_{CRP}$ = 42 ng mL^{-1}, LOD$_{FAK}$ = 208 ng mL^{-1}). The ProLinker strategy was also implemented to a nanoplasmonic-based biosensor displaying advantages for its application in healthcare diagnostic [38]. More examples of CVD detection using biosensors can also be found in recent reviews [39, 40].

4.3.3 SPR Sensors in Infectious Disease Detection

The methods currently employed for the diagnosis of infectious diseases are mostly based on laboratory tests that include microscopy, culture, immunoassays, and nucleic acid amplification. Often these methods result in the undesirable consequences of longer result turnaround times and the need for highly skilled personnel to conduct them [19]. It has been reported that over 95% of deaths from infectious diseases are caused by a lack of proper diagnostics and treatments [20]. The diagnostics of infectious diseases is leaned toward the development and adoption of new technologies like SPR-based bioanalyzers that can potentially enable to facilitate point-of-care diagnostic applications. Infectious diseases are caused by a wide array of pathogens including bacterial and mycobacterial, viruses, fungus, and parasites.

SPR bioanalyzers have been employed for use in the detection of whole cells including pathogenic bacteria [41]. The use of SPR biosensors in the detection of whole cells is still maturing with only a small percentage of all reported SPR biosensors being attributed to the detection of whole-bacterial organisms. The SPR can be used to detect whole *Escherichia coli* [41, 42]. Methicillin-resistant *Staphylococcus aureus* has also been reported using SPR biosensors capable of detecting as little as 103 CFU mL^{-1} with no sample enrichment and without the use of any secondary labels [43]. The detection of *Salmonella* [44] and *Lactobacillus* [45] was also reported by employing SPR technology for the diagnosis of infectious diseases. However the detection of whole bacteria using SPR is criticized for generally producing low sensitivity in some cases due to the limited penetration of bacteria by the electromagnetic field as well as the parity between the refractive index of the bacterial cytoplasm and that of the aqueous medium [46]. An SPR biosensor that can detect *Mycobacterium tuberculosis* has been reported based on DNA sensing [47]. The detection of infectious diseases caused by viruses can also be achieved with SPR biosensors. A multiplex SPR sensor was developed for the detection of nine respiratory viruses: H1N1, influenza A and B, respiratory syncytial virus (RSV), adenovirus, severe acute respiratory syndrome (SARS) coronavirus, and parainfluenza viruses 1–3 (PIV1, 2, 3). For this, the nine respiratory virus-specific oligonucleotides were immobilized on a SPR chip. To achieve high sensitivity, biotin was used to label the PCR primer and further amplify the signal by introducing streptavidin after hybridization. Throat swab specimens representing the nine common respiratory viruses were quantified by the sensor to evaluate the sensitivity, reproducibility, and specificity of the developed SPR method. The output of this research indicates that the sensor has a promising future to simultaneously detect common respiratory viruses [48]. SPR-based biosensors have also been used for detecting human immunodeficiency virus (HIV) [49], human papillomavirus (HPV) [50], and mumps virus [51]. More examples can be found in a recent review paper on biosensor-based virus detection [52].

Novel techniques by utilizing synthetic polymeric receptors on SPR have been developed for the detection of waterborne viruses and endotoxins, leading to severe outbreaks and infectious diseases [52–56]. Among these are MS2 phage [53], adenovirus [54], and *E. coli* endotoxin [55, 56]. In these works, biomimetic receptors were designed and manufactured using supramolecular chemistry. The target-specific receptors were then covalently immobilized onto SPR chips for the detection of viruses and endotoxins. A graphical representation of adenovirus detection in SPR is given in Figure 4.6 as an example. The developed sensor assays offered real-time and label-free detection in a wide concentration range (Table 4.2). Moreover, these in-house synthesized affinity ligands are so cheap, and they can be produced only in a couple of days. It was also confirmed that their performance is comparable to antibodies based on limit of detection and affinity [54].

4.4 Conclusions

SPR technology using biomarkers has a key role in the diagnostic revolution of important diseases such as cancer, cardiac problems, and infectious diseases. Implementations of advanced technologies taking place in microelectronics and nanotechnology industry into the SPR-based sensing devices have enormously improved the assay performances, reliability of the results, and throughput of these systems. Some of the currently available SPR sensors are designed to screen up to 7000 molecular interactions per day, which is significantly important in drug discovery, screening of novel therapeutic

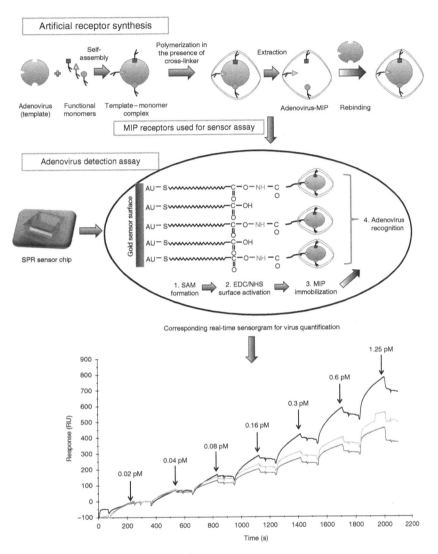

Figure 4.6 Synthesis of an adenovirus-specific biomimetic receptor and its use in an SPR-based biosensor for the real-time detection of viruses.

candidates for cancer cases, and also performing multiplex detection assays for disease biomarkers or discriminating the disease types based on several markers investigated simultaneously. The development of highly specific and stable receptors and also computing the binding interaction between the receptor and target biomarker by applying intensive computational research have also increased the success of the sensing technology in addition to the

Table 4.2 Detection of pathogens using biomimetic SPR sensors.

Pathogen	MS2 phage	Adenovirus	Endotoxin	Endotoxin
Sensor type	Biacore 3000 and SPR-2	Biacore 3000	SPR-2	SPR-2/4
Receptor type	Artificial	Artificial	Artificial	Artificial
Detection range	0.33–27 pM	0.01–20 pM	15.6–500 ng mL^{-1}	0.44–200 ng mL^{-1}
Detection limit	0.3 pM	0.02 pM	15.6 ng mL^{-1}	0.44 ng mL^{-1}
Surface regeneration	Yes	N/A	Yes	Yes
Dissociation constant (K_D) (M)	3.22×10^{-9}	3.10×10^{-11}	$3.24–5.24 \times 10^{-8}$	$4.4–5.3 \times 10^{-10}$
Reference	[53]	[54]	[55]	[56]

use of well-established surface chemistries and smart nanomaterials in health care. In case nonspecific binding problems in real human samples and also the limited penetration of the sensor in some cases are surpassed, SPR-based sensing has been offering a great future in diagnostic with novel implementations as being one of the earliest and well-established techniques in the field of biosensors.

Acknowledgment

Z.A. gratefully acknowledges support from the European Commission, Marie Curie Actions and IPODI as the principle investigator.

References

1 Liedberg, B.; Nylander, C.; Lunstrom, I. *Sensors and Actuators* **1983**, 4, 299–304.
2 Wang, X. P.; Zhan, S. Y.; Huang, Z. H.; Hong, X. Y. *Instrumentation Science and Technology* **2013**, 41, 574–607.
3 Erickson, D.; Mandal, S.; Yang, A. H. J.; Cordovez, B. *Microfluidics and Nanofluidics* **2008**, 4, 33–52.
4 Piliarik, M.; Homola, J. SPR Sensor Instrumentation, in Homola, J. (ed) *Surface Plasmon Resonance Based Sensors*, Springer, Heidelberg, Germany, **2016**, 95–116.
5 Singh, P. *Sensors and Actuators B: Chemical* **2016**, 229, 110–130.
6 Ouellet, E.; Lausted, C.; Lin, T.; Yang, C. W. T.; Hood, L.; Lagally, E. T. *Lab on a Chip* **2010**, 10, 581–588.

7 Peng, W.; Liu, Y.; Fang, P.; Liu, X. X.; Gong, Z. F.; Wang, H. Q.; Cheng, F. *Optics Express* **2014**, 22, 6174–6185.

8 Dostalek, J.; Vaisocherova, H.; Homola, J. *Sensors and Actuators B: Chemical* **2005**, 108, 758–764.

9 Jain, P. K.; El-Sayed, M. A. *Chemical Physics Letters* **2010**, 487, 153–164.

10 Guner, H.; Ozgur, E.; Kokturk, G.; Celik, M.; Esen, E.; Topal, A. E.; Ayas, S.; Uludag, Y.; Elbuken, C.; Dana, A. *Sensors and Actuators B: Chemical* **2017**, 239, 571–577.

11 D'Orazio, P. *Clinica Chimica Acta* **2011**, 412, 1749–1761.

12 Puiu, M.; Bala, C. *Sensors (Switzerland)* **2016**, 16(6), 1–15.

13 Healthcare G. *Biacore Sensor Surface Handbook*, **2003**, AB Biacore and GE Healthcare Bio-Sciences AB, Uppsala, Sweden.

14 Kihm, K. D.; Cheon, S.; Park, J. S.; Kim, H. J.; Lee, J. S.; Kim, I. T.; Yi, H. J. *Optics and Lasers in Engineering* **2012**, 50, 64–73.

15 Tang, Y. J.; Zeng, X. Q.; Liang, J. *Journal of Chemical Education* **2010**, 87, 742–746.

16 Helmerhorst, E.; Chandler, D. J.; Nussio, M.; Mamotte, C. D. *Clinical Biochemist Reviews* **2012**, 33, 4, 161–173.

17 Kabashin, A. V.; Patskovsky, S.; Grigorenko, A. N. *Optics Express* **2009**, 17, 21191–21204.

18 Mariani, S.; Minunni, M. *Analytical and Bioanalytical Chemistry* **2014**, 406, 2303–2323.

19 Uludag, Y.; Tothill, I. E. *Analytical Chemistry* **2012**, 84, 5898–5904.

20 Altintas, Z.; Uludag, Y.; Gurbuz, Y.; Tothill, I. E. *Talanta* **2011**, 86, 377–383.

21 Erturk, G.; Ozen, H.; Tumer, M. A.; Mattiasson, B.; Denizli, A. *Sensors and Actuators B: Chemical* **2016**, 224, 823–832.

22 Yang, M. H.; Yi, X. Y.; Wang, J. X.; Zhou, F. M. *Analyst* **2014**, 139, 1814–1825.

23 Kah, J. C. Y.; Kho, K. W.; Lee, C. G. L.; Sheppard, C. J. R.; Shen, Z. X.; Soo, K. C.; Olivo, M. C. *International Journal of Nanomedicine* **2007**, 2, 785–798.

24 Law, W. C.; Yong, K. T.; Baev, A.; Prasad, P. N. *ACS Nano* **2011**, 5, 4858–4864.

25 Li, Y.; Lee, H. J.; Corn, R. M. *Analytical Chemistry* **2007**, 79, 1082–1088.

26 Jang, H. S.; Park, K. N.; Kang, C. D.; Kim, J. P.; Sim, S. J.; Lee, K. S. *Optics Communications* **2009**, 282, 2827–2830.

27 El-Sayed, I. H.; Huang, X. H.; El-Sayed, M. A. *Nano Letters* **2005**, 5, 829–834.

28 Krishnan, S.; Mani, V.; Wasalathanthri, D.; Kumar, C. V.; Rusling, J. F. *Angewandte Chemie International Edition* **2011**, 50, 1175–1178.

29 Altintas, Z.; Tothill, I. E. *Sensors and Actuators B: Chemical* **2012**, 169, 188–194.

30 Trevino, J.; Calle, A.; Rodriguez-Frade, J. M.; Mellado, M.; Lechuga, L. M. *Clinica Chimica Acta* **2009**, 403, 56–62.

31 Vashist, S. K.; Schneider, E. M.; Luong, J. H. T. *Analyst* **2015**, 140, 4445–4452.

32 Pawula, M.; Altintas, Z.; Tothill, I. E. *Talanta* **2016**, 146, 823–830.

33 Kwon, Y. C.; Kim, M. G.; Kim, E. M.; Shin, Y. B.; Lee, S. K.; Lee, S. D.; Cho, M. J.; Ro, H. S. *Biotechnology Letters* **2011**, 33, 921–927.

34 Jang, H. R.; Wark, A. W.; Baek, S. H.; Chung, B. H.; Lee, H. J. *Analytical Chemistry* **2014**, 86, 814–819.

35 Wu, B.; Jiang, R.; Wang, Q.; Huang, J.; Yang, X. H.; Wang, K. M.; Li, W. S.; Chen, N. D.; Li, Q. *Chemical Communications* **2016**, 52, 3568–3571.

36 Choi, Y. H.; Ko, H.; Lee, G. Y.; Chang, S. Y.; Chang, Y. W.; Kang, M. J.; Pyun, J. C. *Sensors and Actuators B: Chemical* **2015**, 207, 133–138.

37 Aray, A.; Chiavaioli, F.; Arjmand, M.; Trono, C.; Tombelli, S.; Giannetti, A.; Cennamo, N.; Soltanolkotabi, M.; Zeni, L.; Baldini, F. *Journal of Biophotonics* **2016**, 9, 1077–1084.

38 Soler, M.; Estevez, M. C.; Alvarez, M.; Otte, M. A.; Sepulveda, B.; Lechuga, L. M. *Sensors* **2014**, 14, 2239–2258.

39 Altintas, Z.; Fakanya, W. M.; Tothill, I. E. *Talanta* **2014**, 128, 177–186.

40 Fakanya, W. M.; Altintas, Z.; Tothill, I. E. Biosensors for the Diagnosis of Heart Disease, in Ozkan-Ariksoysal, D. (ed) *Biosensors and Their Applications in Healthcare*, Future Medicine, London, UK **2013**, 128–143.

41 Ahmed, A.; Rushworth, J. V.; Hirst, N. A.; Millner, P. A. *Clinical Microbiology Reviews* **2014**, 27, 631–646.

42 Tripathi, S. M.; Bock, W. J.; Mikulic, P.; Chinnappan, R.; Ng, A.; Tolba, M.; Zourob, M. *Biosensors and Bioelectronics* **2012**, 35, 308–312.

43 Tawil, N.; Sacher, E.; Mandeville, R.; Meunier, M. *Biosensors and Bioelectronics* **2012**, 37, 24–29.

44 Mazumdar, S. D.; Barlen, B.; Kampfer, P.; Keusgen, M. *Biosensors and Bioelectronics* **2010**, 25, 967–971.

45 Baccar, H.; Mejri, M. B.; Hafaiedh, I.; Ktari, T.; Aouni, M.; Abdelghani, A. *Talanta* **2010**, 82, 810–814.

46 Torun, O.; Boyaci, I. H.; Temur, E.; Tamer, U. *Biosensors and Bioelectronics* **2012**, 37, 53–60.

47 Hsu, S. H.; Lin, Y. Y.; Lu, S. H.; Tsai, I. F.; Lu, Y. T.; Ho, H. T. *Sensors* **2014**, 14, 458–467.

48 Shi, L.; Sun, Q. X.; He, J. A.; Xu, H.; Liu, C. X.; Zhao, C. Z.; Xu, Y. Q.; Wu, C. L.; Xiang, J. J.; Gu, D. Y.; Long, J.; Lan, H. K. *Bio-Medical Materials and Engineering* **2015**, 26, S2207–S2216.

49 Valizadeh, A. *Artificial Cells Nanomedicine and Biotechnology* **2016**, 44, 1383–1390.

50 Frias, I. A. M.; Avelino, K.; Silva, R. R.; Andrade, C. A. S.; Oliveira, M. D. L. *Journal of Sensors* **2015**, 2015, 1–16.

51 Yuk, J. S.; Ha, K. S. *Methods in Molecular Biology* **2009**, 503, 37–47.

52 Altintas, Z.; Gittens, M.; Pocock, J.; Tothill, I. E. *Biochimie* **2015**, 115, 144–154.

53 Altintas, Z.; Gittens, M.; Guerreiro, A.; Thompson, K.-A.; Walker, J.; Piletsky, S.; Tothill, I. E. *Analytical Chemistry* **2015**, 87, 6801–6807.

54 Altintas, Z.; Pocock, J.; Thompson, K.-A.; Tothill, I. E. *Biosensors and Bioelectronics* **2015**, 74, 996–1004.

55 Abdin, M. J.; Altintas, Z.; Tothill, I. E. *Biosensors and Bioelectronics* **2015**, 67, 177–183.

56 Altintas, Z.; Abdin, M. J.; Tothill, A. M.; Karim, K.; Tothill, I. E. *Analytica Chimica Acta* **2016**, 935, 239–248.

5

Piezoelectric-Based Biosensor Technologies in Disease Detection and Diagnostics

Zeynep Altintas[1] and Noor Azlina Masdor[2,3]

[1] *Technical University of Berlin, Berlin, Germany*
[2] *Cranfield University, Cranfield, UK*
[3] *Malaysian Agricultural Research and Development Institute (MARDI), Kuala Lumpur, Malaysia*

5.1 Introduction

Acoustic-based biosensing platforms such as quartz crystal microbalance (QCM) have shown a promising future for the rapid and sensitive detection of diseases in diagnostic applications. A major improvement to the QCM device is the inclusion of microfluidic systems that has resulted in the developments of microfluidic-based QCM that have many advantages: (i) having throughput, (ii) reducing sample/reagent consumption, and (iii) providing faster and more reliable results. Their construction as close systems has further improved their performance by avoiding environmental effects such as dust on the air and the uncontrolled temperature during the detection assays. It is proved that such a system can provide the detection of single-point mutation on a p53 gene that requires special attention to achieve highly sensitive and stable results [1]. The last decade has witnessed significant improvements in the development of QCM technology, resulting in many commercially available QCM sensors that are successfully utilized in research laboratories. Advances in nanotechnology also improve the entire portfolio of these devices, leading to the reduction of the unit cost for each reaction [2]. Although QCM can be operated label-free, a label-like system in the form of the mass amplifier such as gold nanoparticles (AuNPs) has already been proven to increase the sensitivity of QCM-based detection of disease with dramatic improvement on limits of detection (LODs) over standard sensing methods including direct or normal sandwich assays. With this in mind, the following sections will cover QCM biosensors, their use in disease diagnosis with current trends, and future prospects.

Biosensors and Nanotechnology: Applications in Health Care Diagnostics, First Edition.
Edited by Zeynep Altintas.
© 2018 John Wiley & Sons, Inc. Published 2018 by John Wiley & Sons, Inc.

5.2 QCM Biosensors

A QCM device is based on the phenomenon of piezoelectricity, occurring as a result of the electric charge that builds up in a few solid materials upon being applied with mechanized tension or pressure [3]. The term came from the Greek words piezo or piezein, meaning to squeeze or press, and ēlektron, meaning amber, an old source of electric charge [4]. The famous French physicists Jacques and Pierre Curie discovered piezoelectricity in 1880 [5]. The phenomenon is a reversible process. For instance, the crystals of lead zirconate titanate produce considerable piezoelectricity anytime their static framework is misshaped by way of about 0.1% from the initial dimension. Alternatively, the identical crystals can change about 0.1% of the static dimension anytime the material is applied outside the electric field [6].

Piezoelectric devices generate vibrational waves, which are classified as acoustic waves and include vibration, sound, ultrasound, and infrasound waves [3, 7, 8]. These waves can propagate as either a bulk acoustic wave (BAW) propagating through the interior of the substrate as observed in QCM-based sensors or a surface acoustic wave (SAW) propagating on the surface of the substrate. The majority of commercial devices hinges upon BAW phenomenon since the disadvantage of the SAW sensors is that it is hard to construct a sturdy device because the frequency change is affected by many factors such as the dielectric, conductance and elastic constants of the adsorbent, and conductance of the liquid [3].

The microbalance portion of QCM originated from its classical sensing application in micro-gravimetry or measurement of very small mass [7]. The QCM features limits of detection lower than $1\,\mathrm{ng\,cm}^{-2}$ and covers adsorbents having several hundred nanometers of thickness. Due to this wide, high dynamic range, QCM is employed for the detection of various compounds from small molecules to cells [9, 10]. Despite this, BAW devices need to take into account the viscosity of the liquid according to the Sauerbrey equation as this could affect the measurement, although in modern commercial devices, these drawbacks have been addressed.

The altered resonance frequency could be predicted in Equation 5.1 in accordance with the Sauerbrey formula.

$$\Delta f = -\frac{2 f_0^2}{A\sqrt{\rho_q \mu_q}} \Delta m \tag{5.1}$$

where

f_0—resonant frequency (Hz)
Δf—frequency change (Hz)
Δm—mass change (g)

A—piezoelectrically active crystal area (area between electrodes, cm^2)
ρ_q—density of quartz ($\rho_q = 2.648\,\text{g}\,\text{cm}^{-3}$)
μ_q—shear modulus of quartz for AT-cut crystal ($\mu_q = 2.947 \times 10^{11}\,\text{g}\,\text{cm}^{-1}\,\text{s}^{-2}$)

Sauerbrey initially developed the equation for oscillation in air. In addition, it is only applied to rigid masses attached to the crystal. Kanazawa and coworkers [11] pioneered the work of QCM measurements in the liquid phase, and they showed that the change in resonant frequency taken from air into a liquid is directly proportional to the square root of the liquid's density–viscosity product (Equation 5.2).

$$\Delta f = -f_0^{3/2}\left(\frac{\eta_l \rho_l}{\pi \rho_q \mu_q}\right)^{1/2} \tag{5.2}$$

where ρ_l is the density of the liquid and η_l is the viscosity of the liquid.

In QCM, the standard piezoelectric sensing head is made up of quartz crystal wafer AT-cut and two excitation electrodes that are plated on opposing sides of the crystal. It is manufactured with three parts: frequency counter, an electronic oscillation circuit, and piezoelectric quartz (Figure 5.1) [12–15].

The piezoelectric sensor chip converts the analog results sensed by the sensor molecule into the amplified digital electronic signals. The gold electrodes function to introduce an oscillating electric field that is perpendicular (90°) to

Figure 5.1 Experimental apparatus for a piezoelectric system.

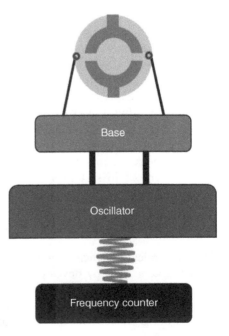

Base

Oscillator

Frequency counter

the chip's and generate a mechanical oscillation because of the piezoelectric effect. The first piezoelectric immunosensor was developed using a QCM immobilized with bovine serum albumin (BSA). The apparatus was applied to detect anti-BSA antibodies [16]. Since then, numerous piezoelectric-based immunosensors for the detection of various analytes have been reported. This ranges from small molecules such as drugs and proteins, to larger units such as viruses and cells [3, 8, 10].

QCM consists of a thin quartz disc with electrodes plated on it. When an oscillating electric field is applied across the disc, an acoustic wave with a certain resonance frequency is induced (Figure 5.2). In QCM, the interaction between the analyte and biomolecule with the sensor surface results in a rapid absorption, which leads to an initial increase in the sensorgram (association phase) followed by a saturation of the surface (dissociation phase) observed as the emergence of a plateau in the sensorgram. The final stage (Figure 5.2) involves the replacement (regeneration phase) of the analyte or biomolecule with regeneration buffer to remove loosely bound materials and regenerate the surface.

The efficient immobilization of the antibody on the surface sensor determines the sensitivity and accuracy of the sensing method when developing a QCM immunosensor [17]. Various immobilization methods have been used over the years including Langmuir–Blodgett film protein A, silanized layer, polymer membrane, and self-assembled monolayer (SAM) [10, 18–21], which possess functional groups that have a strong affinity to the sensor surface [22–27].

One of the superiorities of QCM is that its signal can be further amplified using a label or mass amplifier. Of the mass amplifiers used in amplifying QCM signals, nanoparticles based on gold, magnetic, silica, and polymer have emerged as excellent mass amplifiers. Salam et al. [28] employed AuNPs to detect *Salmonella typhimurium* cells using sandwich assays. The QCM immunosensor-utilizing AuNPs gave a very high sensitivity with the LOD between 10 and 20 CFU mL^{-1} as compared with direct (1.83×10^2 CFU mL^{-1}) and normal sandwich (1.01×10^2 CFU mL^{-1}) assays. In another work for the detection of *Campylobacter jejuni*, the use of AuNPs in conjunction with QCM increases the sensitivity of the system with an LOD of 1.5×10^2 CFU mL^{-1} [10].

5.3 Disease Diagnosis Using QCM Biosensors

Clinical applications including the diagnosis of diseases through the detection of biomarkers, therapy development, and monitoring of diseases related to cancer, cardiovascular, and pathogens are in progress worldwide [29–38]. As the number of biomarkers for these afflictions has increased over the years, the number of clinical tests for the detection of these biomarkers is also on the rise.

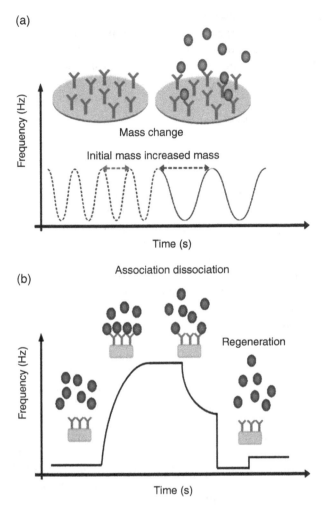

Figure 5.2 The principle of (a) QCM and (b) a sensorgram of an interaction of antibody and antigen showing the establishment of an initial baseline, association, dissociation, and regeneration phases.

Presently, the medical diagnostic market for these major diseases has been dominated by methods based on DNA or protein microarrays [39, 40]. On the other hand, the development of glucose meters for determining the blood glucose levels in the form of a portable analytical instrument that allows patients to monitor diseases themselves has opened the path for many more biosensor development. Of these biosensors, the QCM-based biosensor is an emerging method in the clinical diagnosis field that holds future promises as one of the best candidates for point-of-care applications. As QCM is an extremely sensitive mass sensor, sub-nanogram levels of analyte can be detected via a frequency change from the device. The versatility of QCM in the detection of a wide range of analytes coupled with its sensitivity and selectivity may make it a promising future tool in clinical applications especially for the diagnosis of diseases.

5.3.1 Cancer Detection Using QCM Biosensors

Cancer is one of the major causes of mortality and morbidity cases worldwide. As there are hundreds of cancer and cancer-related diseases, early detection and management are important so that specific treatments can be applied [41, 42]. The current advances in nanotechnology lead to the discovery of hundreds of genomic and proteomic cancer markers that develop novel biosensors [1, 41–44]. The susceptibility of people to certain types of cancers can be screened in a rapid and simple procedure using biosensing devices [45]. Aside from the classical, histological, and morphological methods, the use of biosensors for the detection of these biomarkers is a promising tool for the diagnosis and prognosis of cancer that increases the chance of a successful therapy. Biomarker-based sensing devices for clinical applications are on the rise, particularly the point-of-care tools for early and easy detection [41, 46]. For instance, the advent of the disposable and easy-to-use glucose sensors has allowed the patients and the nontechnical people to use these devices to monitor the progression of the disease themselves. This has opened up new and exciting area for biosensors applications. To date, a number of papers regarding QCM in medical applications were reported. Some cancer-related examples were listed in this section.

One of the most common forms of cancer in men is prostate cancer. In case the level of prostate-specific antigen (PSA) is above $4.0 \, \text{ng mL}^{-1}$, it generally indicates the presence of prostate cancer [47] detected by very well-established biosensor methods. As an example, a QCM-based simple and sensitive sensor method coupled with a nanoparticle amplification system for the detection of PSA and another prostate cancer biomarker, PSA–alpha 1-antichymotrypsin (ACT) complex, was developed and was able to detect both biomarkers in 75% human serum. To overcome the nonspecific binding posed by human serum, a PBS buffer supplemented with 0.5 M NaCl, $200 \, \mu\text{g mL}^{-1}$ BSA, 0.5% Tween 20, and $500 \, \mu\text{g mL}^{-1}$ dextran was utilized. An LOD of $0.29 \, \text{ng mL}^{-1}$ for both PSA and PSA–ACT complex was achieved with a dynamic linear detection ranging up to $150 \, \text{ng mL}^{-1}$ in 75% serum [48]. The LOD value reported in this work is better than the previous gold standard method, which demonstrates LOD value as $4.0 \, \text{ng mL}^{-1}$ [49].

QCM has also been used to investigate the interaction between cancer biomarkers and their respective target receptors to improve detection. For instance, a QCM device with dissipation (QCM-D) was successfully utilized to optimize the response of an electrochemical impedance spectroscopy (EIS) sensor using DNA aptamers against PSA, showing another dimension of the utility of QCM in improving the detection limit for this marker. The biomarker detection was attained by functionalizing the gold sensor surface via thiol chemistry. For this, different ratios of thiolated DNA aptamer and 6-mercapto-1-hexanol (MCH) as spacer molecules were utilized. PSA binding efficiency was monitored by measuring QCM-D signals that provided information about the mass of PSA bound on the surface and aptamer conformation and layer hydration. The ratio of 1:200 for DNA aptamer/spacer molecule (MCH) resulted in the maximum PSA binding. Moreover, by monitoring the QCM-D signal, a dissociation constant (K_D) of 37 nM was achieved for the first time for PSA DNA aptamer. The principle of the assay is given in Figure 5.3 [50].

The up-regulation of α-enolase that is an enzyme of the glycolysis can be used in pancreatic ductal adenocarcinoma (PDAC) as a biomarker. The interactions between unphosphorylated α-enolase peptides and synthetic phosphorylated from the serum of healthy and PDAC patients were quantified by

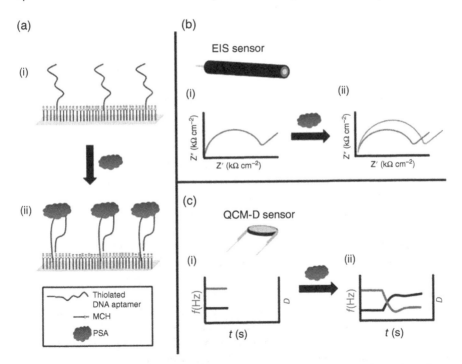

Figure 5.3 Biosensor detection principle with representation of the gold electrode surface (a); EIS measurements showing a signal change in the Nyquist plot before (i) and after (ii) PSA binding (b); QCM-D measurements showing frequency and dissipation responses before (i) and after (ii) PSA binding (c). *Source:* Formisano et al. [50]. Reproduced with permission of Elsevier. (*See insert for color representation of the figure.*)

immobilizing the synthetic peptides to the self-assembled alkanethiol monolayer on the gold surface of a QCM-D sensor [51].

Tumor protein p53, also known as p53, is a crucial compound in multicellular organisms and functions as a tumor suppressor where it avoids cancer formation by preventing genome mutation. The detection of this biomarker and its mutated forms are vital in cancer diagnosis. One of the earliest applications of QCM in detecting a single-nucleotide polymorphism (SNP) in the p53 tumor suppressor gene was carried out by Wang et al. [52]. A toehold-mediated strand displacement reaction (SDR) coupled with QCM biosensor was developed utilizing an hairpin capture probe with an external toehold that was immobilized on the QCM's gold surface. The use of a mass amplifier in the form of a streptavidin-coupled reporter probe enhanced the QCM signal and improved the specificity toward target gene fragment compared with other single-base mutant sequences of p53. A sensitivity assay was obtained with a detection limit of 0.3 nM and a recovery of 84.1% when the target sequence was spiked in HeLa cell lysate [52].

QCM was also utilized to study the hybridization of DNA fragments of the p53 gene near codon 248 with mass amplification using AuNPs as a preliminary study in the development of a sensitive p53 detection assay for clinical application. The results showed that the hybridization sensitivity and the reliability of the QCM were significantly enhanced by three orders of magnitude using AuNPs with a dissociation constant (K_D) value of 5.29×10^{-10} M compared with about 10^{-7} M without the addition of AuNPs [53]. Being a tumor suppressor gene, the p16 is also a very commonly investigated cancer biomarker. The aberrant methylation of the base cytosine, which is the most common epigenetic modification of the p16 gene, was monitored from clinical cholangiocarcinoma tissues using QCM. The developed method was able to successfully discriminate the methylated and unmethylated cytosine bases of the gene, thus presenting a potential POC molecular screening method as being rapid, sensitive, accurate, and cost-effective [14].

As opposed to prostate cancer, which affects mainly men, breast cancer is the most common cancer afflicting women. An assay for the detection of the highly metastatic breast cancer cells using QCM was developed by functionalizing transferrin on the gold sensor surface. Transferrin expression was monitored from MCF7 cells and MDA-MB 231 breast cancer cells with low and high metastatic potentials, respectively. Poly(2-hydroxyethyl methacrylate) (PHEMA) nanoparticles were utilized to enhance the signal. The results showed a good correlation of the number of transferrin receptors on the cells and the QCM signal. The developed QCM biosensor has the potential to detect high metastatic breast cancer cells [54].

In many immortal phenotypes of cancer cells, the telomerase is related to cell senescence and apoptosis of these cells, and hence its detection and monitoring represent an important strategy. More specifically, telomerase inhibitors that bind to telomerase are potential drugs for clinical therapy. Screening of these inhibitors is currently carried out nonquantitatively by the telomerase assay (TRAP assay) with a PCR step. In order to circumvent this problem, a QCM assay was developed to quantify telomerase activity. The developed QCM-based assay was successfully used to investigate the effect and mechanisms of two potential telomerase inhibitors: epigallocatechin gallate (EGCG) and stavudine [55].

Certain virus infections such as human papillomavirus (HPV) infection pose a serious clinical issue since they may lead to cervical cancer. The detection of HPV is currently achieved in real time by employing TaqMan-qPCR. Although this is a sensitive method, it requires skilled personnel and significant operation time. The use of biosensor can present a rapid and simple assay for this virus. In a study, a loop-mediated isothermal amplification LAMP-QCM biosensor was utilized to measure the copy number of HPV16 DNA from cervical samples, and the results were quantitatively compared with the standard TaqMan assay (TaqMan-qPCR) and demonstrated that the LAMP-QCM exhibits 100% specificity and 7.6% imprecision and is 10-fold more sensitive than the TaqMan assay, showing that biosensor development can offer a more sensitive and rapid method for the detection of this virus [56].

5.3.2 Cardiovascular Disease Detection Using Biosensors

Cardiovascular system disorders (CVDs) has become the number one cause of death worldwide with an estimated 17.5 million deaths in 2012. This represents about 31% of deaths globally, of which an estimated 6.7 million deaths were due to stroke, while the 7.4 million were due to coronary heart disease. Low- and middle-income countries pose the greatest number of death with more than three-quarters of CVD deaths. Those who are at high risk of CVD or afflicted with CVD require early and timely diagnostics so that CVD management via medicine or counseling can be administered [57].

Early and rapid diagnosis of CVD is crucial not only for patient survival but also for decreasing the cost and treatment time significantly. Current diagnostic techniques for CVD depend intensely on classical methods that rely on tests carried out in central laboratories. These tests could take many hours to days [34]. Although patients suspected of having CVD should normally exhibit three signs of the disease—chest pain, diagnostic electrocardiogram (ECG) changes, and a raise of sera biochemical markers—the first two criteria are considered poor diagnostic tests for the disease as half of CVD patients do not exhibit these symptoms. This creates a challenge for early diagnosis of the disease [36]. Thus, quantification of cardiac markers becomes the only reliable method for an accurate diagnosis of CVD. One of the most accurate, specific, and rapid methods available or being developed for CVD is through the biosensor platform. More recently, microfluidics-based biosensor technology and lab-on-a-chip technology are in the pipeline for consideration as detection methods for cardiac markers [57, 58].

CVD is a plethora of different disorders, which affect the blood vessels and heart. Atherosclerosis is also a condition affiliated with CVD and occurs through the formation of plaque that builds up in the walls of arteries, narrowing them. This complicates blood flow and leads to stroke or heart attack [59]. A battery of biomarkers is associated with progression and affliction of CVD, including C-reactive protein (CRP) and cardiac troponin I (cTnI) or cardiac troponin T (cTnT). Other less known biomarkers include lipoprotein-associated phospholipase A(2), myoglobin, interlukin-1 (IL-1), interlukin-6 (IL-6), myeloperoxidase (MPO), low-density lipoprotein (LDL), and tumor necrosis factor alpha (TNF-α) [57, 60, 61].

The earliest detection method for CRP carried out in a QCM platform utilized an anti-CRP antibody and its F(ab')2 fragment immobilized on the gold surface of the QCM. The developed method could detect CRP in the range from 10 to 100 μg mL^{-1} [62]. This study evaluated the construction of a high affinity QCM immunosensor using anti-CRP antibody and its fragments for CRP detection. Three types of antibody were immobilized on the surface of a QCM via covalent binding. The affinity between CRP and the antibodies was then determined. The highest affinity was achieved when anti-CRP F(ab')2-IgG antibody (70 μg mL^{-1}) was immobilized on the QCM instead of the normally used anti-CRP IgG antibody. High affinity and selectivity for CRP were achieved in human serum with the treatment of 2-methacryloyloxyethyl phosphorylcholine-co-n-butyl methacrylate. The assay using anti-CRP F(ab')2-IgG antibody demonstrated a linearity of the CRP calibration in the concentration range of 0.01–0.1 ng mL^{-1} [12]. Another attempt in detecting CRP using a batch-type QCM immunosensor was carried out in a direct-binding mode. The developed sensor exhibits a linear detection range of 0.27–106 nM for rat CRP (115 kDa) with a detection limit of 0.53 nM (60.95 μg mL^{-1}) [63].

The development of a microfluidic-based QCM system with an indirect competitive assay approach for CRP is capable to demonstrate a better sensitivity than the batch-type QCM. Such a system was constructed with a micro-dispensing pump, a buffer reservoir, an injector, an oscillator module, a flow-through cell with a sensor chip, and a frequency response analyzer controlled by PC (Figure 5.4). A linear relationship between sensor response and CRP concentration was obtained when plotted in double logarithmic scale showing a wide concentration range between 0.130 ng mL^{-1} and 25 μg mL^{-1} with an LOD of 0.130 ng mL^{-1} [64].

In an attempt to develop a reusable QCM sensor for CRP, a sandwich assay was conducted using 20 nm AuNPs conjugated anti-CRP antibodies. The CRP detection was investigated in the concentration range from 0.02 to 30 μg mL^{-1} that revealed the detection limit as 0.02 μg mL^{-1} [65]. A novel QCM-based assay for CRP was developed using anti-CRP immobilized on ZnO tetrapods,

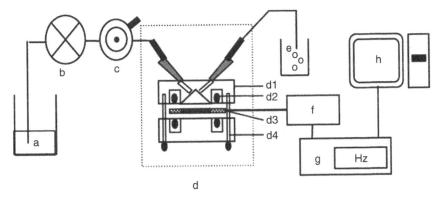

Figure 5.4 Schematic representation of a flow-type QCM immunosensor system. a, Buffer reservoir; b, micro-dispensing pump; c, injector; d, flow-through cell; d1, acryl holder; d2, O-ring; d3, QCM; d4, joint; e, disposal basin; f, oscillator module; g, quartz crystal analyzer; h, PC. Source: Kim et al. [64]. Reproduced with permission of Elsevier.

which were distributed onto the electrode surfaces of QCM, and monoclonal anti-CRP was then immobilized on the ZnO tetrapod film. The QCM assay successfully worked in the linear range from 0.1 to 26 μg mL^{-1} with an acceptable sensitivity of 7 ± 0.17 Hz per decade [66].

QCM can also be built up by incorporating new designs into it to decrease the background noise and increase the performance of the bioassays. The use of inkjet printing equipment in QCM design while developing a detection technique for CRP is a good example for this. The implementation of inkjet printing over the conventional microfluidic system facilitates the commercial development of CRP detection using QCM. The constructed QCM system has achieved highly sensitive detection for this biomarker in the concentration range of 50–1000 ng mL^{-1} and paved the way for commercialization of the device [67].

cTnT detection was also reported using a cost-effective QCM sensor by introducing a carboxylic polyvinyl chloride to the sensing surface as the immobilization layer. A direct detection of cTnT as low as 5 ng mL^{-1} was achieved [58]. In another work, instead of an antibody, a small synthetic peptide identified from a polyvalent phage-displayed library was used as an affinity receptor for the QCM-based quantification of cTnI. The peptide receptor demonstrated a very high affinity for cTnI in the nanomolar levels ($K_D = 66 \pm 4$ nM and 17 ± 8 nM, determined by two independent methods). It was immobilized onto a gold surface, and a sensitivity of 18 ± 1 Hz (μg mL^{-1}) was achieved for the biomarker detection corresponding to an LOD of 0.11 μg mL^{-1} [68].

Fonseca et al. investigated a cTnT detection method by employing a QCM sensor where covalent immobilization of antihuman troponin T (anti-TnT) antibodies on the nanostructured electrode surface by thiol–aldehyde linkages was carried out. The biomarker was quantified in the concentration range of $0.003-0.5\,ng\,mL^{-1}$ in human serum with a very low LOD ($0.0015\,ng\,mL^{-1}$) [69]. Dual QCM technique has also offered a highly sensitive diagnostic method for cTnT analysis with an LOD of $0.008\,ng\,mL^{-1}$ by coupling the technique with monoclonal antibodies (mAb-cTnT). Moreover, this method could measure cTnT in human serum without dilution [61].

5.3.3 Pathogenic Disease Detection Using QCM Biosensors

The detection of pathogens based on the piezoelectric sensor has some advantages compared with another sensor such as SPR. Firstly, the mass change elicited by a big target such as bacteria is greater than smaller targets such as proteins and peptides. Secondly, QCM is as not limited as SPR due to the latter's limited evanescent wave penetration depth [10, 64, 70–74]. The list of pathogens detected using QCM is steadily increasing over the years (Table 5.1).

The detection of *Escherichia coli* O157:H7 and *S. typhimurium* was among the earliest bacterial pathogen detection method carried out with QCM sensors. In the detection of *S. typhimurium* using an anti-salmonella CSA-1 polyclonal goat antibody in a direct assay format, a detection limit of $1 \times 10^2\,CFU\,mL^{-1}$ was achieved after the immunomagnetic separation (IMS) of the bacterium [81]. For the detection of *E. coli* O157:H7, affinity-purified polyclonal antibodies were utilized as the capturing antibodies, and the antibody immobilization method uses protein A. Although both SPR and QCM could detect the bacterium in 1 h, the QCM instrument showed better sensitivity than SPR in terms of both signal-to-noise ratio and detection limit ($10^5–10^8\,CFU\,mL^{-1}$) [80].

Table 5.1 Literature related to QCM immunosensor and the limits of detection achieved for some pathogenic bacteria.

Bacteria	Limit of detection (LOD) (CFU mL^{-1})	Reference
Campylobacter jejuni	1.5×10^2	[10]
Escherichia coli O157:H7	4×10^2	[75]
Salmonella typhimurium	10–20 cells	[28]
Edwardsiella tarda	8×10^2	[76]
E. coli O157:H7	23 cells	[77]
Pseudomonas aeruginosa	1.5×10^2	[78]
Staphylococcus epidermidis	1.3×10^3	[79]
E. coli O157:H7	$10^5–10^8$	[80]
S. typhimurium	1×10^2	[81]

Another label-free QCM sensor for detecting *E. coli* O157:H7 was developed using a specific antimicrobial peptide as a bioreceptor. A detection limit of $400\,\text{CFU}\,\text{mL}^{-1}$ was achieved [75].

Edwardsiella tarda is an opportunistic pathogen in humans, and its detection via QCM was investigated using a peptide ligand from a phage peptide library. The peptide was first labeled with fluorescein or biotin and a mixture of *E. tarda* and the biotinylated peptide was injected into a QCM biosensor with a filter module as a connection. Unbound peptide was separated from the *E. tarda*–peptide complex using the filter and subsequently detected with a streptavidin-coated QCM sensor chip. A very good detection limit (8×10^2) was obtained in this research [76].

Although QCM can be operated label-free, a label-like system in the form of the mass amplifier has already been proven to increase sensitivity. Among the mass amplifiers used in amplifying the sensor signal, nanomaterials such as AuNPs, magnetic, silica, and special polymeric structures act as excellent mass amplifiers [82]. For instance, a QCM-based DNA sensor with streptavidin-conjugated nanoparticles for the detection of *E. coli* O157:H7 improved the LOD from 10^6 to $2.7 \times 10^2\,\text{CFU}\,\text{mL}^{-1}$ [83]. *Staphylococcus epidermidis* is a significant pathogen causing nosocomial infections of the blood. Its detection in clinical samples through QCM is possible by employing nucleic acid biosensor array with streptavidin-coated AuNPs that are conjugated to the PCR-amplified fragments as a mass amplifier. In such a study, thiolated probes with a specific target to the 16S rRNA gene of *S. epidermidis* were utilized. A detection limit of $1.3 \times 10^3\,\text{CFU}\,\text{mL}^{-1}$ was achieved. The developed assay was tested against 55 clinical samples and validated via the conventional clinical microbiological method. The sensitivity and specificity of the developed assay were 97.14 and 100%, respectively [79].

In another work, a QCM-based DNA array for the detection of five bacteria was developed based on the hybridization response of the bacterial 16S–23S rDNA internal transcribed spacer (ITS) region. AuNPs were utilized to amplify the frequency shift signals of the PCR products. The developed sensor was tested against 50 clinical samples and validated by conventional bacterial culture method. A detection limit of $1.5 \times 10^2\,\text{CFU}\,\text{mL}^{-1}$ of *Pseudomonas aeruginosa* was achieved. The detection sensitivity and specificity of the developed biosensor method were 94.12 and 90.91%, respectively [78].

An extensive use of nanoparticles during sample preparation and signal development was carried out for *E. coli* O157:H7 detection using QCM coupled with beacon immunomagnetic (BIMPs) nanoparticles and streptavidin–gold. In the method, *E. coli* O157-BIMPs were coupled with the target polyclonal anti-*E. coli* O157:H7 antibody, the beacon antibody biotin IgG. After capturing *E. coli* O157-BIMPs in a sample, the biotin–avidin system was utilized to conjugate *E. coli* O157-BIMPs to the streptavidin–gold, the AuNPs were subsequently enlarged in solution, and the whole complex was captured on a magnetic plate and then measured via QCM using a monoclonal anti-*E. coli* O157:H7 antibody. A very low detection limit of $23\,CFU\,mL^{-1}$ in phosphate-buffered saline was achieved [77].

Wang et al. enhanced the detection of a SNP in the p53 tumor suppressor gene using a QCM sensor with the aid of mass amplifier based on streptavidin-coupled reporter probe [52]. Salam et al. [28] employed AuNPs to detect *S. typhimurium* cells using indirect, direct, and sandwich assays. The QCM immunosensor-utilizing AuNPs gave a very high sensitivity with the LOD between 10 and $20\,CFU\,mL^{-1}$ as compared with direct and normal sandwich assay at 1.83×10^{2} and $1.01\times10^{2}\,CFU\,mL^{-1}$, respectively. This work indicates that QCM can be used for very sensitive detection of the pathogen at the level that could bypass pre-enrichment steps and decrease assay time. This is a promising achievement for pathogen detection, which normally takes between 5 and 12 days for completion. AuNPs was also utilized in a QCM-based detection of a single nucleotide mutation of the 677TT gene [84] and in the assay for adenosine triphosphate via DNAzyme-activated and aptamer-based target [85]. Various nanoparticles made from silica, magnetic material, and polymer in different sizes with diameters ranging from 30 to 970 nm were explored for the detection of *E. coli* O157:H7 in a QCM immunosensor, and it was discovered that these nanoparticles significantly improved detection [86]. Hence, the nanoparticles could be used to enhance the detection of pathogens. In the most recent publication, the detection of *C. jejuni* on QCM with a sandwich assay using AuNPs mass amplification showed a detection limit of $150\,CFU\,mL^{-1}$, which is among the lowest detection limit available for this bacterium [10].

5.4 Conclusions

The application of biosensors in medical diagnostics is on the rise. Their advantages include rapidity, simple to use, selectivity, and sensitivity that is suitable for point-of-care diagnostics. Of the biosensors being put forward for clinical applications, QCM offers a promising future for rapid diagnostics, particularly with the incorporation of smart nanomaterials. The QCM biosensor has been tested in many areas of medical diagnostics including cancer and cardiovascular diseases as well as pathogen testing, which is the most intensely studied. The label-free and labeled QCM approaches with mass amplifiers are covered in this chapter with their weakness and strengths. The application of QCM in real samples where the biosensor is intended to be used such as sera or other bodily fluids is another area that needs proper and intense attention as this is the ultimate aim of biosensing application in diagnostics and remains the most problematic due to nonspecific binding. An area of future improvement of the QCM sensor is multiparameter analyses, where this involves the integration of several biosensing elements in a single housing. The fabrication of technology such as inkjet printing will propel the use of the sensor further, while efficient and stable microfluidics will allow QCM to be fabricated like the current glucose sensors. The stability of the bioreceptors and repeatability of the measurements will remain important future improvements.

Acknowledgment

Z.A. gratefully acknowledges support from the European Commission, Marie Curie Actions and IPODI as the principle investigator.

References

1 Altintas, Z.; Tothill, I. E. *Sens. Actuators B Chem.* **2012**, 169, 188–194.

2 Huang, C.-J.; Lu, C.-C.; Lin, T.-Y.; Chou, T.-C.; Lee, G.-B. *J. Micromech. Microeng.* **2007**, 17, 835–842.

3 Karaseva, N.; Ermolaeva, T. *Microchim. Acta* **2015**, 182, 1329–1335.

4 Harper, D. **2010**. *Online Etymology Dictionary.* http://www.etymonline.com/index.php?allowed_in_frame=0&search=piezein (accessed on July 24, 2017).

5 Manbachi, A.; Cobbold, R. S. C. *Ultrasound* **2011**, 19, 187–196.

6 Krautkrämer, J.; Krautkrämer, H. *Ultrasonic Testing of Materials* **1990**, 4th ed., Springer-Verlag: Berlin.

7 Lu, C.; Czanderna, A. W. *Applications of Piezoelectric Quartz Crystal Microbalances* **2012**, Elsevier: Amsterdam.

8 Marrazza, G. *Biosensors* **2014**, 4, 301–317.

9 Cooper, M. A.; Singleton, V. T. *J. Mol. Recognit.* **2007**, 20, 154–184.

10 Masdor, N. A.; Altintas, Z.; Tothill, I. E. *Biosens. Bioelectron.* **2016**, 78, 328–336.

11 Kanazawa, K. K.; Gordon, J. G. *Anal. Chem.* **1985**, 57, 1770–1771.

12 Kurosawa, S.; Nakamura, M.; Park, J.-W.; Aizawa, H.; Yamada, K.; Hirata, M. *Biosens. Bioelectron.* **2004**, 20, 1134–1139.

13 Vaughan, R. D.; O'Sullivan, C. K.; Guilbault, G. G. *Enzym. Microb. Technol.* **2001**, 29, 635–638.

14 Prakarnkamanant, P.; Prasongdee, P.; Limpaiboon, T.; Daduang, J.; Promptmas, C.; Jearanaikoon, P. *J. Med. Technol. Phys. Ther.* **2010**, 20, 115–127.

15 Brockman, L.; Wang, R.; Lum, J.; Li, Y. *Open J. Appl. Biosens.* **2013**, 2, 97–103.

16 Shons, A.; Dorman, F.; Najarian, J. *J. Biomed. Mater. Res.* **1972**, 6, 565–570.

17 Benito, J.; Sorribas, S.; Lucas, I.; Coronas, J.; Gascon, I. *ACS Appl. Mater. Interfaces* **2016**, 8, 16486–16492.

18 Nakanishi, K.; Muguruma, H.; Karube, I. *Anal. Chem.* **1996**, 68, 1695–1700.

19 Babacan, S.; Pivarnik, P.; Letcher, S.; Rand, A. *J. Food Sci.* **2002**, 67, 314–320.

20 Alocilja, E. C.; Radke, S. M. *Biosens. Bioelectron.* **2003**, 18, 841–846.
21 Altintas, Z.; Uludag, Y.; Gurbuz, Y.; Tothill, I. E. *Anal. Chim. Acta* **2012**, 712, 138–144.
22 Barlow, S. M.; Raval, R. *Surf. Sci. Rep.* **2003**, 50, 201–341.
23 Love, J. C.; Estroff, L. A.; Kriebel, J. K.; Nuzzo, R. G.; Whitesides, G. M. *Chem. Rev.* **2005**, 105, 1103–1169.
24 Alonso, J. M.; Bielen, A. A. M.; Olthuis, W.; Kengen, S. W. M.; Zuilhof, H.; Franssen, M. C. R. *Appl. Surf. Sci.* **2016**, 383, 283–293.
25 Kushiro, K.; Lee, C.-H.; Takai, M. *Biomater. Sci.* **2016**, 4, 989–997.
26 Yuan, Y. J.; Chen, Q.; Li, J. *RSC Adv.* **2016**, 6, 40336–40342.
27 Phan, H. T. M.; Bartelt-Hunt, S.; Rodenhausen, K. B.; Schubert, M.; Bartz, J. C. *PLoS One* **2015**, 10, e0141282.
28 Salam, F.; Uludag, Y.; Tothill, I. E. *Talanta* **2013**, 115, 761–767.
29 Ranjan, R.; Esimbekova, E. N.; Kratasyuk, V. A. *Biosens. Bioelectron.* **2017**, 87, 918–930.
30 Hong, W.; Lee, S.; Cho, Y. *Biosens. Bioelectron.* **2016**, 86, 920–926.
31 Choi, J. M.; Park, W. S.; Song, K. Y.; Lee, H. J.; Jung, B. H. *Biomed. Chromatogr. BMC* **2016**, 30, 1963–1974.
32 Tan, C.; Qian, X.; Guan, Z.; Yang, B.; Ge, Y.; Wang, F.; Cai, J. *SpringerPlus* **2016**, 5, 467.
33 Manne, U.; Jadhav, T.; Putcha, B.-D. K.; Samuel, T.; Soni, S.; Shanmugam, C.; Suswam, E. A. *Curr. Colorectal Cancer Rep.* **2016**, 12, 332–344.
34 Boutry, C. M.; Nguyen, A.; Lawal, Q. O.; Chortos, A.; Rondeau-Gagné, S.; Bao, Z. *Adv. Mater.* **2015**, 27, 6954–6961.
35 Tang, L.; Liu, J.; Yang, B.; Chen, X.; Yang, C. An impedance wire integrated with flexible flow sensor and FFR sensor for cardiovascular measurements. In Toshiyoshi, H. and Wang, X.E. (eds), *2016 IEEE 29th International Conference on Micro Electro Mechanical Systems (MEMS)*, January 24–28, **2016**, Shanghai, China. IEEE: Piscataway, 345–348.
36 Aragón, A. M.; Hernando-Rydings, M.; Hernando, A.; Marín, P. *AIP Adv.* **2015**, 5, 87132.
37 Lim, J. Y.; Choi, S.-I.; Choi, G.; Hwang, S. W. *Mol. Brain* **2016**, 9, 26.
38 Stephen Inbaraj, B.; Chen, B. H. *J. Food Drug Anal.* **2016**, 24, 15–28.
39 Liew, K. J. L.; Chow, V. T. K. *J. Virol. Methods* **2006**, 131, 47–57.
40 Santiago, G. A.; Vergne, E.; Quiles, Y.; Cosme, J.; Vazquez, J.; Medina, J. F.; Medina, F.; Colón, C.; Margolis, H.; Muñoz-Jordán, J. L. *PLoS Negl. Trop. Dis.* **2013**, 7, e2311.
41 Soper, S. A.; Brown, K.; Ellington, A.; Frazier, B.; Garcia-Manero, G.; Gau, V.; Gutman, S. I.; Hayes, D. F.; Korte, B.; Landers, J. L.; Larson, D.; Ligler, F.; Majumdar, A.; Mascini, M.; Nolte, D.; Rosenzweig, Z.; Wang, J.; Wilson, D. *Biosens. Bioelectron.* **2006**, 21, 1932–1942.
42 Tothill, I. E. *Semin. Cell Dev. Biol.* **2009**, 20, 55–62.
43 Altintas, Z.; Tothill, I. *Sens. Actuators B Chem.* **2013**, 188, 988–998.

44 Zhu, H.; Dale, P. S.; Caldwell, C. W.; Fan, X. *Anal. Chem.* **2009**, 81, 9858–9865.

45 Altintas, Z.; Uludag, Y.; Gurbuz, Y.; Tothill, I. E. *Talanta* **2011**, 86, 377–383.

46 Prakrankamanant, P. *J. Med. Assoc. Thai.* **2014**, 97(Suppl 4), 56–64.

47 Okuno, J.; Maehashi, K.; Kerman, K.; Takamura, Y.; Matsumoto, K.; Tamiya, E. *Biosens. Bioelectron.* **2007**, 22, 2377–2381.

48 Uludağ, Y.; Tothill, I. E. *Talanta* **2010**, 82, 277–282.

49 Morgan, T. O.; Jacobsen, S. J.; McCarthy, W. F.; Jacobson, D. J.; McLeod, D. G.; Moul, J. W. N. *Engl. J. Med.* **1996**, 335, 304–310.

50 Formisano, N.; Jolly, P.; Bhalla, N.; Cromhout, M.; Flanagan, S. P.; Fogel, R.; Limson, J. L.; Estrela, P. *Sens. Actuators B Chem.* **2015**, 220, 369–375.

51 Bianco, M.; Aloisi, A.; Arima, V.; Capello, M.; Ferri-Borgogno, S.; Novelli, F.; Leporatti, S.; Rinaldi, R. *Biosens. Bioelectron.* **2013**, 42, 646–652.

52 Wang, D.; Tang, W.; Wu, X.; Wang, X.; Chen, G.; Chen, Q.; Li, N.; Liu, F. *Anal. Chem.* **2012**, 84, 7008–7014.

53 Li, S.; Li, X.; Zhang, J.; Zhang, Y.; Han, J.; Jiang, L. *Colloids Surf. A Physicochem. Eng. Asp.* **2010**, 364, 158–162.

54 Atay, S.; Pişkin, K.; Yılmaz, F.; Çakır, C.; Yavuz, H.; Denizli, A. *Anal. Methods* **2016**, 8, 153–161.

55 Wang, J.; Liu, L.; Ma, H. *Sens. Actuators B Chem.* **2017**, 239, 943–950.

56 Jearanaikoon, P.; Prakrankamanant, P.; Leelayuwat, C.; Wanram, S.; Limpaiboon, T.; Promptmas, C. *J. Virol. Methods* **2016**, 229, 8–11.

57 Altintas, Z.; Fakanya, W. M.; Tothill, I. E. *Talanta* **2014**, 128, 177–186.

58 Wong-ek, K.; Chailapakul, O.; Nuntawong, N.; Jaruwongrungsee, K.; Tuantranont, A. *Biomed. Tech. (Berl.)* **2010**, 55, 279–284.

59 McDonnell, B.; Hearty, S.; Leonard, P.; O'Kennedy, R. *Clin. Biochem.* **2009**, 42, 549–561.

60 Fakanya, W. M.; Tothill, I. E. *Biosensors* **2014**, 4, 340–357.

61 Mattos, A. B.; Freitas, T. A.; Silva, V. L.; Dutra, R. F. *Sens. Actuators B Chem.* **2012**, 161, 439–446.

62 Kurosawa, S.; Nakamura, M.; Aizawa, H.; Park, J.-W.; Tozuka, M.; Kobayashi, K.; Yamada, K.; Hirata, M. *Proceedings of the Annual IEEE International Frequency Control Symposium*, New Orleans, LA, May 29–31, **2002**, 273–275.

63 Kim, N.; Kim, D. K.; Cho, V. J. *KSBB J.* **2007**, 22, 443–446.

64 Kim, N.; Kim, D.-K.; Cho, Y.-J. *Sens. Actuators B Chem.* **2009**, 143, 444–448.

65 Ding, P.; Liu, R.; Liu, S.; Mao, X.; Hu, R.; Li, G. *Sens. Actuators B Chem.* **2013**, 188, 1277–1283.

66 Wang, X.; Zhang, J.; Zheng, Z. *Sens. Lett.* **2013**, 11, 1617–1621.

67 Fuchiwaki, Y.; Tanaka, M.; Makita, Y.; Ooie, T. *Sensors* **2014**, 14, 20468–20479.

68 Wu, J.; Cropek, D. M.; West, A. C.; Banta, S. *Anal. Chem.* **2010**, 82, 8235–8243.

69 Fonseca, R. A. S.; Ramos-Jesus, J.; Kubota, L. T.; Dutra, R. F. *Sensors* **2011**, 11, 10785–10797.

70 Souto, D. E. P.; Faria, A. R.; deAndrade, H. M.; Kubota, L. T. *Curr. Protein Pept. Sci.* **2015**, 16, 782–790.

71 Taylor, A. D.; Ladd, J.; Yu, Q.; Chen, S.; Homola, J.; Jiang, S. *Biosens. Bioelectron.* **2006**, 22, 752–758.

72 Wei, D.; Oyarzabal, O. A.; Huang, T.-S.; Balasubramanian, S.; Sista, S.; Simonian, A. L. *J. Microbiol. Methods* **2007**, 69, 78–85.

73 Hong, S.-R.; Kim, M.-S.; Jeong, H.-D.; Hong, S. *Aquac. Res.* **2017**, 48 (5), 2055–2063.

74 Farka, Z.; Kovář, D.; Skládal, P. *Sensors (Switzerland)* **2015**, 15, 79–92.

75 Dong, Z.-M.; Zhao, G.-C. *Talanta* **2015**, 137, 55–61.

76 Choi, H.; Choi, S.-J. *Anal. Biochem.* **2012**, 421, 152–157.

77 Shen, Z.-Q.; Wang, J.-F.; Qiu, Z.-G.; Jin, M.; Wang, X.-W.; Chen, Z.-L.; Li, J.-W.; Cao, F.-H. *Biosens. Bioelectron.* **2011**, 26, 3376–3381.

78 Cai, J.; Yao, C.; Xia, J.; Wang, J.; Chen, M.; Huang, J.; Chang, K.; Liu, C.; Pan, H.; Fu, W. *Sens. Actuators B Chem.* **2011**, 155, 500–504.

79 Xia, H.; Wang, F.; Huang, Q.; Huang, J.; Chen, M.; Wang, J.; Yao, C.; Chen, Q.; Cai, G.; Fu, W. *Sensors* **2008**, 8, 6453–6470.

80 Su, X. L.; Li, Y. *Trans. Am. Soc. Agric. Eng.* **2005**, 48, 405–413.

81 Su, X. L.; Li, Y. *Biosens. Bioelectron.* **2005**, 21, 840–848.

82 Altintas, Z.; Tothill, I. E. Nanomaterial applications in health care diagnostics. In Kharisov, B. I., Kharissova, O. V., Ortiz-Mendez, U. (eds), *CRC Concise Encyclopaedia of Nanotechnology* **2015**, CRC Press, Boca Raton, 317–329.

83 Mao, X.; Yang, L.; Su, X.-L.; Li, Y. *Biosens. Bioelectron.* **2006**, 21, 1178–1185.

84 Nie, L. B.; Yang, Y.; Li, S.; He, N. Y. *Nanotechnology* **2007**, 18, 305501–305505.

85 Song, W.; Zhu, Z.; Mao, Y.; Zhang, S. *Biosens. Bioelectron.* **2014**, 53, 288–294.

86 Jiang, X.; Wang, R.; Wang, Y.; Su, X.; Ying, Y.; Wang, J.; Li, Y. *Biosens. Bioelectron.* **2011**, 29, 23–28.

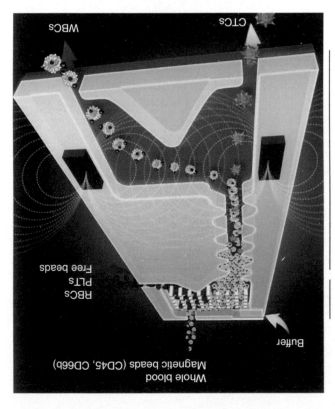

Figure 8.2 CTC isolation chip. Whole blood mixed with antibody-tagged magnetic beads is sent to the channel in parallel with buffer solution. Deterministic lateral displacement compartment of the chip enables the separation of white blood cells (WBCs) and CTCs from the whole blood. Then, WBCs and CTCs are brought on a single line after passing asymmetric focusing elements utilizing inertial microfluidics. Finally, magnetic force is applied over centrally aligned CTCs to a different force magnetically labeled CTCs to a different outlet, completing the isolation process. *Source:* Karabacak et al. [22]. Reproduced with permission of Nature Publishing Group.

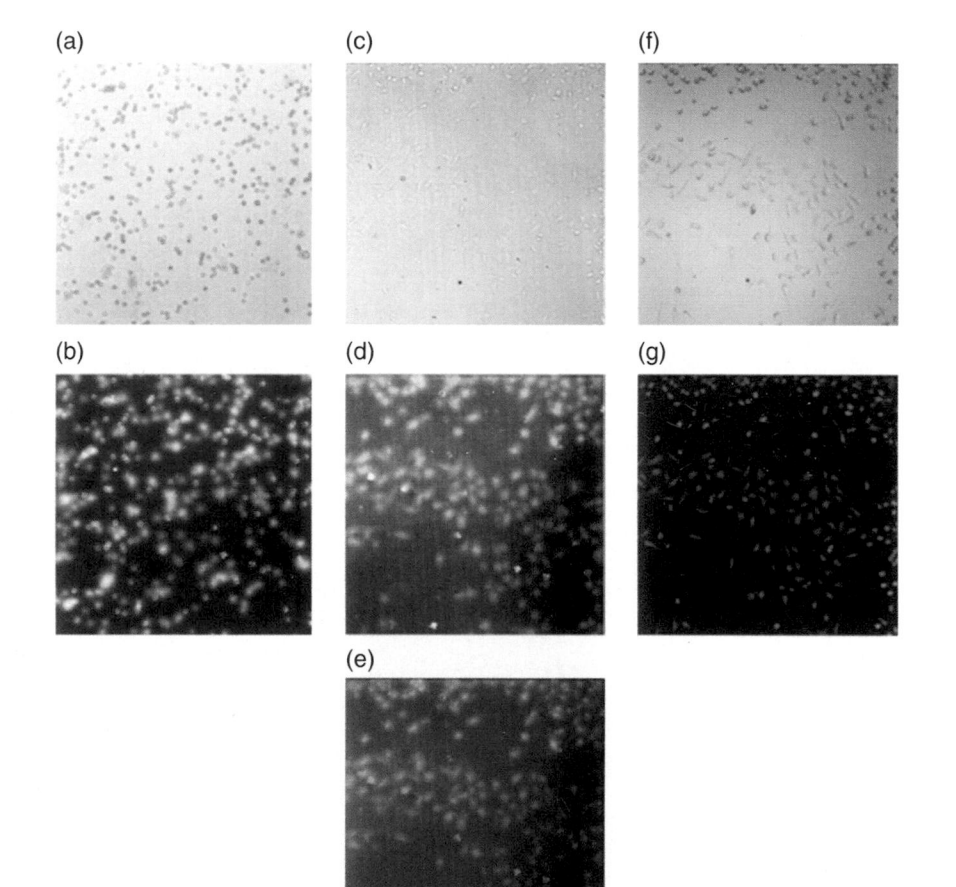

(a)

(c)

(f)

(b)

(d)

(g)

(e)

100 μm

Figure 7.3 Lensless shadow image (a) and fluorescence image of the fluorescent microspheres (b); lensless shadow image (c) and fluorescence image of L929 cells (d), as well the composition (e) of both type pictures; and the bright filed (f) and fluorescent microscopy images for the same L929 sample (g). *Source:* Li et al. [9]. Reproduced with permission of SPIE.

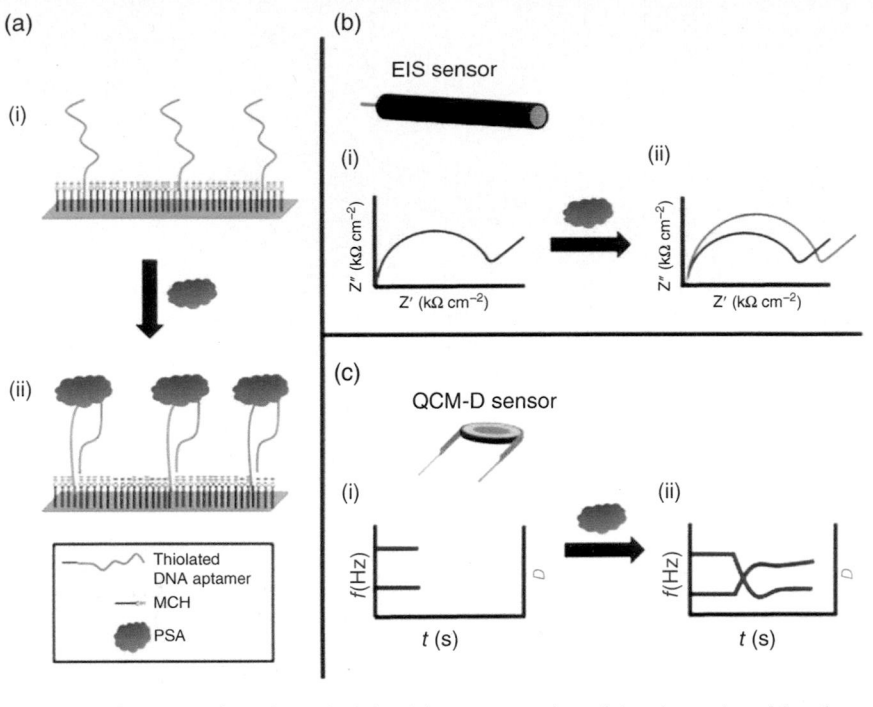

Figure 5.3 Biosensor detection principle with representation of the electrode gold surface (a); EIS measurements showing a signal change in the Nyquist plot before (i) and after (ii) PSA binding (b); QCM-D measurements showing frequency and dissipation responses before (i) and after (ii) PSA binding (c). *Source:* Formisano et al. [50]. Reproduced with permission of Elsevier.

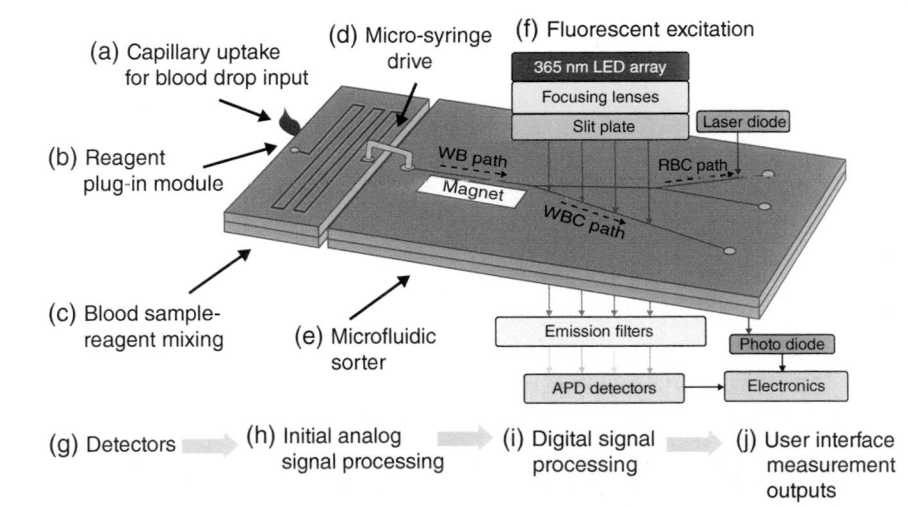

Figure 7.2 Schematic of the subcomponents and processes that constitute an integrated portable micro-cytometer reader, substance packs, and disposable microfluidic chip. *Source:* Grafton et al. [8]. Reproduced with permission of SPIE.

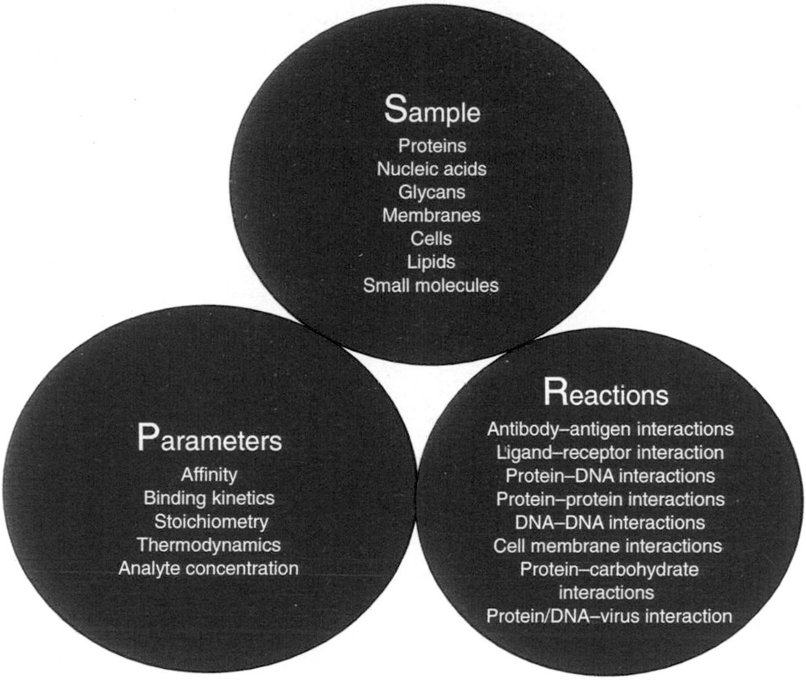

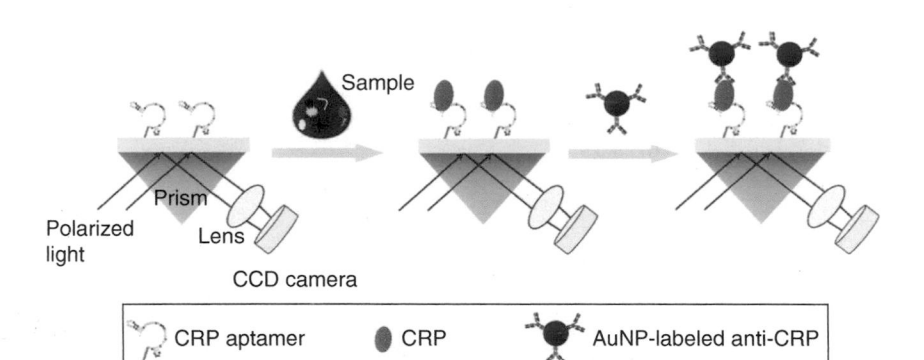

Figure 4.1 Applications of SPR including sample targets, measurable parameters, and reactions that facilitate measurement.

Figure 4.5 Schematic illustration of the AuNP-enhanced SPR biosensor with an aptamer–antibody sandwich assay. *Source:* Wu et al. [35]. Reproduced with permission of Royal Society of Chemistry.

Biosensors and Nanotechnology: Applications in Health Care Diagnostics, First Edition.
Edited by Zeynep Altintas.
© 2018 John Wiley & Sons, Inc. Published 2018 by John Wiley & Sons, Inc.

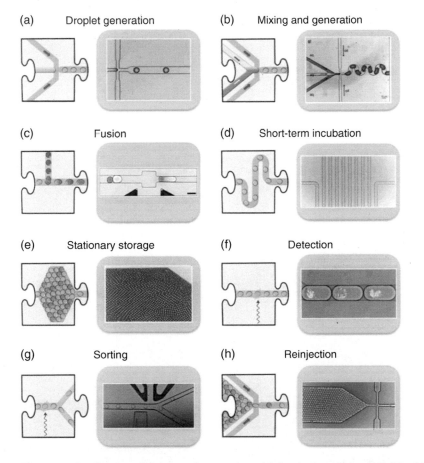

Figure 8.7 A collection of droplet unit operations: (a) droplet generation [46], (b) mixing and generation [47], (c) fusion [48], (d) incubation, (e) storage [49], (f) detection, (g) sorting, and (h) re-injection [46]. *Source:* From Kintses et al. [50]. Adapted with permission from Elsevier. (a and h) *Source:* Schaerli and Hollfelder [46]. Reproduced with permission of Royal Society of Chemistry. (b) *Source:* Huebner et al. [47]. Reproduced with permission of The Royal Society of Chemistry. (e) *Source:* Courtois et al. [49]. Adapted with permission of John Wiley Sons, Inc.

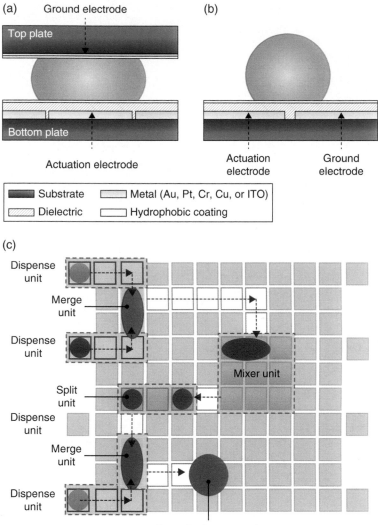

Figure 8.10 Digital microfluidic platforms: (a) Close system and (b) open system. (c) 2D schematic illustration of digital microfluidic chip illustrating dispensing, merging, splitting, and mixing units. *Source:* Malic et al. [67]. Reproduced with permission of Royal Society of Chemistry.

(a)

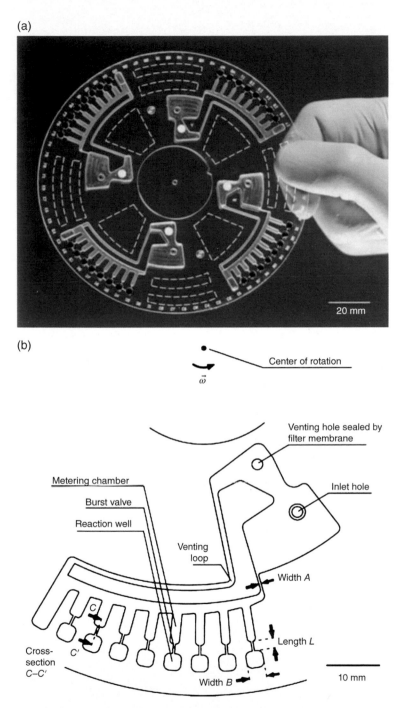

(b)

Center of rotation

$\vec{\omega}$

Venting hole sealed by
filter membrane

Metering chamber

Inlet hole

Burst valve

Reaction well

Venting
loop

Width *A*

C

Length *L*

Cross-
section
C–C'

C'

10 mm

Width *B*

Figure 8.11 (a) Photograph of a CD microfluidic cartridge unit, reaction wells are filled with dye solution. (b) Schematic of the microfluidic design showing some critical components such as inlet, metering chambers, burst valves, and reaction wells. *Source:* Focke et al. [76]. Reproduced with permission of Royal Society of Chemistry.

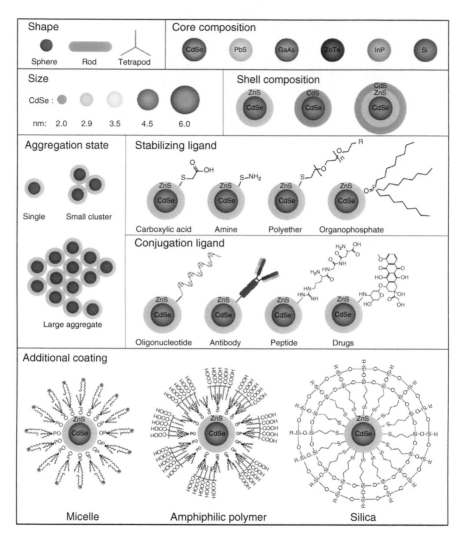

Figure 9.2 Quantum dots (QDs) are a heterogeneous group of materials. Biological fate and toxicity depend on QD physicochemical properties. Shape, core composition, size, and shell composition can be manipulated during QD synthesis. Post-synthesis surface ligands are added to solubilize and target the particles. An additional coating can further protect the QD core from oxidation. Surface chemistry influences the quantum dot's propensity to aggregate, particularly in biological solutions. *Source:* Tsoi et al. [14]. Reproduced with permission of American Chemical Society.

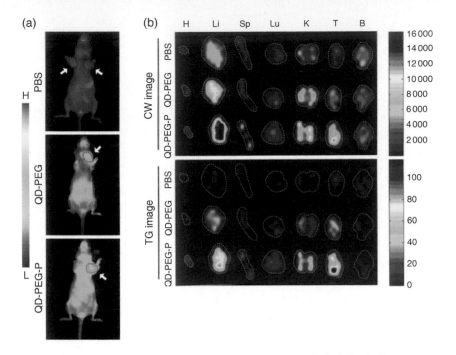

Figure 9.5 Breast tumor targeting with QD-PEG-P. (a) Intravital whole-body fluorescent imaging (ventral view) of MCF10CA1a breast tumor-bearing nude mice at 28 h after intravenous injection of PBS, QD-PEG, or QD-PEG-P. Images were taken under 700 nm channel of LI-COR Pearl Impulse small animal imaging system. White arrows show the position of MCF10CA1a tumors (circled in black). (b) *Ex vivo* CW and TG imaging of the tissues harvested from the mice in (a) after the *in vivo* imaging. B, brain; H, heart; K, kidney; Li, liver; Lu, lung; Sp, spleen; T, tumor. *Source:* Liu et al. [12]. Reproduced with permission of John Wiley & Sons.

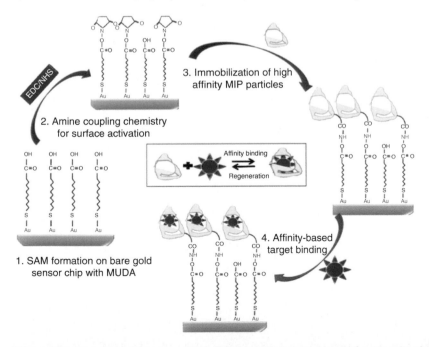

Figure 10.2 Virus detection using molecularly imprinted nanopolymer-functionalized SPR biosensor. *Source:* Altintas et al. [73]. Reproduced with permission of American Chemical Society.

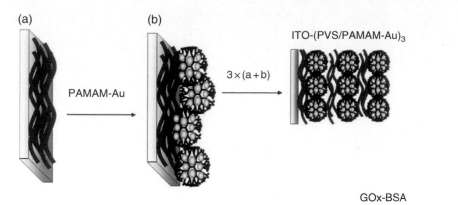

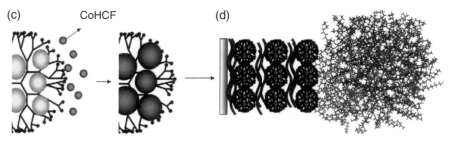

Figure 11.2 Glucose biosensor containing AuNPs. (a, b) Deposition of LbL multilayers containing polyvinyl sulfonate/PAMAM-Au. (c) After deposition of three bilayers, an ITO-(PVS/PAMAM-Au)$_3$@CoHCF electrode was prepared by potential cycling. (d) Immobilizing enzyme using bovine serum albumin (BSA), glutaraldehyde, and glucose oxidase (GOx). *Source:* Crespilho et al. [42]. Reproduced with permission of Elsevier.

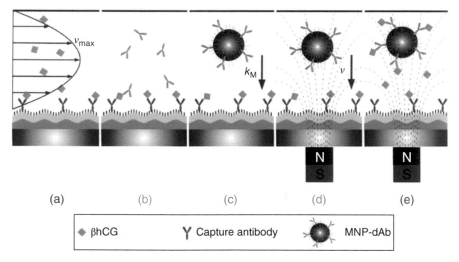

Figure 11.3 Schematics of used detection formats: direct detection (a), sandwich assays with amplification by detection antibody (b), and MNP-dAb without (c) and with (d) applied magnetic field. Detection format consisting of preincubating MNP-dAb with βhCG followed by sandwich assay upon applied magnetic field gradient (e). *Source:* Wang et al. [84]. Reproduced with permission of American Chemical Society.

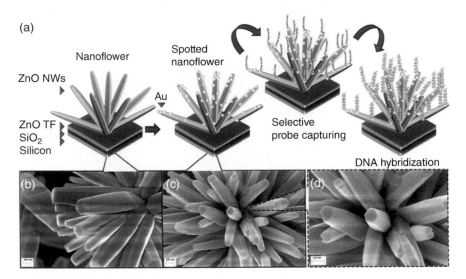

Figure 11.11 (a) Schematic of the steps involved in the synthesis of the spotted nanoflower (NF) DNA bioelectrode. (b) FESEM image of low magnification revealing the flower-like ZnO nanostructure possessing hexagonally shaped tips, which demonstrate the high crystallinity of the prepared ZnO nanowire ends. (c and d) Low- and high-magnification images of spotted NFs indicate that radially oriented NFs have an average length of 2–3 μm and a diameter of approximately 100 nm. *Source:* Perumal et al. [225]. https://www.nature.com/articles/srep12231#comments. Used under CC BY 4.0 http://creativecommons.org/licenses/by/4.0/.

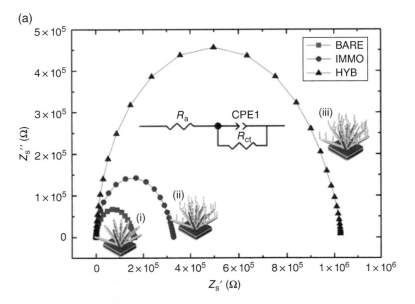

Figure 11.12 (a) Impedance spectra of (i) spotted NF, (ii) spotted NF/p-DNA (probe), and (iii) spotted NF/p-DNA/t-DNA (duplex) bioelectrode; the inset shows the Randles equivalent circuit, where the parameters R_a, R_{ct}, and CPE represent the bulk solution resistance, charge transfer resistance, and constant phase element, respectively.

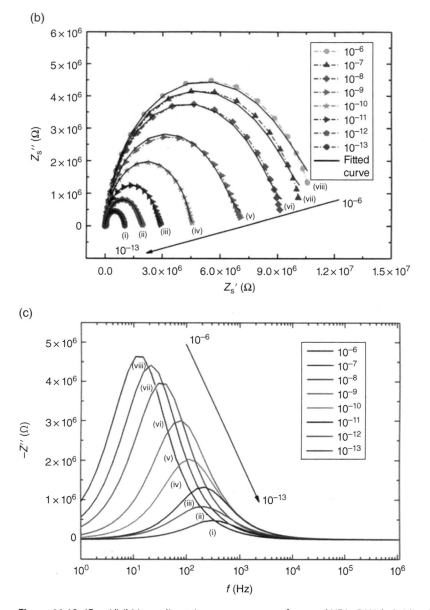

Figure 11.12 (Cont'd) (b) Impedimetric response curve of spotted NF/p-DNA hybridized with different concentrations of complementary target DNA (i–viii), 10 μM to 100 fm. (c) Imaginary part showing the overall impedance, which decreases, and the peak frequency, which is shifted toward the higher frequencies as the concentration of complementary DNA decreases.

(d)

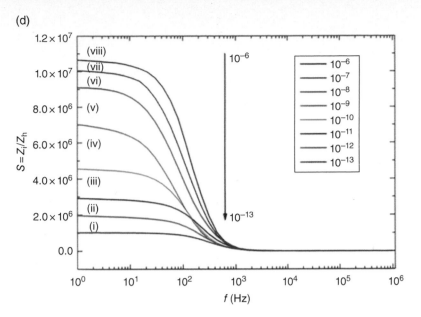

Figure 11.12 (Cont'd) (d) The gain curve of spotted NF/p-DNA hybridized at different concentrations. *Source:* Perumal et al. [225]. https://www.nature.com/articles/srep12231#comments. Used under CC BY 4.0 http://creativecommons.org/licenses/by/4.0/.

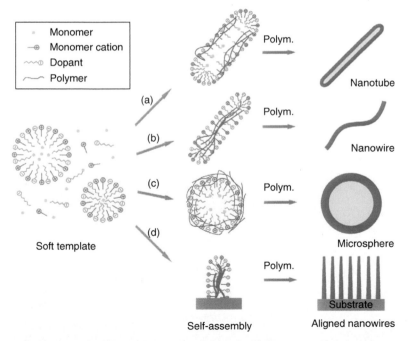

Figure 11.19 Schematic of the mechanism of the soft-template synthesis of different conducting polymer nanostructures: (a) micelles acted as soft templates in the formation of nanotubes. Micelles were formed by the self-assembly of dopants, and polymerization was carried out on the surface of the micelles; (b) nanowires formed by the protection of dopants. The polymerization was carried out inside the micelles; (c) monomer droplets acted as soft templates in the formation of microsphere; and (d) polymerization on the substrate producing aligned nanowire arrays. Nanowires were protected by the dopants, and polymerization was carried out on the tips of nanowires. *Source:* Xia et al. [292]. Reproduced with permission of Elsevier.

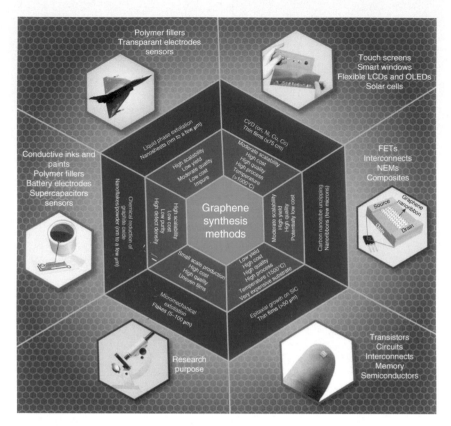

Figure 13.2 A schematic showing the conventional methods commonly used for the synthesis of graphene along with their key features and the possible applications. *Source:* Image CKMNT, http://www.nanowerk.com/spotlight/spotid=25744.php

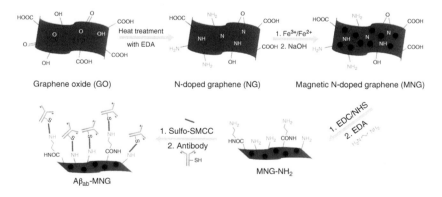

Figure 13.7 Schematic representation of the Aβab-MNG platform construction. *Source:* Li et al. [77]. https://www.ncbi.nlm.nih.gov/pmc/articles/PMC4844990/figure/f1/. Used under CC BY 4.0 http://creativecommons.org/licenses/by/4.0/.

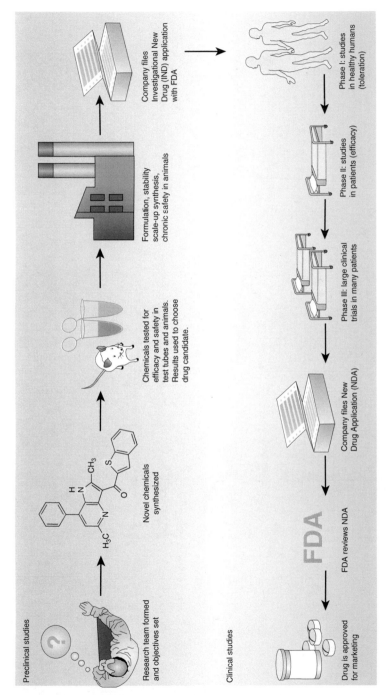

Figure 14.1 Stages of drug discovery process. *Source:* Lombardino and Lowe [41]. Reproduced with permission of Nature Publishing Group.

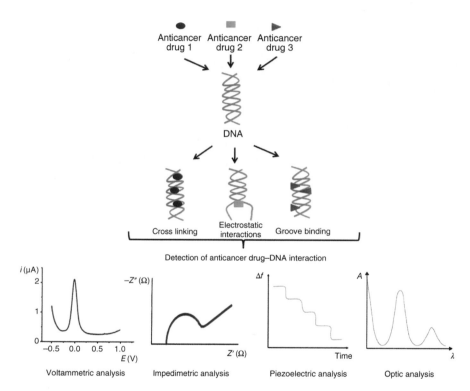

Figure 15.1 Schematic presentation of biosensing of anticancer drug–DNA interactions.

6

Electrochemical-Based Biosensor Technologies in Disease Detection and Diagnostics

Andrea Ravalli and Giovanna Marrazza

Department of Chemistry "Ugo Schiff", University of Florence, Florence, Italy

6.1 Introduction

The field of molecular diagnostics has expanded rapidly over the past decade. Applications range from the detection of mutations responsible for human-inherited disorders, disease-causing and food-contaminating viruses, and research into bacteria and forensics. In order to improve patient care, molecular diagnostics laboratories have been challenged to develop new tests that are reliable, cost-effective, and accurate and to optimize existing protocols by making them faster and more economical.

A rapid and accurate screening of health conditions represents a key step in order to identify the sign of symptoms of a disease or of an altered physiological process although their effect did not come forward. Although few tests can be performed directly by a doctor (i.e., blood pressure analysis for hypertensions diagnosis), other tests require the use of sophisticated instrumentations such as computerized axial tomography (CAT) scan or nuclear magnetic resonance imaging (NMRI), which are expensive, performed by trained personnel with long waiting time before the analysis response is completed and delivered to the patient.

Patient conditions can also be detected by the analysis of the so-called biomarker, or rather, as defined by the United States and European Union, National Institute of Health, a "characteristic that is objectively measured and evaluated as an indicator of normal biological processes, pathogenic processes, or pharmacologic responses to therapeutic intervention." The biomarkers act as indicators of a normal or a pathogenic biological process. They allow assessing the pharmacological response to a therapeutic intervention. A biomarker shows a specific physical trait or a measurable biologically produced change in the

Biosensors and Nanotechnology: Applications in Health Care Diagnostics, First Edition.
Edited by Zeynep Altintas.
© 2018 John Wiley & Sons, Inc. Published 2018 by John Wiley & Sons, Inc.

body that is linked to a disease or a particular health condition. Briefly, biomarker (i.e., enzymes, proteins, DNA, RNA, etc.) concentration increases in biological fluids (urine, blood, serum, etc.) over a threshold level in the presence of a particular disease (cancer, heart infraction, diabetes, etc.) or physiological condition (i.e., pregnancy) and decreases after the patient's condition has returned to normal.

Based on their properties, biomarkers can be classified into different inter-connected groups [1–3]:

- Disease biomarkers, correlated with the history of the disease, which should be referred with known clinical indicators
- Diagnostic biomarkers, which help to find a specific disease in a population of patients
- Biomarkers for drug discovery, which define the interaction of the drug with its target
- Predictive biomarkers, which can be used to evaluate the possibility to develop a specific disease in a population of patients
- Drug activity biomarkers, which can be used to detect the effectiveness of a given drug
- Translational biomarkers, which can be applied in preclinical and clinical test
- Surrogate biomarkers, which can be used as a substitute measure for a clinical endpoint to evaluate the benefit or lack of clinical treatments

Analyses of fluids, tissue, and so on in equipped laboratory with qualified staff routinely address biomarker analysis [4]. The use of electrochemical biosensors for biomarker determination represents an ideal tool considering the low cost of the device, the short time required for the analysis, and the possibility to detect two or more biomarkers at the same time. Moreover, the biosensors can be outside the laboratory or at the bedside of the patient used as a point-of-care-testing (POCT) device by untrained personnel [5–7].

Focusing on the recent activity of researchers worldwide, without pretending to being exhaustive, the aim of this chapter is to give emphasis on the recent advances on the biosensors for cancer, cardiovascular, and autoimmune diseases. An overview of biorecognition elements and transduction technology will be presented as well as the biomarkers and biosensing systems currently used to detect the onset and monitor the progression of selected diseases.

6.2 Electrochemical Biosensors: Definitions, Principles, and Classifications

As defined by the International Union of Pure and Applied Chemistry (IUPAC),

> biosensor is a self-contained integrated device which is capable of providing specific quantitative or semi-quantitative analytical information

using a biological recognition element (biochemical receptor) which is in direct spatial contact with a transducer. The transducer is used to convert (bio)chemical signal resulting from the interaction of the analyte with the bioreceptor into an electronic one. The intensity of generated signal is directly or inversely proportional to analyte concentration. [8]

In accordance with transduction method, biosensors can be classified as electrochemical [8], piezoelectric [9], optical [10, 11], thermometric [12], and magnetic [13].

The analyte detection can be achieved by electrochemical biosensor through two main approaches: catalysis and affinity reaction (Figure 6.1).

Catalytic biosensors employ enzymes and microorganisms as the biorecognition molecule that catalyzes a reaction involving the analyte to give a product. Various types of biosensors have been developed over the years, since 1962, when Clark and Lyons reported the first biosensor for glucose in blood measurement. The enzyme-based sensor was the first generation of biosensors and in the subsequent years a variety of biosensors for other clinically important substances were developed. In enzymatic biosensor an electrochemical signal (i.e., redox-related current) is measured in relation to the conversion of the analyte in an electroactive species by the bioreceptor (i.e., transducer-integrated enzymes, cellular organelles, tissues, or whole microorganisms) (Figure 6.1a) [14, 15].

In affinity biosensors the transducer incorporates biological or biomimetic receptor molecules that can reversibly bind the target analyte with high selectivity and specificity in a nondestructive way [16]. Commonly used

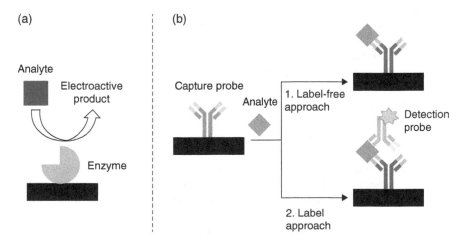

Figure 6.1 Schematic representation of electrochemical biosensors development: (a) catalytic biosensor; (b) affinity biosensor: (1) label-free approaches and (2) label approach.

biorecognition molecules include antibodies, acid nucleics, molecular-imprinted polymers, peptides, and lectins. Then, electrochemical detection of the captured analyte can be achieved following either of the two main routes: labeled or label-free assay (Figure 6.1b). In labeled assay, similar to well-known sandwich ELISA test, once the capture probe was immobilized on the surface of the transducer followed by the affinity reaction with the analyte, a detection probe was introduced. Detection probe can be directly labeled or can be detected by the introduction of a tertiary-labeled molecules. Labeled molecules can include enzyme, magnetic beads, nanostructures, and so on. Electrical signal obtained in the presence of the label (i.e., enzymatic substrate conversion into an electroactive compound) is proportional to the number of analyte molecules bound to the electrode surface and allows the evaluation of analyte concentration both in standard and in real samples. The use of a labeled detection probe allows an increase of the sensibility (coupled with a decreasing of nonspecific signal) for the detection of analytes in a complex matrix but with an increase of costs on assay-working time and with limitation for the direct use on field [17–28].

In label-free biosensors, the changing of electrode surface properties (i.e., capacitance, resistance, conductibility, etc.) can be directly evaluated after the bioreceptor–analyte affinity reaction. In comparison with labeled assays, label-free techniques allow a real-time evaluation of the analyte–bioreceptor binding with reduced costs of analysis and assay time. In biosensor development, in order to increase the sensibility and decrease the nonspecific signal, in particular using real samples, a blocking agent (such as bovine serum albumin or casein) needs to be introduced [29–35].

Immunoassays are currently the predominant analytical technique for the quantitative determination of a broad variety of analytes in clinical analysis. The recognition elements are immunochemical antibody–antigen (Ab–Ag) interactions. The quality of the designed immunosensor depends on the affinity and selectivity of the selected antibody to its antigen, as well as the proper immobilization of the antibody, with an optimum density and adjusted orientation for the antigen binding [36].

As recognition elements, the nucleic acids (DNA and RNA) are chemically more stable than antibodies. A genosensor is a biosensor that employs an immobilized oligonucleotide as the biorecognition element. Specifically, electrochemical genosensors rely upon the conversion of the base-pair recognition event into a useful electrical signal. Typically, the design of an electrochemical genosensor involves immobilization of the DNA probe, hybridization with the target sequence, labeling, and electrochemical investigation of the surface. Optimization of each step is required to improve the overall performance of the devices [37].

Recently recognition elements, such as aptamers and molecularly imprinted polymers (MIPs), attracted the attention of the scientific community in order to develop chemical sensors and biosensors with improved analytical performance [38]. Aptamers are short and single-stranded DNA or RNA sequences, selected *in vitro* using a technique called *s*election *e*volution of *l*igands by *ex*ponential (SELEX) enrichment from synthetic oligonucleotides libraries that are able to bind different kinds of target molecules (e.g., other oligonucleotides, biomarkers, or small synthetic molecules such pesticides) with high selectivity and specificity. Compared to antibodies, aptamers show higher detection range, a higher stability under different chemical and physical conditions, a prolonged shelf life, and acceptable cross-reactivity and can be obtained using efficient and cost-effective processes. They can be also easily modified, giving the possibility to obtain various labeled probe elements [39–41].

MIPs are realized generating specific recognition sites on polymeric nanoparticles in order to mimic a biological receptor. They are synthesized using a template-assisted approach; functional monomers form a complex with the template (that will be the analytical target) and then the polymerization is started, using an appropriate solvent. Removing then the template by extensive washing steps allows the polymer to maintain specific recognition sites, complementary to the template in size, shape, and position of interacting functional groups [42–44].

Bioreceptor immobilization represents a key step in biosensor development, which drastically affects the analytical performance of the device in terms of sensibility, stability, response time, and reproducibility. Immobilized biomolecules have to retain unchanged structure, to maintain their biological properties, and to remain bound to the surface of the transducer during the use of the biosensor [45]. The immobilization strategies most generally employed are physical or chemical methods. The choice of the immobilization method is dependent from assay format.

Enzymes are usually entrapped in 3D matrix by different strategies such as polymeric film (obtained by electropolymerization or by photopolymerization) or in polysaccharide-, graphite-, or carbon-based nanostructures [46–50].

Other enzyme immobilization procedures involve the use of cross-linking agents (such as glutaraldehyde), the direct adsorption, or the covalent binding (i.e., *N*-hydroxysuccinimide (NHS) with 1-ethyl-3-(3-dimethylaminopropyl) carbodiimide (EDAC) coupling chemistry) on the electrode surface [15].

Antibody immobilization must take into account the two main effects: antibodies density (in order to avoid nonspecific binding) and optimal orientation (to maximize antigen binding). Surface modification strategies may exploit the specific binding of Fc region (i.e., with the use of protein A- or G-modified surface), the binding of specific molecule attached to the Fc part (i.e., biotinylated Ab/streptavidin-modified electrode), or the covalent attachment to the transducer surface by covalent coupling (generally with the use of NHS/EDAC reagents) [51]. For gold or gold nanoparticle (AuNP)-modified surfaces, electrode formation of self-assembled monolayer is generally used. In this case, a carboxyl of amine-modified thiols is first immobilized on the surface of the transducer (exploiting the SH/Au bound) followed by antibody immobilization through amidic reaction or affinity reaction of Fc-modified biomolecules [52, 53].

The possibility to insert various functional groups directly from synthesis into oligonucleotides probes (such as DNA and RNA) or into engineered-scaffold protein (i.e., affibody, nanobody) facilitates electrode surface modification procedure. Biotin/streptavidin affinity and thiols/gold surface are mostly used [54, 55].

The use of micro- and nano-magnetic particles either as immobilization platforms or as labels has attracted major attention lately. Their use as a solid support for the immobilization of the recognition element has many advantages such as fast and specific immobilization of a wide amount of bioelements, easy separation after the washing and reaction steps, easy manipulation, and reduction of the analysis time and reagents consumption [56].

Catalytic and affinity electrochemical biosensors can also be easily coupled with microfluidic devices. In particular, these systems allow the development of automatic devices with easy data analysis, easy handling of chemicals (solution replacing and washing steps), and use of low volume of solutions, which results in an increase of the precision and the accuracy of the measurements [57–60].

Recently, miniaturization advances in electronic and fluidic field lead to the development of the so-called lab-on-a-chip devices (also named as micro total analysis, µTAS) in which one or more laboratories' features are integrated in a chip of dimensions between one millimeter and few centimeters. However, technological advances are still necessary to confer the proper sensitivity and selectivity (with reduced false signal errors due to nonspecific adsorption) to these devices necessary for the detection of analytes in complex matrices [61–64].

The common electrochemical techniques used for the detection of the analyte include potentiometry, amperometry, voltammetry, conductometry, and electrochemical impedance spectroscopy (EIS). Potentiometric devices measure the potential difference between the working and the reference electrodes when zero or no significant current flows between them. An example of potentiometric biosensor is represented by ion-selective electrode (ISE) in which the measured potential is obtained by the accumulation of ions at the surface of ion-selective membrane [65–67]. Nowadays the development of field-effect-based biosensors (mainly ion-selective field-effect transistor (ISFET) and enzyme field-effect transistor (EnFET)) is becoming common. ISFET is a classical metal/oxide/semiconductor (MOS) field-effect transistor (FET) with a gate formed by a separated reference electrode and attached to the gate area via a solution. These semiconductor FETs have an ion-sensitive surface. The surface electrical potential changes due to the interaction between ions and the semiconductor and can be subsequently measured [30, 68–71].

Amperometric devices measure the current related to the redox reaction of an electroactive species in a biochemical reaction. The measured current is obtained by maintaining a fixed voltage between the working and the reference electrodes [72]. Amperometry is the electrochemical technique usually applied in commercially available biosensors for clinical analyses that detect redox reactions.

If the current is measured in relation to a potential variation, biosensor devices can be referred to as voltammetric. Some examples of voltammetric techniques used in electrochemical biosensor include cyclic, linear, differential pulse, and square wave voltammetry [39, 73–76].

In conductometric devices the reaction of the biorecognition process produces a change in the ionic species concentration, resulting in a variation of the electrical conductivity of the solution or of the current flow, which is measured between two metal electrodes placed at a certain distance [77, 78].

EIS-based biosensor has attracted great attention due to the possibility to monitor the biorecognition reaction occurring on the surface of modified electrode without the use of a label. In this technique, a sinusoidal current in relation to a sinusoidal voltage application is produced. By the variation of the frequency of the applied potentials, the impedance of the modified surface can be calculated [79, 80]. Each measurement is generally reported in a graph (called Nyquist plot) in which the real component of the impedance is plotted against its imaginary part. Simulated circuit (i.e., Randles equivalent circuit) can be used to retrieve information about the bio-reaction taking place at the modified electrode surface, which are commonly expressed in terms of electrolyte resistance (R_{el}), charge transfer resistance (R_{ct}), double-layer capacitance (C_{dl}), mass transfer resistance (R_{mt}), and Warburg impedance (W) [53, 81, 82].

Like many other technologies, electrochemical sensors and biosensors have benefited from the growing power of new materials, design, and processing tools; thus, many technologies are available to fabricate miniaturized, simple-to-operate, and low-cost devices. Among these, thick-film technology is one of the most used since the equipment needed is less complex and costly than others. Moreover, thick-film electrochemical transducers can be easily mass-produced at low cost and thus used as disposable. The use of disposable strips that obviate the need for a regeneration step would appear to be the most promising approach, since it meets the needs of decentralized testing. Additionally they are suitable for working with micro-volumes and for point-of-care test.

Recently, nanomaterials are finding wide applications in biomedical and bioanalytical devices because they can greatly facilitate the miniaturization of sensors and instrumentation and improve the analytical performance or due to their unique electrochemical properties. In fact, nanostructures increase both the electron transfer between the redox center and the electrode surface and the number of immobilized bioreceptor because of their high surface/volume ratio [83–87].

The most commonly used nanomaterials are carbon, gold, silver, and nano-hybrid materials [88]. Several carbon-based surfaces have been investigated as electrochemical transducers. The unique physical and chemical properties of carbon nanomaterials, such as high mechanical strength, excellent electrical and thermal conductivity, highly ordered structure, and high surface area, are responsible for the increasing interest in biosensor design. Carbon nanotubes, carbon nanospheres, carbon nanohorns, graphene oxide, and graphene nanoribbons have been immobilized onto electrodes, sometimes the carbon nanomaterial being modified with biomolecules prior or after the immobilization step, to amplify the electrochemical signal and decrease the limit of detection in biosensing systems [89, 90].

Metal nanoparticles, silver, gold, copper, palladium, platinum, and titanium have been incorporated in electrochemical biosensor configurations with two main purposes: the fabrication of nanostructured supports and their use as signal enhancers. They possess high electrocatalytic activity, stability, and biocompatibility, and they can be easily functionalized. They favor electron transfer, and, as a result, higher sensitivities and lower limit of detections are attained by metal nanomaterial-modified electrochemical biosensors.

The embedding of metal nanoparticles into polymer matrices represents a simple way to use the advantages of nanoparticles. This technique is one of the most efficient strategies to avoid the aggregation of nanosized metal and conserve their properties. Therefore, polymer/metal nanocomposite in which polymer phase acts as a stabilizer, template, or protecting agent shows many important attributes [91]. Nanomaterials coupled to sensors offer versatile opportunities for signal amplification in (bio)analytical systems; how it can be observed from publications is reported in the following sections.

6.3 Biomarkers in Clinical Applications

In the following sections, some particular examples of electrochemical biosensors developed for detecting cancer and cardiovascular and autoimmune diseases are briefly discussed.

6.3.1 Electrochemical Biosensors for Tumor Markers

Cancer must be defined as a disease that is characterized by abnormal growth and development of normal cells beyond their natural boundaries. In the last 50 years despite the global efforts to limit the incident of this disease, cancer has become the leading cause of death [92]. A central issue of the immunosensors applied in clinical diagnosis or in screening and monitoring is that the biomarker used should be expressed in detecting limits in biological fluids. A list of cancer biomarkers (Table 6.1) approved by the Food and Drug

Table 6.1 FDA-approved serum and urine biomarkers for cancer diagnosis, prognosis, and therapy selection.

Type of cancer	Biomarker	Source	Cutoff value
Breast	CA15-3	Serum	$25\,U\,mL^{-1}$
	CA27-29	Serum	$36.4\,U\,mL^{-1}$
	HER2/NEU	Serum	$15\,ng\,mL^{-1}$
Ovarian	CA125	Serum	$35\,U\,mL^{-1}$
Colon	CEA	Serum	$5\,\mu g\,L^{-1}$
Prostate	PSA total	Serum	$4\,ng\,mL^{-1}$
	PSA free/PSA total ratio	Serum	0.10–0.25% age dependant
Thyroid	Thyroglobulin	Serum	$>1\,ng\,mL^{-1}$ (after thyroid surgery)
	TSH-stimulated thyroglobulin	Serum	$>2\,ng\,mL^{-1}$ (after thyroid surgery)
Bladder	NMP22	Urine	$10\,U\,mL^{-1}$
	Fibrin/FDP	Urine	$10\,mg\,L^{-1}$
	BTA	Urine	$14\,U\,mL^{-1}$
	High molecular weight CEA	Urine	$5\,\mu g\,L^{-1}$
Testicular	α-Fetoprotein	Serum	$6\,ng\,mL^{-1}$
	Human gonadotropin-β	Serum	$5\,U\,mL^{-1}$
Pancreatic	CA19-9	Serum	$55\,U\,mL^{-1}$

BTA, bladder tumor-associated antigen; CA, cancer antigen; CEA, carcinoembryonic antigen; FDA, US Food and Drug Administration; FDP, fibrin degradation product; HER2, human epidermal growth factor receptor 2; NMP22, nuclear matrix protein No. 22; PSA, prostate-specific antigen.

Administration (FDA) suggests that some biomarkers for breast, ovarian, and prostate cancers could be used in monitoring (such as CA15-3, CA27-29, and CA125), screening and monitoring (PSA, PSA total, and PSA complex), prognosis (cytokeratins in breast tumor), staging (human chorionic gonadotropin-β in testicular cancer), or drug development [93].

Taking into account that cancer is one of the leading causes of mortality, the early clinical diagnosis is crucial for the successful treatment of the disease. The combination between two or several biomarkers could lead to the correct identification of cancer presence in early stages instead of using one biomarker. Several examples of biosensors for the detection of these biomarkers can be found in the literature (Table 6.2).

Carbohydrate antigen 125 (CA125), also known as mucin 16 (MUC16), is a high-molecular glycoprotein and is the main cancer biomarker screened for the evaluation of ovarian cancer [95]. Ren et al. proposed an immunosensor for

Table 6.2 Biosensor for cancer biomarkers detection.

Biomarker	Electrochemical technique	Linear range	Detection limit	Reference
CA125	SWV	$0.001-25\,ng\,mL^{-1}$	$0.25\,pg\,mL^{-1}$	[35]
	EIS	$0-100\,U\,mL^{-1}$	$6.7\,U\,mL^{-1}$	[53]
	EIS	$0-100\,U\,mL^{-1}$	$0.1\,U\,mL^{-1}$	[30]
	EIS	$0-150\,U\,mL^{-1}$	$1.03\,U\,mL^{-1}$	[34]
	EIS	$0-0.1\,U\,mL^{-1}$	$0.0016\,U\,mL^{-1}$	[31]
	Amperometry	$5-1000\,U\,mL^{-1}$	$40\,U\,mL^{-1}$	[17]
	DPV	$0-25\,U\,mL^{-1}$	$2\,U\,mL^{-1}$	[82]
	DPV	$0.05-20\,U\,mL^{-1}$	$0.002\,U\,mL^{-1}$	[21]
	Amperometry	$0.002-20\,U\,mL^{-1}$	$0.001\,U\,mL^{-1}$	[24]
HE4	SWV	$0-400\,pM$	$6.8\,fM$	[19]
HER2	Amperometry	$1-200\,\mu g\,mL^{-1}$	$1\,\mu g\,mL^{-1}$	[54]
	EIS	10^{-5} to $10^{2}\,ng\,mL^{-1}$	$5\,ng\,mL^{-1}$	[94]
	EIS	$0-40\,\mu g\,mL^{-1}$	$6\,\mu g\,mL^{-1}$	[55]
PSA	CV	$0-80\,ng\,mL^{-1}$	$1\,pg\,mL^{-1}$	[25]
	EIS	$0-25\,ng\,mL^{-1}$	$0.5\,ng\,mL^{-1}$	
CEA	DPV	$0.02-20\,ng\,mL^{-1}$	$7.0\,pg\,mL^{-1}$	[21]

CV, cyclic voltammetry; DPV, differential pulse voltammetry; EIS, electrochemical impedance spectroscopy; SWV, square wave voltammetry.

CA125 detection based on the use of an acid site compound-modified glassy carbon electrode (GCE) [35]. The acid site is composed of ferrocenecarboxylic acid (FA), HCl-doped polyaniline, and chitosan hydrochloride (CS-HCl) that generate analytical signal and increase specific surface area. Anti-CA125 antibody was incubated on the surface of $Ag-Co_3O_4$ nanosheet and functionalized with the acid site-modified GCE. After the affinity reaction with the antigen, redox behavior of FA was used to construct the calibration curve in a linear range between 0.001 and $25\,ng\,mL^{-1}$ with a limit of detection of $0.25\,pg\,mL^{-1}$. The proposed immunosensor showed also good selectivity and good performance for CA125 detection in CA125-spiked human serum samples.

In our previous work, we developed a label-free immunosensor for CA125 detection based on gold nanostructured screen-printed electrodes [53]. First, AuNPs were electrodeposed on the surface of graphite screen-printed electrode by means of cyclic voltammetry technique starting from $0.6\,mM$ $HAuCl_4$ solution prepared in $0.5\,M$ H_2SO_4. Mixed self-assembly monolayer (SAM) was then formed on the surface of gold-nanostructured electrodes using

11-mercaptoundecanoic acid (MUDA) and 1-mercaptohexanol (MCH). Afterward, anti-CA125 antibody was covalently immobilized on the electrode surface by means of EDAC/NHS coupling. Affinity reaction with antigen was the carried out followed by label-free CA125 detection by EIS in the presence of $[Fe(CN)_6]^{3-/4-}$ redox mediator.

A linear calibration curve was obtained between 0 and $100\,U\,mL^{-1}$ CA125 with a detection limit of $6.7\,U\,mL^{-1}$. The proposed biosensor was also applied for the detection of CA125 in fortified serum samples.

Label-free immunosensor for CA125 detection was reported by Das et al. [30]. In this work, gold nanostructures were electrodeposited on the surface of a microchip (composed of a layer of SiO_2 covered by Au and by another layer of SiO_2 with a circular aperture of $5\,\mu m$) and after functionalized with cystamine in order to form an SAM. Cystamine residues were then coupled with glutaraldehyde and with anti-CA125 antibody. Affinity reaction with antigen was evaluated analyzing the reduction of $[Fe(CN)_6]^{3-/4-}$ oxidation current due to the formation of a layer of protein that blocks the electron transfer from the solution to the surface of the chip. This sensor allowed the detection of CA125 in serum and blood with good detection limit ($0.1\,U\,mL^{-1}$) and reproducibility.

Another CA125 immunosensor was reported by Raghav et al. [34]. In this work, gold screen-printed electrodes were modified with cysteamine hydrochloride followed by incubation with Au/AgNPs. Anti-CA125 antibody was immobilized by two different routes: (i) by direct adsorption (Figure 6.2, ISA route) and (ii) by covalent amide bond formation (Figure 6.2, ISB route). Affinity reaction with CA125 was evaluated by EIS measurements. A linear correlation between R_{ct} difference ($\Delta R_{ct} = R_{ct, CA125} - R_{ct, blank}$) and antigen concentration in the range between 1 and $250\,U\,mL^{-1}$ CA125 and between 1 and $150\,U\,mL^{-1}$ were observed, respectively, for ISA (sensitivity $35\,U\,mL\,U^{-1}\,cm^{-2}$) and ISB (sensitivity $190\,U\,mL\,U^{-1}\,cm^{-2}$). Good correlation with reference method (ELISA) was also found.

Impedimetric biosensor was also proposed by Johari-Ahar et al. [31]. In this work, a gold electrode was modified with mercaptopropionic acid (MPA) and then with silica-coated gold nanoparticles ($AuNP/SiO_2$), CdSe quantum dots (QDs), and anti-CA125 monoclonal antibody. The affinity reaction with the cancer biomarker was evaluated by means of EIS in a linear range of $0–0.1\,U\,mL^{-1}$ with a limit of detection of $0.0016\,U\,mL^{-1}$.

An enzyme-linked sandwich immunoassay for the analysis of CA125 cancer biomarker was reported by Al-Ogaidi et al. [17] based on the use of nanoelectrode array chip (NEA), which was functionalized with a monoclonal primary anti-CA125 antibody. Affinity reactions with antigen and with secondary alkaline phosphatase-labeled anti-CA125 antibody were carried out. After the incubation with the substrate (*para*-aminophenylphosphate (PAPP)) redox behavior of enzymatic product (*para*-aminophenol (PAP)) was evaluated

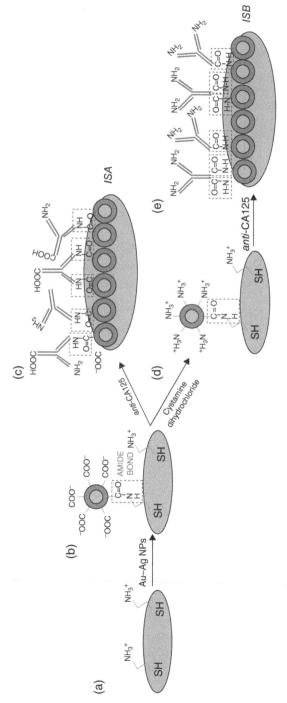

Figure 6.2 Schematic representation of immunosensor development for CA125 detection: direct adsorption (ISA) and covalent amine bond formation (ISB). (a) Cystamine-modified gold electrode, (b) covalent immobilization of Au–Ag NPs, (c) covalent immobilization of *anti*-CA125 on citrate stabilized Au–Ag NPs for immunosensor A (*ISA*), (d) amine functionalization of immobilized Au–Ag NPs, and (e) covalent immobilization of *anti*-CA125 on amine functionalized Au–Ag NPs for immunosensor B (*ISB*). Source: Raghav and Srivastava [34]. Reproduced with permission of Elsevier.

between -0.1 and 0.3 V. The proposed immunoassay showed a linear range for CA125 detection between 5 and $1000\,U\,mL^{-1}$ with a detection limit of $40\,U\,mL^{-1}$ and $0.15\,nA\,mL\,U^{-1}$ as sensitivity. Optical detection was also used using *para*-nitrophenylphosphate as substrate. The enzymatic product *para*-nitrophenol was detected at $405\,nm$ and allows the detection of CA125 in the same linear range in respect to the electrochemical method with a detection limit of $1.3\,U\,mL^{-1}$ and a sensitivity of $0.0005\,OD_{405}\,mL\,U^{-1}$.

Another sandwich immunoassay for CA125 detection was proposed by Taleat et al. [82]. In this work, the surface of graphite screen-printed electrodes was modified with anthranilic acid (AA) by means of cyclic voltammetry. Polyanthranilic acid (PAA) carboxyl groups were then covalently functionalized with a primary monoclonal anti-CA125 antibody followed by affinity reaction with the cancer protein. The assay was completed by incubation with a secondary anti-CA125 antibody labeled with AuNPs, which induced silver deposition from a silver enhancer solution. The formed Ag/AuNPs were then treated with HNO_3, and the dissolved Ag^+ was detected by anodic stripping voltammetry technique. Using this approach CA125 cancer biomarker was detected in a range between 0 and $25\,U\,mL^{-1}$ with a detection limit of $2\,U\,mL^{-1}$.

CA125 was detected together with cancer antigen 153 (CA153) and carcinoembryonic antigen (CEA) using a screen-printed array composed of three graphene-modified graphite working electrodes, a single screen-printed Ag/AgCl pseudo reference, and a single graphite auxiliary electrode [21]. Briefly each working electrode was covalently functionalized with anti-CA125, anti-CA153, and anti-CEA antibodies followed by incubation with a solution containing three cancer biomarkers. Afterward, the immunoassay was completed by reaction with mesoporous platinum nanoparticles (M-Pt NPs) functionalized with anti-CA125 or CA153 or anti-CEA antibodies. Electroreduction of hydrogen peroxide allowed the determination of CA125 (linear range 0.05–$20\,U\,mL^{-1}$, detection limit $0.002\,U\,mL^{-1}$), CA153 (linear range 0.008–$24\,U\,mL^{-1}$, detection limit $0.001\,U\,mL^{-1}$), and CEA (linear range 0.02–$20\,ng\,mL^{-1}$, detection limit $7.0\,pg\,mL^{-1}$) both in buffered solutions and in protein-spiked serum samples (Figure 6.3).

Guo et al. [24] evaluated the use of Au/Pd core–shell nanoparticles in a sandwich immunoassay for CA125 analysis both for electrode surface nanostructuration and as secondary antibody label. CA125 protein was sandwiched between a primary antibody immobilized on the surface of Au/Pd nanoparticles-modified GCE and a secondary anti-CA125 antibody labeled with Au/Pd core–shell nanoparticles. Electrocatalytic reduction of hydrogen peroxide (mediated by AuNPs surface) was used to obtain a linear calibration curve for CA125 detection between 0.002 and $20\,U\,mL^{-1}$ with a detection limit of $0.001\,U\,mL^{-1}$.

Epididymis protein 4 (HE4), a tumor marker for ovarian carcinoma, has come to the fore of interest mainly due to the possibility for its detection in early stages of the disease when the sensing of other biomarkers is limited [96].

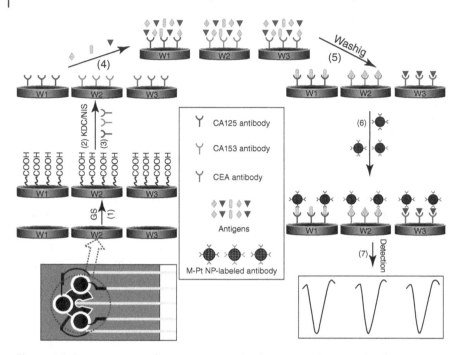

Figure 6.3 Representation of immunosensors development and approaches for CA125 and CEA determination. *Source:* Cui et al. [21]. Reproduced with permission of Elsevier.

Èadková et al. [19] proposed an immunoassay for HE4 detection based on a sandwich format using magnetic beads as support. The assay was assembled by immobilization of a primary anti-HE4 antibody on the surface of the magnetic particles, followed by affinity reaction with the antigen and by the incubation with a secondary anti-HE4 antibody and with a tertiary alkaline phosphatase-labeled antibody. The analysis of redox behavior of *para*-aminophenol (produced by the enzyme from *para*-aminophenylphosphate) obtained through square wave voltammetry techniques allowed the determination of HE4 protein in a range of concentration between 0 and 400 pM, with a detection limit of 6.8 fM and a limit of quantification of 23 fM. The immunosensor was also applied for the detection of HE4 in complex matrices with a recoveries ranging from 87 to 93%.

In the case of breast cancer, the main tumoral marker is represented by human epidermal growth factor receptor 2 (HER2 or HER2/new), which is a member of ErbB protein family and is a transmembrane glycoprotein composed of three distinct regions: an N-terminal extracellular domain (ECD), a single á-helix transmembrane domain (TM), and an intracellular tyrosine kinase domain. Overexpression of HER2 usually results in malignant transformation of cells accounting for 25% of all breast cancer cases (clinical cutoff: 15 ng mL^{-1}), and it is always associated with more aggressive tumor phenotypes [97]. HER2 electrochemical biosensor development has been intensively studied and includes the use of antibody, aptamer, and affibody molecules in label-free or labeled approaches [54, 55, 94, 98–101].

Patris et al. [54] reported the development of a sandwich-type immunoassay for detection of HER2 cancer biomarker, based on the use of Nanobodies® (Nb) as bioreceptor. A primary anti-HER2 Nb was immobilized on the surface of H_2SO_4-activated graphite screen-printed electrode through EDAC/NHS coupling chemistry. After the surface-blocking step with BSA, the affinity reaction with HER2 was carried out, followed by incubation with an HRP-labeled secondary Nb. Amperometric detection of H_2O_2 allowed the construction of a calibration curve in the range between 1 and $200\,\mu g\,mL^{-1}$ with a limit of detection and quantification of 1 and $4\,\mu g\,mL^{-1}$, respectively.

An aptamer-based label-free biosensor was reported by Chun et al. [94]. Thiolated anti-HER2 aptamer was immobilized on the surface of gold electrode followed by surface blocking with PBS/BSA solution. Affinity reaction with HER2 protein was then evaluated by means of EIS in a linear range between 10^{-5} and $10^2\,ng\,mL^{-1}$ HER2 (LOD = $5\,ng\,mL^{-1}$).

In our recent work, we reported the use of affibody molecules as bioreceptor for the development of a label-free electrochemical biosensor for HER2 detection [55]. The biosensor was based on the immobilization of a terminal cysteine-modified affibody on the surface of gold-nanostructured graphite screen-printed electrode. After mixing SAM formation with 6-mercapto-1-hexanol (MCH) and surface-blocking step with BSA, affinity reaction with HER2 was performed and evaluated by EIS technique. The proposed biosensor showed a linear response between 0 and $40\,\mu g\,L^{-1}$ HER2 with a limit of detection of $6\,\mu g\,L^{-1}$. The developed affisensor showed also good response in HER2-fortified serum samples. Kinetic and thermodynamic parameters of affibody-HER2 affinity reaction were furthermore evaluated by surface plasmon resonance experiment.

Kavosi et al. [25] reported the development of an electrochemical immunosensor for the detection of prostate-specific antigen (PSA) based on the use of AuNPs–incorporated polyamidoamine dendrimer (AuNPs–PAMAM) and multiwalled carbon nanotubes/ionic liquid/chitosan nanocomposite (MWCNTs/IL/Chit). In particular GCE was modified with MWCNTs/IL/Chit nanocomposites followed by covalent attachment of PAMAM-modified AuNPs through the use of phthaloyl chloride and linker. PSA cancer biomarker was then sandwiched between a primary and an HRP-modified secondary antibody. DPV analysis of H_2O_2 enzymatic reduction allowed the determination of PSA in a linear range between 0 and $80\,ng\,mL^{-1}$ with a detection limit of $1\,pg\,mL^{-1}$. In addition label-free determination of PSA by EIS was carried out in a linear range up to $25\,ng\,mL^{-1}$ with a detection limit of $0.5\,ng\,mL^{-1}$.

6.3.2 Electrochemical Biosensors for Cardiac Markers

The early and quick diagnosis of cardiovascular disease (CVD) is extremely important not only for patient survival but also for saving cost and a great deal of time in successful prognosis of diseases [102]. Existing methods of diagnosis for CVD rely heavily on classical methods that are based on tests conducted in central laboratories that may take several hours or even days from when tests are ordered to when results are received. Laboratory tests are an important part in diagnosing heart infarction and fast and cost-effective diagnostics is needed. At present three of the most informative markers are cardiac troponin I or T (cTnI/T), myoglobin, and natriuretic peptide, particularly of the B type (BNP). Cardiac biomarkers based on disease type are given in Table 6.3. These markers are determined by different immunoassay methods, such as ELISA, radioimmunoassay, and immunochromatographic tests [103].

As reported in Table 6.4, different biosensors have been developed for cardiac biomarker detection [104, 105].

Moreira et al. [33] reported the development of an MIP-based electrochemical biosensor for the detection of myoglobin, a cardiac marker expressed in the presence of myocardial infarction. PVC-COOH layer was assembled on the surface of gold screen-printed working electrode followed by the covalent

Table 6.3 Biomarkers for cardiovascular diseases.

Type of cardiac disease	Biomarker	Cutoff value
AMI	cTnI	$0.1\,ng\,mL^{-1}$
	cTnT	$0.1\,ng\,mL^{-1}$
	Myoglobin	$70\,ng\,mL^{-1}$
	CK-MB	$10\,ng\,mL^{-1}$
Inflammation cardiac risk factor	CRP	$1-15\,mg\,L^{-1}$
	MPO	$350\,ng\,mL^{-1}$
	TNF-α	$0.0036\,ng\,mL^{-1}$
	IL-6	$0.00138-0.002\,ng\,mL^{-1}$
Myocardial necrosis	11-FABP	$6\,ng\,mL^{-1}$
Acute coronary/heart failure syndrome	BNP	$200\,pg\,mL^{-1}$ (age and sex dependence)
	NT-proBNP	$300\,pg\,mL^{-1}$ (age and sex dependence)
	Fibrogen	$0.002-4.2\,g\,L^{-1}$

AMI, acute myocardial infarction; BNP, B-type natriuretic peptide; CK-MB, creatine kinase MB subform; CRP, C-reactive protein; cTnI/T, cardiac troponin I/T; 11-FABP, fatty acid-binding protein; IL-6, interlukin-6; MPO, myeloperoxidase; NT-proBNP, N-terminal pro-B-type natriuretic peptide; TNF-α, tumor necrosis factor.

Table 6.4 Electrochemical biosensor for cardiac biomarkers detection.

Biomarker	Electrochemical technique	Linear range	Detection limit	Ref.
Myoglobin	EIS	$0.582-4.24\,\mu g\,mL^{-1}$	$2.25\,\mu g\,mL^{-1}$	[33]
	SWV	$1.1-2.98\,\mu g\,mL^{-1}$	—	
Cardiac Troponin T	CV	$0.009-0.8\,ng\,mL^{-1}$	$9\,pg\,mL^{-1}$	[32]
	OCP	10^6 magnitude	$<5\,pg\,mL^{-1}$	[28]
	Amperometry	$0.1-10\,ng\,mL^{-1}$	$0.2\,ng\,mL^{-1}$	[27]
C-reactive protein	EIS	$0.5-5\,nM$	$176\,pM$	[29]
	Chronoamperometry	$2.2-200\,ng\,mL^{-1}$	$2.6\,ng\,mL^{-1}$	[22]
MPO_{active}	Amperometry	$0.200\,ng\,mL^{-1}$	$0-16\,ng\,mL^{-1}$	[26]
MPO_{mass}		$0.004\,ng\,mL^{-1}$	$0-16\,ng\,mL^{-1}$	
miRNA 499	CV	10^{-6} to $1\,\mu M$	$0.3\,pM$	[20]

CV, cyclic voltammetry; EIS, electrochemical impedance spectroscopy; OCP, open circuit potential; MPO, myeloperoxidase; SWV, square wave voltammetry.

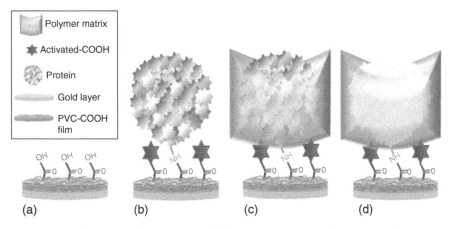

Figure 6.4 Molecular-imprinted polymer (MIP)-based electrochemical biosensor for myoglobin determination. (a) Formation of a polymer layer on bare gold chip with carboxylated poly(vinyl chloride) (PVC-COOH). (b) Covalent attachment of myoglobin protein on the activated surface. (c) Imprinting of the protein template. (d) Removal of the template from polymer matrix and obtaining MIP for myoglobin detection. *Source:* Moreira et al. [33]. Reproduced with permission of Elsevier.

attachment of myoglobin protein (Figure 6.4). Subsequently, imprinted polymer was formed through incubation with acrylamide and *N,N*-methylen-ebisacrylamide monomer in the presence of ammonium persulphate. Once the MIP was assembled, myoblogin protein was removed by washing step with oxalic acid. The analytical performance of biosensor was evaluated both by EIS (linear range: $0.852–4.26\,\mu g\,mL^{-1}$; detection limit: $2.25\,\mu g\,mL^{-1}$) and square wave voltammetry (linear range: $1.1–2.98\,\mu g\,mL^{-1}$) using $5\,mM$ $[Fe(CN)_6]^{3-/4-}$ redox probe solution. Selectivity and sensor response in synthetic serum sample were also evaluated.

Another MIP-based electrochemical biosensor for the detection of cardiac troponin T (TnT) was reported by Karimian et al. [32]. In this case, the imprinted template was obtained on the surface of a gold electrode by electropolymerization (performed through cyclic voltammetry) of o-phenylenediamine (PPD) in the presence of TnT protein, which can be removed by washing the modified gold electrode in basic solution and water. The sensor surface was characterized by EIS and CV electrochemical techniques and atomic force microscopy (AFM). CV was also used to evaluate the analytical performance of the developed biosensor using $1\,mM$ $[Fe(CN)_6]^{3-/4-}$ solution as redox probe. The sensor response was linear to the TnT concentration between 0.009 and $0.8\,ng\,mL^{-1}$ with a detection limit of $9\,pg\,mL^{-1}$. The sensor also showed good sensitivity and selectivity in the human blood serum.

Potentiometric immunosensor for cardiac troponin I-T-C complex protein was reported by Zhang et al. [28]. In this work, a GCE was coated using a mixture containing polymerized polyaniline and dinonylnaphthylsulfonic acid (PANI/DNNSA). Subsequently, a typical ELISA sandwich assay was assembled on the surface of modified GCE using a primary and an HRP-modified secondary antibody. Evaluation of cardiac troponin I-T-C complex was evaluated by the changing of open circuit potential (OCP) obtained after the addition of the enzymatic substrate. The proposed immunosensor exhibited an excellent detection limit ($<5\,pg\,mL^{-1}$) with a wide dynamic linear range (>6 order of magnitude).

Another immunosensor for cTnT detection was proposed by Silva et al. [27] that was based on the use of a polyethylene terephthalate (PTE) wafer (functionalized with an epoxy–graphite composite) as screen-printed electrode and on the use of streptavidin-modified microspheres as immobilization platform for the immunosandwich assay (performed using a biotin-modified primary and a HRP-modified secondary antibody). Amperometric determination of H_2O_2 allowed the construction of a calibration curve for cTnT in a linear range between 0.1 and $10\,ng\,mL^{-1}$ with a detection limit of $0.2\,ng\,mL^{-1}$.

The importance of biomarkers for inflammation in the early diagnosis of heart failure is subject to intense inquiry since inflammation has a high relevance in heart failure pathogenesis and progression. Several clinical studies demonstrated a correlation between inflammatory biomarkers, such as C-reactive protein, cytokines, interleukin-6, and tumor necrosis factor α, and high risk of the future development of heart failure in asymptomatic older subjects.

Impedimetric label-free biosensor for the detection in blood serum of C-reactive protein was reported by Bryan et al. [29]. The biosensor was based on the surface modification of polycrystalline gold electrode with a carboxylated polyethylene glycol (PEG)-thiol. After SAM formation, anti-C-reactive protein antibody was attached to the electrode surface by EDAC/NHS covalently coupling. EIS was used for surface characterization and for the development of C-reactive protein calibration curve. A linear correlation between charge transfer resistance (R_{ct}) and C-reactive protein was obtained in the range between 0.5 and 5.0 nM with a detection limit of 176 pM. The developed immunosensor showed good analytical performance for C-reactive protein detection in blood serum samples with limited aspecific adsorption of matrix proteins. Surface regeneration was also evaluated in buffered and serum sample solutions.

Another electrochemical immunosensor for C-reactive protein determination in blood serum samples was developed by Fakanya et al. [22]. Different gold screen-printed electrodes were developed and optimized in order to find the best analytical performance for C-reactive protein detection. The selected screen-printed gold electrode was directly functionalized with an anti-C reactive protein antibody followed by the affinity reaction with cardiac biomarker protein and HRP-labeled secondary anti-C-reactive protein antibody (Figure 6.5a). CRP evaluation was performed by chronoamperometric measurement at −200 mV versus Ag/AgCl pseudo-reference using

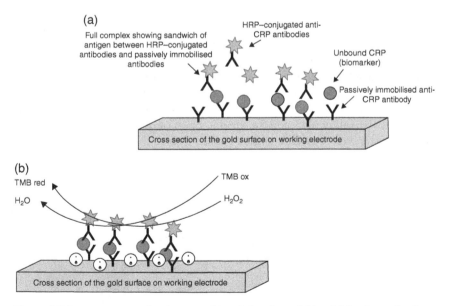

Figure 6.5 Immunosensor development (a) and signal acquisition (b) for determination C-reacted protein in serum samples. *Source:* Fakanya and Tothill [22]. http://www.mdpi.com/2079-6374/4/4/340/htm. Used under CC BY 4.0 http://creativecommons.org/licenses/by/4.0/.

3,3',5,5'-tetramethylbenzidine hydrochloride (TMB) as enzymatic substrate (Figure 6.5b). The proposed immunosensor exhibited a linear range from 2.2 to $100 \, \text{ng mL}^{-1}$ with a limit of detection of $2.6 \, \text{ng mL}^{-1}$. The analytical performances of the screen-printed-based immunosensor were found in good agreement with commercially available ELISA test for CRP detection in human serum samples.

A microfluidic device for the detection of endogenous (MPO_{active}) and total myeloperoxidase (MPO_{mass}) was reported by Moral-Vico et al. [26]. In both cases streptavidin-modified magnetic beads were functionalized with a biotinylated primary anti-MPO antibody, followed by a blocking step with biotin and affinity reaction with MPO protein. Endogenous peroxidase activity (MPO_{active}) was then detected in flow by the addition of 2,2'-azino-bis(3-ethylbenzothiazoline-6-sulfonicacid) (ABTS) liquid and the evaluation of the amperometric signal obtained after the enzymatic reaction at $-0.1 \, \text{V}$ versus gold reference electrode (Figure 6.6, route 3a). For the detection of MPO_{mass}, the MPO/Ab/Mb biocomplex was modified with an HRP-modified secondary anti-MPO antibody and injected in a secondary microfluidic channel (Figure 6.6, route 3b). Amperometric determination of MPO_{mass} was conducted under the same condition reported earlier. The proposed microfluidic biosensor showed good analytical performance for the detection of MPO_{active} (limit of detection: $0.200 \, \text{ng mL}^{-1}$) and MPO_{mass} (limit of detection: $0.004 \, \text{ng mL}^{-1}$) proteins both in buffered solutions and human plasma samples.

An electrochemical biosensor for the detection of acute myocardial infarction (AMI)-related circulating microRNA 499 (miRNA-499) was reported by Chen et al. [20]. In this work, a gold electrode was first modified with a miRNA complementary thiolated DNA probe followed by mixed SAM formation with 6-mercapto-1-hexanol. The DNA-modified gold electrode was then incubated with the target miRNA and with methylene blue as intercalate. Due to the insufficient sensibility obtained for the detection of miRNA, HRP enzyme and H_2O_2 substrate were introduced into the electrolyte solution, which work like

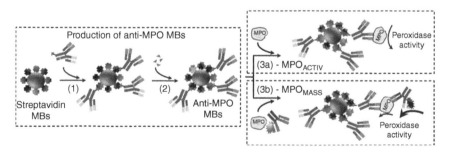

Figure 6.6 Scheme of immunosensor development for MPO_{active} (route 3a) and MPO_{mass} (route 3b) detections. *Source:* Moral-Vico et al. [26]. Reproduced with permission of Elsevier.

a second-generation biosensors; in particular MB works as an electron mediator and amplified in H_2O_2 enzymatic reaction. The current response (evaluated by cyclic voltammetry measurements) increases linearly with miRNA concentration in the range 1 pM to 1 µM with a detection limit lower than 0.3 pM.

6.3.3 Electrochemical Biosensors for Autoimmune Disease

Autoimmune diseases such as type 1 diabetes, coeliac, multiple sclerosis, rheumatoid arthritis, systemic lupus erythematosus (SLE) arise when the body's immune system attacks its own substances and tissues. These diseases are characterized by the production of high affinity autoantibodies that can be detected and correlated to the presence of the disease [106]. However, until now, no reference methods exist for the detection of autoantibodies due to lack of standardization and low negative predicted values. Therefore, electrochemical biosensors represent a valid alternative because they can give important information regarding antigen–autoantibody affinity reaction.

Some biosensors for autoimmune and infectious disease biomarker detection are reported in Table 6.5.

Konstantinov et al. reported the development of a portable electrochemical biosensor for the detection of anti-chromatin autoantibodies in human serum [107]. In particular a chromatin-modified membrane was placed on the surface of 3-electrode screen-printed electrode strip and assembled in a homemade flow cell. Subsequently, after the introduction of the serum containing the

Table 6.5 Electrochemical biosensor for autoimmune and infectious disease biomarkers detection.

Biomarker	Electrochemical technique	Linear range	Detection limit	Reference
Anti-chromatin auto-Ab	Pulse amperometry	Correlation with ELISA		[107]
Anti-DNA auto-Ab	Amperometry	—	0.04 µg IgG	[108]
Tranglutaminase auto-Ab	CV	0–40 U mL^{-1}	—	[109]
Anti-citrullinated auto-Ab	Amperometry	0–0.005 serum dilution	—	[110]
Hepatitis C virus	DPV	0.01–8 µM	1.0 pM	[111]
Hepatitis B virus	Amperometry	Single-nucleotide polymorphism		[112]
	CV	10^{-6} to 10 µM	2.0 pM	[113]
	DPV	0.4–4 pM	0.3 pM	[114]

CV, cyclic voltammetry; DPV, differential pulse voltammetry.

autoantibodies and a secondary HRP-labeled antibody, pulsed amperometric detection of H_2O_2 was performed. Autoantibody determination is completed in 20 min, and the obtained results show good correlation with ELISA test.

Similarly, another electrochemical biosensor for detection of anti-DNA autoantibodies in human serum was proposed by Rubin et al. [108]. In this work, 8 screen-printed array electrodes were modified with target DNA and placed in 8 well-methacrylate supports. The assay was completed by interaction with human serum and a secondary HRP-labeled antihuman IgG. After the addition of H_2O_2 and TMB electrochemical substrates, amperometric measurements were performed. Good correlation with ELISA test was found and no interaction with control serum samples was highlighted.

Electrochemical biosensor for detection of tranglutaminase (tTG) autoantibodies was reported by Neves et al. [109]. tTG tissue was firstly immobilized on the surface of a multiwalled carbon nanotube/AuNPs-modified graphite screen-printed electrode, followed by a surface-blocking step with BSA solution. After incubation with serum sample, the tTG autoantibodies bound to the surface of the sensors were labeled with an alkaline phosphatase-modified human IgG. The addition of the substrate (3-indoxyl phosphate with silver ions) induces the formation of an indoxyl intermediated that reduces the Ag^+ to Ag^0 subsequently detected by cyclic voltammetry measurements. The proposed biosensor showed a linear correlation between 0 and $40\,U\,mL^{-1}$ autoantibodies concentration with a relative standard deviation (RSD) of 9.32% for the negative samples and 2.01% for the positive samples. Results were then compared with ELISA test with good accordance.

Similar approach (exploiting multiwalled carbon nanotubes/polystyrene-modified electrode) was described by Villa et al. for the detection of anti-citrullinated peptide antibodies (ACPAs) in serum that are specific for rheumatoid arthritis (RA) autoimmune disease [110]. Other interesting electrochemical biosensors for autoimmune disease detection can be found in Refs. [115–119].

6.3.4 Electrochemical Biosensors for Autoimmune Infectious Disease

In addition to genetic factors, autoimmune diseases can be also triggered by environmental factors, in particular by interaction of the host with viruses, bacteria, or other infectious pathogens. In this case, the derived pathologies can be named as autoimmune infectious diseases [106]. The most important autoimmune disorder caused by interaction with a virus is hepatitis.

Tang et al. reported the development of novel electrochemical biosensor for the detection of hepatitis C virus (HCV) with the use of BamHI (a type-II restriction endonuclease extracted from *Bacillus amyloliquefaciens*) coupled with enzymatic signal enhancement by thionine and HRP-encapsulated nanogold hollow spheres [111]. 21-mer oligonucleotide relative to HCV was immobilized on the surface of activated GCE followed by reaction with the gold nano-compound. After the hybridization, BamHI cleavage was performed. BamHI catalyzes the cleavage of the formed double-stranded DNA between particular duplex symmetrical sequences, resulting in the detachment of the nano-compound from the GCE. As a result, the electrochemical signal of H_2O_2 reduction was decreased or disappeared. The proposed biosensor showed a linear correlation in the range 0.01–8 µM HCV-DNA sequence with a limit of detection of 1.0 pM.

Electrochemical biosensor for the detection of hepatitis B virus (HBV) was reported by Liu et al. [112]. The biosensors were developed using 16-electrode array modified with a complementary and a one mismatched DNA capture probe. In order to reduce the nonspecific adsorption on the electrode surface, thiolated oligo(ethylene glycol) was used as non-fouling agent. After the hybridization reaction, biotin-modified DNA detection probe was used followed by ligase and denaturation steps. Only when the sequence of the target DNA is fully complementary to the capture probe, ligation between the tandem capture and the detection probe can be performed. The assay was finally completed by incubation with avidin-HRP enzyme followed by amperometric measurement. The proposed biosensor was then successfully applied for the detection of pre-core mutation in the HBV genome at G1896A and for the two adjacent polymorphisms in the human CYP2C19 genome at C680T and G681A.

Shakoori et al. proposed the development of gold nanorods and nanostructured electrochemical biosensor for the detection of HBV [113]. Bare gold electrode was first nanostructured using 1,6-hexanedithiol (as linker) and gold nanorods. Then, ss-DNA capture probe was immobilized on the surface of the nanostructured sensor followed by hybridization reaction HBV-related ss-DNA target. Cyclic voltammetry studies of $[Co(phen)_3]^{3+}$ redox probe allowed the construction of a calibration curve with a linear range between 1.0 pM and 10 µM and a limit of detection of 2.0 pM. In addition, the proposed biosensor showed good specificity for one-point mismatched and noncomplementary DNA from DNA sequences.

Another HBV biosensor was reported by Zheng et al. [114]. In this case, a DNA molecular beacon was immobilized into electrode surface by the biotin at 3′-end, while the 5′-end of the probe was modified with 4-(4-dimethyl aminophenylazo) benzoic acid (dabcyl) as label. In the presence of target DNA, the hybridization reaction led to a change in the stem-loop conformation, allowing the reaction of 5′-end label with β-cyclodextrins-modified AuNPs. Immobilized AuNPs were then electrochemically dissolved in HCl solution and detected by differential pulse voltammetry (Figure 6.7). The proposed method showed high sensitivity and specificity with a linear range from 0.4 to 4 pM detection limit of 0.3 pM for HBV-DNA sequence.

Further interesting electrochemical biosensors for autoimmune infection disease-associated biomarkers detection can be found in Refs. [120–125].

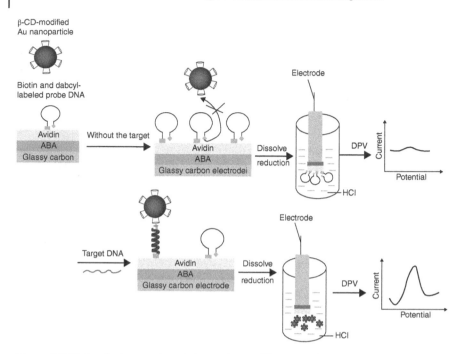

Figure 6.7 Molecular beacon-based electrochemical biosensor for hepatitis B virus determination. *Source:* Zheng et al. [114]. Reproduced with permission of Elsevier.

6.4 Conclusions

This chapter has presented some recent biosensors based on various biomolecular receptors for the detection of different biomarkers. The analytical characteristics of the biosensors together with their innovative aspects have been briefly described. Innovative biosensor-based strategies could allow a biomarker testing reliably in a decentralized setting due to their attractive characteristics such as reduced size, cost, and required time of analysis.

References

1 Biomarkers Definitions Working, Group. *Clin. Pharmacol. Ther.* **2001**, 69, 89–95.

2 Davis, J.; Maes, M.; Andreazza, A.; McGrath, J. J.; Tye, S. J.; Berk, M. *Mol. Psychiatry* **2015**, 20, 152–153.

3 Perera, F. P.; Weinstein, I. B. *Carcinogenesis* **2000**, 21, 517–524.

4 Rifai, N.; Gillette, M. A.; Carr, S. A. *Nat. Biotechnol.* **2006**, 24, 971–983.

5 Ahmed, M. U.; Hossain, M. M.; Safavieh, M.; Wong, Y. L.; Rahman, I. A.; Zourob, M.; Tamiya, E. *Crit. Rev. Biotechnol.* **2016**, 36 (3), 495–505.

6 Justino, C. I. L.; Rocha-Santos, T. A. P.; Duarte, A. C. *TrAc Trends Anal. Chem.* **2013**, 45, 24–36.

7 Turner, A. P. *Chem. Soc. Rev.* **2013**, 42, 3184–3196.

8 Thevenot, D. R.; Toth, K.; Durst, R. A.; Wilson, G. S. *Biosens. Bioelectron.* **2001**, 16, 121–131.

9 Marrazza, G. *Biosensors* **2014**, 4, 301.

10 Mariani, S.; Minunni, M. *Anal. Bioanal. Chem.* **2014**, 406, 2303–2323.

11 Parolo, C.; Merkoci, A. *Chem. Soc. Rev.* **2013**, 42, 450–457.

12 Ramanathan, K.; Danielsson, B. *Biosens. Bioelectron.* **2001**, 16, 417–423.

13 Llandro, J.; Palfreyman, J. J.; Ionescu, A.; Barnes, C. H. W. *Med. Biol. Eng. Comput.* **2010**, 48, 977–998.

14 Antuña-Jiménez, D.; Blanco-López, M. C.; Miranda-Ordieres, A. J.; Lobo-Castañón, M. J. *Sens. Actuators B Chem.* **2015**, 220, 688–694.

15 Sassolas, A.; Blum, L. J.; Leca-Bouvier, B. D. *Biotechnol. Adv.* **2012**, 30, 489–511.

16 Rogers, K. R. *Mol. Biotechnol.* **2000**, 14, 109–129.

17 Al-Ogaidi, I.; Aguilar, Z. P.; Suri, S.; Gou, H.; Wu, N. *Analyst* **2013**, 138, 5647–5653.

18 Barreda-García, S.; González-Álvarez, M. J.; de-los-Santos-Álvarez, N.; Palacios-Gutiérrez, J. J.; Miranda-Ordieres, A. J.; Lobo-Castañón, M. J. *Biosens. Bioelectron.* **2015**, 68, 122–128.

19 Èadková, M.; Dvoøáková, V.; Metelka, R.; Bílková, Z.; Korecká, L. *Electrochem. Commun.* **2015**, 59, 1–4.

20 Chen, G.; Shen, Y.; Xu, T.; Ban, F.; Yin, L.; Xiao, J.; Shu, Y. *Biosens. Bioelectron.* **2016**, 77, 1020–1025.

21 Cui, Z.; Wu, D.; Zhang, Y.; Ma, H.; Li, H.; Du, B.; Wei, Q.; Ju, H. *Anal. Chim. Acta* **2014**, 807, 44–50.

22 Fakanya, W. M.; Tothill, I. E. *Biosensors* **2014**, 4, 340–357.

23 Florea, A.; Ravalli, A.; Cristea, C.; Sãndulescu, R.; Marrazza, G. *Electroanalysis* **2015**, 27, 1594–1601.

24 Guo, A. P.; Wu, D.; Ma, H. M.; Zhang, Y.; Li, H.; Du, B.; Wei, Q. *J. Mater. Chem. B* **2013**, 1, 4052–4058.

25 Kavosi, B.; Salimi, A.; Hallaj, R.; Amani, K. *Biosens. Bioelectron.* **2014**, 52, 20–28.

26 Moral-Vico, J.; Barallat, J.; Abad, L.; Olive-Monllau, R.; Munoz-Pascual, F. X.; Galan Ortega, A.; del Campo, F. J.; Baldrich, E. *Biosens. Bioelectron.* **2015**, 69, 328–336.

27 Silva, B. V.; Cavalcanti, I. T.; Mattos, A. B.; Moura, P.; Sotomayor Mdel, P.; Dutra, R. F. *Biosens. Bioelectron.* **2010**, 26, 1062–1067.

28 Zhang, Q.; Prabhu, A.; San, A.; Al-Sharab, J. F.; Levon, K. *Biosens. Bioelectron.* **2015**, 72, 100–106.

29 Bryan, T.; Luo, X.; Bueno, P. R.; Davis, J. J. *Biosens. Bioelectron.* **2013**, 39, 94–98.

30 Das, J.; Kelley, S. O. *Anal. Chem.* **2011**, 83, 1167–1172.

31 Johari-Ahar, M.; Rashidi, M. R.; Barar, J.; Aghaie, M.; Mohammadnejad, D.; Ramazani, A.; Karami, P.; Coukos, G.; Omidi, Y. *Nanoscale* **2015**, 7, 3768–3779.

32 Karimian, N.; Vagin, M.; Zavar, M. H.; Chamsaz, M.; Turner, A. P.; Tiwari, A. *Biosens. Bioelectron.* **2013**, 50, 492–498.

33 Moreira, F. T. C.; Dutra, R. A. F.; Noronha, J. P. C.; Sales, M. G. F. *Electrochim. Acta* **2013**, 107, 481–487.

34 Raghav, R.; Srivastava, S. *Sens. Actuators B Chem.* **2015**, 220, 557–564.

35 Ren, X.; Wang, H.; Wu, D.; Fan, D.; Zhang, Y.; Du, B.; Wei, Q. *Talanta* **2015**, 144, 535–541.

36 Justino, C. I. L.; Freitas, A. C.; Pereira, R.; Duarte, A. C.; Santos, T. A. P. R. *TrAc Trends Anal. Chem.* **2015**, 68, 2–17.

37 Berti, F.; Lozzi, L.; Palchetti, I.; Santucci, S.; Marrazza, G. *Electrochim. Acta* **2009**, 54, 5035–5041.

38 Chen, A.; Yang, S. *Biosens. Bioelectron.* **2015**, 71, 230–242.

39 Ravalli, A.; Rivas, L.; De La Escosura-Muñiz, A.; Pons, J.; Merkoçi, A.; Marrazza, G. *J. Nanosci. Nanotechnol.* **2015**, 15, 3411–3416.

40 Zhou, W.; Huang, P. J.; Ding, J.; Liu, J. *Analyst* **2014**, 139, 2627–2640.

41 Blind, M.; Blank, M. *Mol. Ther-Nucl. Acids* **2015**, 4, e223.

42 Panasyuk, T.; Dall'Orto, V. C.; Marrazza, G.; El'skaya, A.; Piletsky, S.; Rezzano, I.; Mascini, M. *Anal. Lett.* **1998**, 31, 1809–1824.

43 Berti, F.; Todros, S.; Lakshmi, D.; Whitcombe, M. J.; Chianella, I.; Ferroni, M.; Piletsky, S. A.; Turner, A. P.; Marrazza, G. *Biosens. Bioelectron.* **2010**, 26, 497–503.

44 Poma, A.; Guerreiro, A.; Whitcombe, M. J.; Piletska, E. V.; Turner, A. P. F.; Piletsky, S. A. *Adv. Funct. Mater.* **2013**, 23, 2821–2827.

45 Ricci, F.; Adornetto, G.; Palleschi, G. *Electrochim. Acta* **2012**, 84, 74–83.

46 Hwa, K.-Y.; Subramani, B. *Biosens. Bioelectron.* **2014**, 62, 127–133.

47 Cernat, A.; Le Goff, A.; Holzinger, M.; Sandulescu, R.; Cosnier, S. *Anal. Bioanal. Chem.* **2014**, 406, 1141–1147.

48 Lanzellotto, C.; Favero, G.; Antonelli, M. L.; Tortolini, C.; Cannistraro, S.; Coppari, E.; Mazzei, F. *Biosens. Bioelectron.* **2014**, 55, 430–437.

49 Pérez, S.; Sánchez, S.; Fàbregas, E. *Electroanalysis* **2012**, 24, 967–974.

50 Biscay, J.; Rama, E. C.; García, M. B. G.; Carrazón, J. M. P.; García, A. C. *Electroanalysis* **2011**, 23, 209–214.

51 Diaconu, I.; Cristea, C.; Harceaga, V.; Marrazza, G.; Berindan-Neagoe, I.; Sandulescu, R. *Clin. Chim. Acta* **2013**, 425, 128–138.

52 Pan, Y.; Sonn, G. A.; Sin, M. L. Y.; Mach, K. E.; Shih, M.-C.; Gau, V.; Wong, P. K.; Liao, J. C. *Biosens. Bioelectron.* **2010**, 26, 649–654.

53 Ravalli, A.; dos Santos, G. P.; Ferroni, M.; Faglia, G.; Yamanaka, H.; Marrazza, G. *Sens. Actuators B Chem.* **2013**, 179, 194–200.

54 Patris, S.; De Pauw, P.; Vandeput, M.; Huet, J.; Van Antwerpen, P.; Muyldermans, S.; Kauffmann, J. M. *Talanta* **2014**, 130, 164–170.

55 Ravalli, A.; da Rocha, C. G.; Yamanaka, H.; Marrazza, G. *Bioelectrochemistry* **2015**, 106, Part B, 268–275.

56 Xu, Y. H.; Wang, E. K. *Electrochim. Acta* **2012**, 84, 62–73.

57 Laschi, S.; Miranda-Castro, R.; Gonzalez-Fernandez, E.; Palchetti, I.; Reymond, F.; Rossier, J. S.; Marrazza, G. *Electrophoresis* **2010**, 31, 3727–3736.

58 Rackus, D. G.; Shamsi, M. H.; Wheeler, A. R. *Chem. Soc. Rev.* **2015**, 44, 5320–5340.

59 Zani, A.; Laschi, S.; Mascini, M.; Marrazza, G. *Electroanalysis* **2011**, 23, 91–99.

60 Berti, F.; Laschi, S.; Palchetti, I.; Rossier, J. S.; Reymond, F.; Mascini, M.; Marrazza, G. *Talanta* **2009**, 77, 971–978.

61 Lafleur, J. P.; Jonsson, A.; Senkbeil, S.; Kutter, J. P. *Biosens. Bioelectron.* **2016**, 76, 213–233.

62 Miserere, S.; Merkoçi, A. Castillo-León, J.; Svendsen, W.E. (eds), *Lab-on-a-Chip Devices and Micro-Total Analysis Systems*; Springer International Publishing, Cham, **2015**, p. 141–160.

63 Rios, A.; Zougagh, M.; Avila, M. *Anal. Chim. Acta* **2012**, 740, 1–11.

64 Temiz, Y.; Lovchik, R. D.; Kaigala, G. V.; Delamarche, E. *Microelectron Eng* **2015**, 132, 156–175.

65 Gupta, V. K.; Ganjali, M. R.; Norouzi, P.; Khani, H.; Nayak, A.; Agarwal, S. *Critical Reviews in Analytical Chemistry* **2011**, 41, 282–313.

66 Mascini, M.; Marrazza, G. *Anal. Chim. Acta* **1990**, 231, 125–128.

67 Xie, X.; Zhai, J.; Bakker, E. *J. Am. Chem. Soc.* **2014**, 136, 16465–16468.

68 Barik, M. A.; Sarma, M.; Sarkar, C. R.; Dutta, J. *Appl. Biochem. Biotechnol.* **2014**, 174, 1104–1114.

69 Dou, Y. H.; Haswell, S. J.; Greenman, J.; Wadhawan, J. *Electroanalysis* **2012**, 24, 264–272.

70 Müntze, G. M.; Baur, B.; Schäfer, W.; Sasse, A.; Howgate, J.; Röth, K.; Eickhoff, M. *Biosens. Bioelectron.* **2015**, 64, 605–610.

71 Sarkar, D.; Liu, W.; Xie, X.; Anselmo, A. C.; Mitragotri, S.; Banerjee, K. *ACS Nano* **2014**, 8, 3992–4003.

72 Yoo, E. H.; Lee, S. Y. *Sensors* **2010**, 10, 4558–4576.

73 Karimi-Maleh, H.; Sanati, A. L.; Gupta, V. K.; Yoosefian, M.; Asif, M.; Bahari, A. *Sens. Actuator B Chem.* **2014**, 204, 647–654.

74 Kumar, J.; D'Souza, S. F. *Biosens. Bioelectron.* **2011**, 26, 4289–4293.

75 Lotierzo, M.; Abuknesha, R.; Davis, F.; Tothill, I. E. *Environ. Sci. Technol.* **2012**, 46, 5504–5510.

76 Saberi, R.-S.; Shahrokhian, S.; Marrazza, G. *Electroanalysis* **2013**, 25, 1373–1380.

77 Nguyen-Boisse, T. T.; Saulnier, J.; Jaffrezic-Renault, N.; Lagarde, F. *Sens. Actuators B Chem.* **2013**, 179, 232–239.

78 Soldatkin, O. O.; Peshkova, V. M.; Saiapina, O. Y.; Kucherenko, I. S.; Dudchenko, O. Y.; Melnyk, V. G.; Vasylenko, O. D.; Semenycheva, L. M.; Soldatkin, A. P.; Dzyadevych, S. V. *Talanta* **2013**, 115, 200–207.

79 Lisdat, F.; Schäfer, D. *Anal. Bioanal. Chem.* **2008**, 391, 1555–1567.
80 Randviir, E. P.; Banks, C. E. *Anal. Methods* **2013**, 5, 1098–1115.
81 Sonuç, M. N.; Sezgintürk, M. K. *Talanta* **2014**, 120, 355–361.
82 Taleat, Z.; Ravalli, A.; Mazloum-Ardakani, M.; Marrazza, G. *Electroanalysis* **2013**, 25, 269–277.
83 Pérez-López, B.; Merkoçi, A. *Microchim. Acta* **2012**, 179, 1–16.
84 Pingarrón, J. M.; Yáñez-Sedeño, P.; González-Cortés, A. *Electrochim. Acta* **2008**, 53, 5848–5866.
85 Ravalli, A.; Marrazza, G. *J. Nanosci. Nanotechnol.* **2015**, 15, 3307–3319.
86 Zhu, C.; Yang, G.; Li, H.; Du, D.; Lin, Y. *Anal. Chem.* **2015**, 87, 230–249.
87 Fenzl, C.; Hirsch, T.; Baeumner, A. J. *TrAC Trends Anal. Chem.* **2016**, 79, 306–316.
88 Zhu, X.; Li, J.; He, H.; Huang, M.; Zhang, X.; Wang, S. *Biosens. Bioelectron.* **2015**, 74, 113–133.
89 Reverté, L.; Prieto-Simón, B.; Campàs, M. *Anal. Chim. Acta* **2016**, 908, 8–21.
90 Lawal, A. T. *Mater. Res. Bull.* **2016**, 73, 308–350.
91 Zare, Y.; Shabani, I. *Mater. Sci. Eng. C* **2016**, 60, 195–203.
92 Malvezzi, M.; Bertuccio, P.; Rosso, T.; Rota, M.; Levi, F.; La Vecchia, C.; Negri, E. *Ann. Oncol.* **2015**, 26, 779–786.
93 Ludwig, J. A.; Weinstein, J. N. *Nat. Rev. Cancer* **2005**, 5, 845–856.
94 Chun, L.; Kim, S. E.; Cho, M.; Choe, W. S.; Nam, J.; Lee, D. W.; Lee, Y. *Sens. Actuator B Chem.* **2013**, 186, 446–450.
95 Bast, R. C., Jr.; Hennessy, B.; Mills, G. B. *Nat. Rev. Cancer* **2009**, 9, 415–428.
96 Karlan, B. Y.; Thorpe, J.; Watabayashi, K.; Drescher, C. W.; Palomares, M.; Daly, M. B.; Paley, P.; Hillard, P.; Andersen, M. R.; Anderson, G.; Drapkin, R.; Urban, N. *Cancer Epidemiol. Biomark. Prev.* **2014**, 23, 1383–1393.
97 Tse, C.; Gauchez, A. S.; Jacot, W.; Lamy, P. J. *Cancer Treat. Rev.* **2012**, 38, 133–142.
98 Marques, R. C.; Viswanathan, S.; Nouws, H. P.; Delerue-Matos, C.; Gonzalez-Garcia, M. B. *Talanta* **2014**, 129, 594–599.
99 Al-Khafaji, Q. A. M.; Harris, M.; Tombelli, S.; Laschi, S.; Turner, A. P. F.; Mascini, M.; Marrazza, G. *Electroanalysis* **2012**, 24, 735–742.
100 Emami, M.; Shamsipur, M.; Saber, R.; Irajirad, R. *Analyst* **2014**, 139, 2858–2866.
101 Qureshi, A.; Gurbuz, Y.; Niazi, J. H. *Sens. Actuator B Chem.* **2015**, 220, 1145–1151.
102 McGill, H. C., Jr.; McMahan, C. A.; Gidding, S. S. *Circulation* **2008**, 117, 1216–1227.
103 Gerszten, R. E.; Wang, T. J. *Nature* **2008**, 451, 949–952.
104 Qureshi, A.; Gurbuz, Y.; Niazi, J. H. *Sens. Actuator B Chem.* **2012**, 171, 62–76.
105 Rezaei, B.; Ghani, M.; Shoushtari, A. M.; Rabiee, M. *Biosens. Bioelectron.* **2016**, 78, 513–523.

106 Delogu, L. G.; Deidda, S.; Delitala, G.; Manetti, R. *J. Infect. Dev. Ctries.* **2011**, 5(10), 679–687.

107 Konstantinov, K. N.; Sitdikov, R. A.; Lopez, G. P.; Atanassov, P.; Rubin, R. L. *Biosens. Bioelectron.* **2009**, 24, 1949–1954.

108 Rubin, R. L.; Wall, D.; Konstantinov, K. N. *Biosens. Bioelectron.* **2014**, 51, 177–183.

109 Neves, M. M. P. S.; González-García, M. B.; Nouws, H. P. A.; Costa-García, A. *Biosens. Bioelectron.* **2012**, 31, 95–100.

110 de Gracia Villa, M.; Jiménez-Jorquera, C.; Haro, I.; Gomara, M. J.; Sanmartí, R.; Fernández-Sánchez, C.; Mendoza, E. *Biosens. Bioelectron.* **2011**, 27, 113–118.

111 Tang, D.; Tang, J.; Su, B.; Li, Q.; Chen, G. *Chem. Commun.* **2011**, 47, 9477–9479.

112 Liu, G.; Lao, R.; Xu, L.; Xu, Q.; Li, L.; Zhang, M.; Song, S.; Fan, C. *Biosens. Bioelectron.* **2013**, 42, 516–521.

113 Shakoori, Z.; Salimian, S.; Kharrazi, S.; Adabi, M.; Saber, R. *Anal. Bioanal. Chem.* **2015**, 407, 455–461.

114 Zheng, J.; Chen, C.; Wang, X.; Zhang, F.; He, P. *Sens. Actuators B Chem.* **2014**, 199, 168–174.

115 Neves, M. M. P. S.; González-García, M. B.; Delerue-Matos, C.; Costa-García, A. *Sens. Actuators B Chem.* **2013**, 187, 33–39.

116 Wei, W.; Zhang, L.; Ni, Q.; Pu, Y.; Yin, L.; Liu, S. *Anal. Chim. Acta* **2014**, 845, 38–44.

117 Centi, S.; Tombelli, S.; Puntoni, M.; Domenici, C.; Franek, M.; Palchetti, I. *Talanta* **2015**, 134, 48–53.

118 Neves, M. M. P. S.; González-García, M. B.; Santos-Silva, A.; Costa-García, A. *Sens. Actuators B Chem.* **2012**, 163, 253–259.

119 Martín-Yerga, D.; González-García, M. B.; Costa-García, A. *Talanta* **2014**, 130, 598–602.

120 Zhang, S.; Tan, Q.; Li, F.; Zhang, X. *Sens. Actuators B Chem.* **2007**, 124, 290–296.

121 Hassen, W. M.; Chaix, C.; Abdelghani, A.; Bessueille, F.; Leonard, D.; Jaffrezic-Renault, N. *Sens. Actuator B Chem.* **2008**, 134, 755–760.

122 Ding, C.; Zhao, F.; Zhang, M.; Zhang, S. *Bioelectrochemistry* **2008**, 72, 28–33.

123 Hong, S. A.; Kwon, J.; Kim, D.; Yang, S. *Biosens. Bioelectron.* **2015**, 64, 338–344.

124 Chen, C. C.; Lai, Z. L.; Wang, G. J.; Wu, C. Y. *Biosens. Bioelectron.* **2015**, 77, 603–608.

125 Li, X.; Scida, K.; Crooks, R. M. *Anal. Chem.* **2015**, 87, 9009–9015.

7

MEMS-Based Cell Counting Methods

Mustafa Kangül, Eren Aydın, Furkan Gökçe, Özge Zorlu*, Ebru Özgür*, and Haluk Külah

Department of Electrical and Electronics Engineering, Middle East Technical University, Ankara, Turkey
*Current address: Mikro Biyosistemler Inc, Ankara, Turkey

7.1 Introduction

Micro-electro-mechanical systems (MEMS) can be defined as miniature, multifunctional microsystems consisting of sensors, actuators, and electronics. Development in IC fabrication techniques led the making of invisibly small machines to be possible. Micromachining, which uses many of standard IC technology, is an enabling technology allowing the formation of physical and electronic devices in microscale. Apart from being portable and accurate, MEMS devices are low cost due to mass fabrication techniques. MEMS market reaches over $10B with devices such as inkjet cartridge heads, pressure sensors, inertial sensors, and projection sensors. MEMS products are commonly used in many applications basically in military, communication, and automotive industries [1].

Developments in MEMS technology have grabbed great attention from the researchers in biomedical field, too. Since the dimensions of the MEMS and the biological and chemical particles are comparable, a new research area has been established: biomedical micro-electro-mechanical systems (BioMEMS). BioMEMS research focuses on the development of microscale devices with abilities of detection, separation, and any kind of manipulation of the biological or chemical units.

Laboratory diagnostic tools, individualized treatments, tissue scaffolding devices, drug delivery devices, and minimally invasive operations can be considered as leading application areas of BioMEMS. Most of these applications

Biosensors and Nanotechnology: Applications in Health Care Diagnostics, First Edition.
Edited by Zeynep Altintas.

are traditionally performed using conventional macroscale devices. However, BioMEMS brings some crucial advantages over its conventional counterparts, such as

- Decreased reagent consumption (in the order of microliters or nanoliters)
- Enhanced sensitivity
- Enhanced analysis speed
- Point-of-care detection
- Portability due to low weight and low power consumption
- *In vivo* and *in vitro* usage possibility
- Biocompatibility

Selective cell quantification is one of the most exciting and promising application areas of diagnostic studies in the scope of the BioMEMS. For example, detection of rare target cells among millions of blood cells can enlighten the road of the early diagnosis of cancer.

In this chapter, MEMS-based cell counting techniques will be summarized under three subtopics: optical, electrical and electrochemical, and gravimetric methods.

7.2   MEMS-Based Cell Counting Methods

7.2.1   Optical Cell Counting Methods

Optical detection is widely used in cell quantification in the macroscale systems due to its accuracy and sensitivity. However, such optical cell counting systems are not widely accessible because of their expensive setups. Achievements in MEMS, on the other hand, made it possible to miniaturize these optical detection systems by decreasing both fabrication and maintenance costs significantly. Using semiconductors' optical properties such as interactions with photons at specific wavelengths and their suitability for micromachining, many studies have shown that micro-optical cell counting can be successfully practiced indeed.

In the micro-optical cell counting systems, there are still some issues on rare cell counting related with smaller sample volumes because the less the sample volume is, the less the cells are in the sample to detect [2]. In other words, it is more difficult to detect rare cells in microscale systems when compared with the macroscale counterparts. As a result, one should consider two main factors while choosing the micro-optical detection systems over the macroscale counterparts: sensitivity and scalability to smaller dimensions.

The optical detection methods can be grouped into two, according to the optical detection technique used. The first is the detection of the optical characteristics of scattered and/or stimulated photons (luminescence), and the second is the detection of cells via high-resolution imaging techniques.

7.2.1.1 Quantification of the Cells by Detecting Luminescence

Detecting luminescence of the cells is a popular method on which a lot of research was conducted and practical results were reported. Characteristics like fluorescence, chemiluminescence, and bioluminescence are some of the luminescence forms that are detected in microscale quantification of the chemical or biological entities such as cells.

Luminescence is the emission of light, and it is independent of heat. In other words, luminescence occurs at normal temperatures, which is important for cell quantification as high temperatures may damage cells. For it to occur, the atoms are excited to higher energy states from their ground state by an excitation source. The excited atoms, then, return to their ground state by emitting photons. Generally, the emitted photons have less energy than the excitation energy. This process is illustrated in the Jablonski diagram in Figure 7.1.

If the excitation energy results from a chemical reaction, the process is called chemiluminescence. It is highly sensitive; nonetheless, because there is not much reagents, microchip designs are more complex in chemiluminescence detection [2]. If the light is emitted by a living thing itself, it is named bioluminescence. It is also a form of chemiluminescence as the excitation energy comes from the chemical reactions inside the living organism. If the process receives its excitation energy from the absorption of light, fluorescence takes place. The absorption causes the atoms to shift to a higher energy level, which is unstable. As a result, the atoms return to their ground state by emitting light. During this process, emitted light can be detected for

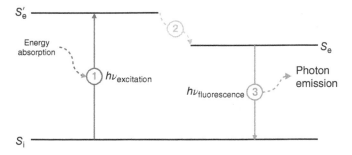

Figure 7.1 The fluorescence Jablonski diagram. (1) Excitation of atoms to higher energy states S_1'. (2) Dissipation of some energy as heat or other background processes. (3) Emission of photons (luminescence) and return of the atoms to the initial state S_i.

quantification aims. The absorbed photon has shorter wavelength when compared with the emitted light. Hence, in order to reject the excitation light, that is, the absorbed wavelength, filters can be used, leading to better selectivity [3]. Due to its selectivity, fluorescence detection is widely preferred in the microscale optical detection systems. Selectivity is important as the presence or a change in the optical output of a detector is an indication of a possible binding reaction [4]. Fluorescent detection can be laser induced, lamp based, or LED induced. Lasers are preferred due to ease of integration; however, lamps are both inexpensive and more flexible in terms of frequency. For example, in some microscope-based systems, mercury or xenon light sources are being used [2]. Moreover, LEDs are also good choices for excitation of the atoms since they consume much less power. However, additional filters may be required due to their broad spectrum [2]. There are also examples that use OLEDS and blue LED as excitation sources [5–7].

Miniaturized flow cytometric and spectroscopic cell counting techniques that use fluorescent detection have been reported in the literature. Grafton et al. have successfully miniaturized the flow cytometer utilizing avalanche breakdown photodiodes to detect fluorescently labeled cells [8]. Figure 7.2 shows the schematic of the detection system. It consists of a custom-designed microfluidic channel for separation of white blood cells from human blood by using electromagnetic properties of the cells. Then, the cells are counted on-chip via fluorescent excitation using the photodiodes. In another study,

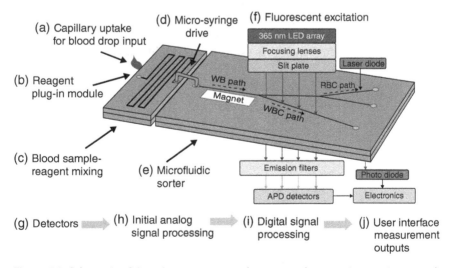

Figure 7.2 Schematic of the subcomponents and processes that constitute an integrated portable micro-cytometer reader, substance packs, and disposable microfluidic chip. *Source:* Grafton et al. [8]. Reproduced with permission of SPIE. (*See insert for color representation of the figure.*)

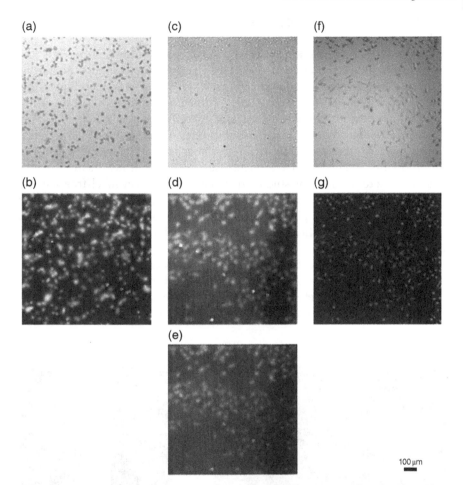

Figure 7.3 Lensless shadow image (a) and fluorescence image of the fluorescent microspheres (b); lensless shadow image (c) and fluorescence image of L929 cells (d), as well the composition (e) of both type pictures; and the bright filed (f) and fluorescent microscopy images for the same L929 sample (g). *Source:* Li et al. [9]. Reproduced with permission of SPIE. (*See insert for color representation of the figure.*)

Li et al. developed a lensless spectroscopy module for cell detection, which is also based on the fluorescence method [9]. The module uses high-resolution CMOS sensors to track the fluorescence activities of labeled cells. They coated the surface of the sensor with an additional layer of interference filter to construct a band-pass filter that enhances the resolution of analyses. In the experiments, L929 cells were observed and the images captured via the sensor with and without the coating. The results were comparable with optical microscopy images, which are shown in Figure 7.3.

7.2.1.2 Quantification of the Cells via High-Resolution Imaging Techniques

Although detecting the emission/absorption of photons, especially the fluorescence, is very popular, there is an increasing interest in direct counting of the cells by using high-resolution and/or wide field-of-view techniques, too. Direct counting of the cells such as shadow imaging requires almost no or less preparation of the analyte when compared with the fluorescent detection [3]. However, in some cases, further image processing is required to get clear images that are free from diffraction and interference of the light. Lensless high-resolution imaging not only decreases the cost but also fastens up the analyses. There are several studies that have focused on label-free lensless imaging of the cells using complementary metal–oxide–semiconductor (CMOS) and charge-coupled device (CCD) sensors. Moreover, some have shown that lensless wide field-of-view imaging results in accurate quantization of the cells.

Talbot self-imaging effect, in which the light is focused on a plane on different points forming a periodic grid, is used for cell culture monitoring [10]. Although the aim is to detect the fluorescently labeled particles/cells on chip, since the method is based on scanning a large area with high resolution, it is a

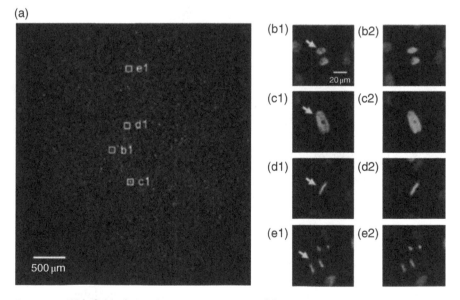

Figure 7.4 Wide field-of-view fluorescence imaging of the GFP cells. (a) The 3.7 mm × 3.5 mm image. (b1, c1, d1, and e1) Cropped images of typical cells in (a), including G1 (b1), G2 (c1), metaphase (d1), and anaphase (e1) (arrows) (b2, c2, d2, and e2). The same cells as imaged by a conventional microscope with a 20×/0.4 NA objective. *Source:* Han et al. [10]. Reproduced with permission of American Chemical Society.

good example of how high-resolution imaging techniques can enhance the sensitivity of such systems. The trajectories of the cells over 24 h were tracked, and the division of individual cells and also the cells in metaphase, anaphase, G1, and G2 were captured as shown in Figure 7.4.

Demircan *et al.* developed a label-free lensless lab-on-a-chip system to detect leukemia cells by integrating a dielectrophoretic chip with a high-resolution CMOS sensor [11]. The cells were detected inside a parylene-based microfluidic channel, and the captured images were compared with microscope observations for the verification of the system. It is reported that the module is capable of sensing cells with 3 μm diameter, with a noise level 28.3 e—rms and a multiplexing rate up to 400 kHz. In another study, a lensless cell monitor named LUCAS (lensless, ultra-wide-field cell monitoring array platform based on shadow imaging), which uses an optoelectronic sensor array to capture shadow image of cells, was developed and tested by Ozcan and Demirci [12]. The module can be combined with microchannels for parallel on-chip cell quantification. Images of monocytes, fibroblasts, and red blood cells were captured by using LUCAS and analyzed, which are shown in Figure 7.5. CCD chips were used for monitoring due to their light sensitivity. Quantification was completed via a computer-assisted auto-detection system.

The abovementioned studies prove that it is possible to construct accurate and inexpensive cell quantification systems by using high-resolution imaging techniques and by detecting the luminescence of the cells. Apparently, as the studies in the field continue, there will be more accurate and less expensive microscale detection systems that are not only more sensitive but also cheaper when compared with the macroscale counterparts. For example, sensitivities around a few cells in whole blood samples will pave the early detection of several diseases and conditions, too. As a result of low cost, less sample volume, and increased sensitivity, microscale cell detection can provide instant point-of-care solutions in healthcare.

7.3 Electrical and Electrochemical Cell Counting Methods

Electrical/electrochemical sensors can be categorized according to measured electrical entities. Impedimetric sensing depends on the variation of the resistivity between two electrodes with the existence of the bio-particles. Voltammetric and amperometric sensing techniques relate the chemical change occurring due to redox reactions at the electrode–electrolyte interface to electrical quantities such as potential or current. These methods are investigated further with the applications in the following sections.

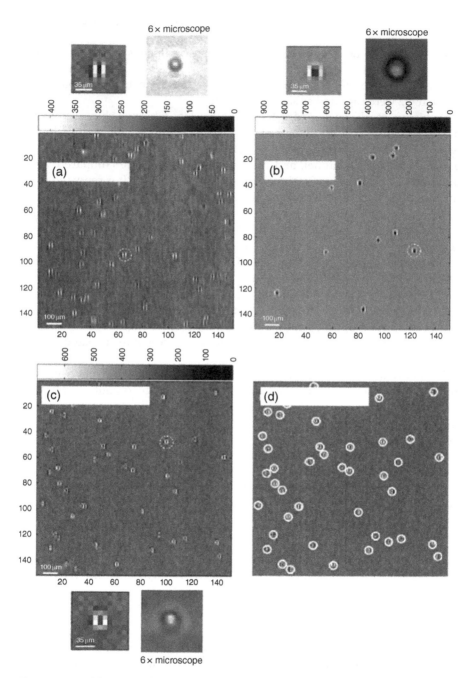

Figure 7.5 LUCAS images for (a) monocytes, (b) NIH-3T3 fibroblasts, (c) red blood cells. Each image has next to it a zoomed version of a cell and its comparison with a low NA microscope image. (d) Computer-assisted automated detection of the dynamic location of the red blood cells. A total of 41 cells are counted within a field of view of 1.4 mm × 1.4 mm. *Source:* Ozcan and Demirci [12]. Reproduced with permission of Royal Society of Chemistry.

7.3.1 Impedimetric Cell Quantification

The basic principle in the impedimetric cell sensing is the measurement of electric charge of the accumulated cells on the electrode surface. The working electrode is functionalized via target-specific probes, such as antibodies or aptamers, and the impedance response of the electrochemical system is measured. Then, the captured cell number is approximated based on the relation of cell concentration with the change in electrical response measured.

Initially developed impedimetric cell sensors aimed for the real-time measurement of cell proliferation, growth inhibition, or apoptosis events in cell culture media. This was achieved by implementing small working electrodes at the bottom of culture vessels/microwell plates. The presence of cells affects the local ionic environment at the electrode–solution interface, leading to an increase in the electrode impedance. The number and the degree of cell adherence to the electrode, as well as the biological status of the cells including cell viability, morphology, and so on, affect the impedance change. This approach has been implemented to measure cell growth rate, cytotoxicity tests, apoptosis events, and so on in real time, utilizing microwell plates with bottom electrodes, in a variety of studies [13]. Similarly, microbial sensors have been developed for the detection of pathogens based on the same principle. Yang et al. [14] developed an electrochemical impedance immunosensor for the detection of *E. coli* utilizing an immobilized anti-*E. coli* antibodies onto an indium–tin oxide interdigitated array (IDA) microelectrodes. The increased electron transfer resistance upon binding of *E. coli* cells onto antibody functionalized IDA microelectrodes was measured via electrochemical impedance spectroscopy, in the presence of $[Fe(CN)6]^{3-/4-}$ as a redox probe. A detection limit of 106 colony forming units (CFU) mL^{-1} was achieved with this method. An increased sensitivity has been reported by Santos et al. [15], who utilized a similar technique for the detection of pathogenic *E. coli* O157:H7. A detection limit of 2 CFU mL^{-1} was achieved via electrochemical impedance spectroscopy measurement.

To increase the precision of cell counting, the size of the sensor surface can be reduced to the size of single cells. Jiang and Spencer [16] have developed such an impedance-based electrochemical sensor for the precise counting of CD4+ helper cells for the detection of HIV infection. Instead of a single large working electrode, they have constructed densely packed working electrode pixels with size comparable with that of single CD4+ cell. The working electrodes were modified with antihuman CD4+ antibody for specific capture of CD4+ cells. The pixelated impedance sensor was reported to have single-cell sensitivity and an accuracy that is linearly correlated with the optical counting techniques. Another impedance-based cell detection sensor has been successfully developed by inspiring the famous Coulter

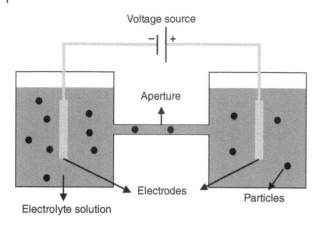

Figure 7.6 Illustration of a Coulter counter.

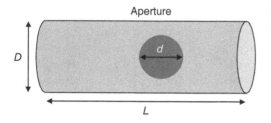

Figure 7.7 Dimensions of a cell in an aperture.

counter. This method, which is named after Wallace H. Coulter, utilizes two electrodes inside two chambers. Electrolyte-filled chambers are connected with a narrow aperture that determines the impedance between the two electrodes (Figure 7.6).

Particle flow through the aperture changes the local ionic concentration. As the particle enters the aperture, electrolyte solution of the same volume exits. Therefore, motion of particle through aperture creates resistive pulses (Figure 7.7).

Change in resistivity can be calculated with the help of R_p, the initial aperture resistance, aperture diameter D, length L, and cell diameter d:

$$\delta R_p = R_p \frac{d^3}{D^2 L} \tag{7.1}$$

Enhanced MEMS technologies enable to fabricate a variety of apertures and electrodes on a microchannel. Therefore, the miniaturization of macro-Coulter counter devices has been done by several groups in different aspects. Dynamic

(a)

(b)

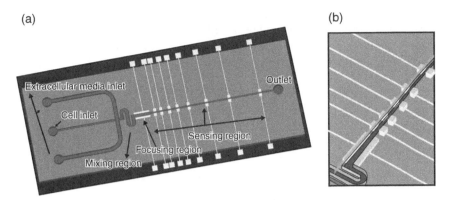

Figure 7.8 (a) Schematics of the MEMS Coulter counter. (b) A magnified view of the focusing and detection electrodes, and the Coulter channel. *Source:* Wu et al. [17]. Reproduced with permission of Springer.

quantization of cells [17] and blood cell counting [18] has been achieved with Coulter counters. Figure 7.8 illustrates a MEMS-based Coulter counter design.

Although Coulter counter is easy to utilize with MEMS, there is one major design drawback. The occurrence of multiple cells inside the aperture causes two close but different peaks or one combined peak. The problem is originated from a size-based separation, and it is a threat on reliability of the system. In order to avoid this problematic case, aperture size should be minimized to allow only single-cell passage at a time instant, which limits the throughput under continuous flow.

7.3.2 Voltammetric and Amperometric Cell Quantification

Voltammetric and amperometric electrochemical sensors have also been developed for cell counting. In these sensors, current resulting from redox reactions of electroactive species is measured at a fixed potential (amperometric) or at varying potentials of different types (linear sweep, differential staircase, normal pulse, reverse pulse, differential pulse, etc.), and the information about the rate of the redox reaction or concentration of analyte can be gathered. For cell sensing applications, a redox reporter (e.g., $[Fe(CN)6]^{3-/4-}$, methylene blue, etc.) should be used for current generation upon excitation. As in the case of impedimetric methods, decreasing the sensing electrode size to single-cell levels can increase sensing sensitivity of voltammetric sensors for cell counting applications. Moscovici et al. [19] have designed an electrochemical sensor for the detection of prostate cancer cells using an aperture sensor array. They utilized antibody-modified gold apertures with sizes ranging from 50 to 300 μm for the cell-specific capturing and counted the number of cells based on the current response gathered via differential pulse voltammetry. The binding of prostate cancer cells onto antibody-modified gold surface alters the interfacial electron transfer reaction of a redox reporter and allows cell populations as low as $125\,cells\,mL^{-1}$ to be read out.

Although these methods give an estimation of the cell count in the sample, they do not provide a high sensitivity measurement, which is necessary for the counting of rare cells, especially CTCs, where detection of even one cell in the milieu of billons of blood cells is necessary. Even though the decreased aperture size used in microelectrode arrays electrochemical sensors, studies demonstrated so far do not meet the requirements for CTC counting.

7.4 Gravimetric Cell Counting Methods

MEMS-based gravimetric cell counting devices track the changes created by the mass load of the captured or placed particles. The detection of the bending angle of a cantilever structure and resonance frequency change of a resonant sensor are two main methods for the mass measurement with MEMS. Developments in MEMS technologies have enabled producing mass sensors with the capability of measuring extremely small masses in the range of picograms [20].

7.4.1 Deflection-Based Cell Quantification

Loading a particle on the surface of a cantilever beam causes a deflection. The maximum deflection length, y_{max}, and the deflection angle θ_{max} are determined by the loaded mass, as well as the physical and mechanical characteristics of the cantilever beam. In Figure 7.9, the effect of mass loading on the cantilever surface is explained.

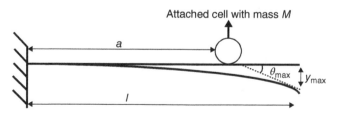

Figure 7.9 Effect of mass loading on cantilever.

In Equations 7.2 and 7.3, maximum deflection (y_{max}) and maximum angle θ_{max} are defined, respectively:

$$y_{max} = \frac{Mga^2}{6EI}(3l - a) \tag{7.2}$$

$$\theta_{max} = \frac{Mga^2}{2EI} \tag{7.3}$$

$$I = \frac{w \times h^3}{12} \tag{7.4}$$

where g is gravitational acceleration, E is Young's Modulus of the material, I is the moment of inertia, w is the width, and h is the thickness of the cantilever. Mass sensitivity of such sensors can be defined as

$$\frac{y_{max}}{M} \quad \text{or} \quad \frac{\theta_{max}}{M} \tag{7.5}$$

Sensitivity of the bending-based cantilever beam depends on the location of the captured cell. Simply, deflection increases as the captured cell moves away from the fixed point. Although activating only a certain region of the cantilever ensures the same deflection with the same loaded mass, this decreases the sensor efficiency since the cell capturing area of the sensor also decreases.

There is also another deflection-based sensing method that depends on the surface stress difference of the cantilever. In the surface stress method, deflection is caused by the interaction of the analyte with the active sensor layer. One surface of the cantilever is activated and the analyte flows from the activated surface. If the surface interaction generates ionic groups, electrostatic repulsion causes compressive stress and cantilever bends down. The bending radius of cantilever (r) depends on the surface stress difference according to Stoney's equation [21]:

$$\frac{1}{r} = \frac{6(1-v)}{Eh^2}\Delta\sigma \tag{7.6}$$

where E is Young's Modulus, v is Poisson's ratio, and $\Delta\sigma$ is the difference in surface stress of the top and bottom plates of the cantilever. For $r \gg l$ deflection can be calculated as in Equation 7.7:

$$\Delta x = \frac{3(1-v)}{E}\left(\frac{l}{h}\right)^2 \Delta\sigma \tag{7.7}$$

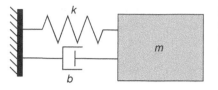

Figure 7.10 Illustration of mass, spring, and damping system. m denotes the resonating mass, B is damping constant, and k is spring constant.

In such sensors sensitivity can be defined as $\Delta x/\Delta\sigma$ and can be increased by increasing the l/h ratio and elasticity of the material used.

In a surface stress sensor based on bending illustrated in Ref. [22], the analyte is caught by activated cantilever beam, and deflection is measured using laser and photodetectors.

7.4.2 Resonant-Based Cell Quantification

Resonant-based sensors have many application areas in MEMS technology like accelerometers, gyros, thermal sensors, and so on. They are also promising sensors for gravimetric cell detection applications. Relation between the resonance frequency and the mass of the sensor raises the opportunity of mass measurement since the resonance frequency is a precisely measurable quantity with different techniques. The basic operation principle of resonance-based detection relies on measuring the shift in resonance frequency due to the added mass coming from the captured particles. The theory of the resonant-based sensor systems is investigated in the following part.

7.4.2.1 Theory of the Resonant-Based Sensors

A resonator can be modeled as a second-order mass, spring, and damper system as illustrated in Figure 7.10.

Motion equilibrium under the applied force (F) can be expressed as in Equation 7.8, where m, B, and k denote mass, damping constant, and spring constant, respectively:

$$m\ddot{X} + B\dot{X} + kx = F \tag{7.8}$$

Transfer function expression of such system where the input is the applied force and the output is the displacement is given in Laplace domain in Equation 7.9:

$$\frac{X(s)}{F(s)} = \frac{1}{ms^2 + Bs + k} \tag{7.9}$$

Two different parameters are used to define a resonating system. The natural frequency of the system, w_n, refers to the undamped free oscillation frequency, while the damping ratio, ζ, refers to how much the system is damped:

$$w_n = \sqrt{\frac{k}{m}} \tag{7.10}$$

$$\zeta = \frac{B}{2\sqrt{km}} \tag{7.11}$$

The systems whose damping ratio is smaller than unity are referred to as underdamped systems. In an underdamped resonance system, energy dissipation is sufficiently small, and the free vibration response of the system is oscillatory. Therefore, resonance-based cell detection sensor systems must be designed as underdamped, and energy loss must be sustained back to the system to ensure a closed-loop oscillation.

Free vibration equations of underdamped systems are given in (7.12):

$$x(t) = e^{-\zeta w_n t} \left(A\cos(w_d t) + B\cos(w_d t) \right) \tag{7.12}$$

$$w_d = w_n \sqrt{1-\zeta^2} \tag{7.13}$$

A and B are constants and depend on initial values of $x(t)$ and $\dot{x}(t)$. w_d is the damped natural frequency that corresponds to the oscillation frequency of free vibration.

In weakly damped systems, damped natural frequency is almost equal to natural frequency ($\zeta < 0.2$):

$$w_d \cong w_n = \sqrt{\frac{k}{m}} \tag{7.14}$$

In mass sensing applications, the most important parameter of a resonating system is its quality factor (Q), which is an alternative way to express how much the system is damped. In other words, as quality factor gets higher, oscillation of the resonator decays more slowly due to damping:

$$Q = \frac{1}{2\zeta} = \frac{\sqrt{km}}{B} \tag{7.15}$$

Resonance frequency of loaded resonator changes as in Equation 7.16, and the shift in resonance frequency can be expressed as in Equation 7.17, after mass change of Δm is introduced to the system:

$$w_{n\_loaded} = \sqrt{\frac{k}{M + \Delta m}} \tag{7.16}$$

$$\Delta w = w_n - w_{n\_loaded} \tag{7.17}$$

The load mass can be calculated by using Equation 7.18:

$$\Delta m \cong \frac{2M\Delta w}{w_{res}} \tag{7.18}$$

The sensitivity of such system can be calculated as shown in Equation 7.19:

$$S = \frac{\Delta f}{\Delta m} = \frac{1}{2\pi} \frac{\Delta w}{\Delta m} = \frac{f_{res}}{2M} \tag{7.19}$$

These are the basic mechanical aspects of resonating structures. The electronic part of a MEMS system generally includes the actuation and sensing mechanisms of a transducer. In the following section, generation of the actuation force and sensing output of the resonator-based MEMS sensor are described.

7.4.2.2 Actuation and Sensing Methods of Resonators in MEMS Applications

In this section, mechanisms that are necessary to create a displacement at resonance frequency and sense the displacement are investigated.

Actuation Methods To make a resonator system resonating, a harmonic force should be applied. Details of harmonically forced vibration are reported in Ref. [23]. In order to generate the force precisely and easily in the desired frequency and amplitude, an appropriate actuation mechanism, which converts electrical energy into kinetic energy, is essential. Electrostatic, piezoelectric, and electrothermal actuation methods are the ones that are most commonly used in the gravimetric MEMS sensors. The generation of electrostatic force is based on changing the stored energy between parallel plates. In Figure 7.11, generation of electrostatic force is illustrated. Proof mass is charged using a DC power supply, and AC signal is applied to change the stored energy on the electrode. Generated force is related with the gradient of the energy, and Equations 7.20–7.25 show the related force equation:

$$\nabla E = F = \frac{\partial E}{\partial x}\hat{a}_x + \frac{\partial E}{\partial y}\hat{a}_y + \frac{\partial E}{\partial z}\hat{a}_z \tag{7.20}$$

$$E = \frac{1}{2}CV^2 \tag{7.21}$$

Figure 7.11 Generation of drive force of resonator.

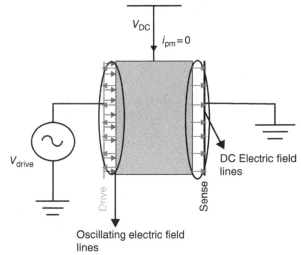

For the case in Figure 7.11, $(\partial E/\partial y)$ and $(\partial E/\partial z)$ are zero. Assuming that the resonator is only able to move in the x direction,

$$F = \frac{1}{2}\frac{\partial}{\partial x}\left(CV^2\right) = \frac{V^2}{2}\frac{\partial C}{\partial x} \tag{7.22}$$

$$F_{\text{drive}} = \frac{\left(V_{\text{DC}} - V_{\text{drive}}\right)^2}{2}\left(\frac{\partial C}{\partial x}\right) \tag{7.23}$$

$$F_{\text{sense}} = \frac{V_{\text{DC}}^2}{2}\frac{\partial C}{\partial x} \tag{7.24}$$

$$F_{\text{net}} = F_{\text{drive}} - F_{\text{sense}} \cong -V_{\text{DC}} \times V_{\text{drive}} \times \frac{\partial C}{\partial x} \quad \text{for} \quad V_{\text{DC}} \gg V_{\text{drive}} \tag{7.25}$$

Negative sign in F_{net} expression indicates 180° phase difference between applied signal and actuation force. Increasing the DC voltage will increase the magnitude of the applied force. $(\partial C/\partial x)$ ratio should be as large as possible to increase the applied F_{net} force.

Piezoelectric actuation is another method to generate the harmonic force required. Piezoelectricity is the coupling between mechanical and electrical properties of materials, which exhibit a change in their polarization due to an applied strain. The degree of polarization is proportional to the applied strain as stated in Ref. [24]. Normally, piezoelectric materials have random orientation of polar domains; however, this random polarization can be oriented under high electric fields. In addition, the materials are usually anisotropic, that is, the properties are dependent on the direction of applied strain and the orientation of polar domains or electrodes.

The standard form of the piezoelectric constitutive equations is given in Equations 7.26 and 7.27, which are known as the strain-charge form [25]:

$$S = sT + eE \tag{7.26}$$

$$D = e^T S + \varepsilon E \tag{7.27}$$

Piezoelectricity can also be expressed with equations as in (7.28) and (7.29), which are known as the stress-charge form [25]:

$$T = cS + dE \tag{7.28}$$

$$D = d^T T + \varepsilon E \tag{7.29}$$

In these equations S represents the 6×1 strain matrix, T represents the 6×1 stress matrix, E is the 3×1 electric field matrix, D is the 3×1 electric displacement matrix, s is the 6×6 compliance matrix, e is the 6×3 direct piezoelectric matrix, ε is the 3×3 dielectric constant, d is the piezoelectric matrix, and c is the stiffness matrix.

Electrothermal actuation relies on converting electrical energy into heat and heat energy into mechanical energy. Basically, current flowing in a resistor causes temperature increase and increase in temperature results in expansion in the material and the structure.

Assume that an AC voltage having frequency w with a DC offset is applied to a resistor:

$$P = P_{\text{static}} + P_{\text{dyn}} = \frac{V_{\text{DC}}^2}{R} + \frac{V_{\text{AC}}^2}{2R} + \frac{2V_{\text{DC}}V_{\text{AC}}}{R}\cos(wt) + \frac{V_{\text{AC}}^2}{2R}\cos(2wt) \tag{7.30}$$

Obviously, dynamic power dissipation has two different frequency components; for $V_{\text{DC}} \gg V_{\text{AC}}$ case, the second harmonic term can be neglected:

$$\frac{\partial T}{\partial t} = \frac{\kappa}{\rho C}\nabla^2 T + \frac{P}{\rho C} \tag{7.31}$$

$\dfrac{\kappa}{\rho C}$ is thermal diffusivity, ρ is mass density, and C is specific heat per mass. Since Equation 7.31 is linear, it can be solved separately for static and dynamic terms.

Static heat term causes static heat change, ΔT_{stat}, which can cause compressive stress in clamped–clamped beams, and it can affect resonance characteristics significantly. However, in clamped-free beams, ΔT_{stat} will not affect resonance characteristics [26]. The dynamic heat term will cause dynamic temperature change in the resistors. The temperature change will generate strain [26]:

$$\varepsilon(T) = \varepsilon(T_0) + \alpha_T(T - T_0) \tag{7.32}$$

Figure 7.12 Electrical model of thermal actuators.

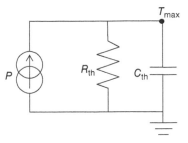

The periodic strain will cause mechanical force and moment, which will actuate the mechanical structure.

Thermal domain can be roughly modeled using lumped electrical elements. Low pass RC filter, as in Figure 7.12, can be used as system model in electrical domain:

$$\frac{T_{max}}{P} = \frac{1}{1 + jwR_{th}C_{th}} \tag{7.33}$$

$$\tau_{th} = R_{th}C_{th} \tag{7.34}$$

R_{th} and C_{th} are usually distributed elements and frequency dependent. That is why static and dynamic response of the system will differ from each other.

Due to the low pass characteristics, it takes time for the capacitor to reach its steady-state voltage value, which refers to the thermal equilibrium temperature of the system. In the macroworld, thermal phenomena have long time constants and are considered as slow actuators. However in micro- and nanoscale resonators, it is possible to actuate an actuator thermally at high frequencies even in GHz ranges [27]:

$$R_{th} = \frac{L}{\kappa A} \tag{7.35}$$

$$C_{th} = mass \times C \tag{7.36}$$

Scaling an actuator with a factor K will result in an increase in R_{th} by the factor K and a decrease in C_{th} by the factor K^3. Therefore, thermal time constant decreases K^2 times. Moreover, if a mechanical structure is scaled with a factor K, its resonance frequency will increase K times. As a result, it is possible to make a mechanical structure resonate with an electrothermal actuation while scaling it.

Sensing Methods Converting the microscale displacement of a MEMS device into a detectable quantity with sufficient gain is one of the most important issues of designing a mechanical sensor. There are expensive optical systems for sensing the movement. However, such systems bring complexity and cost to the designed sensor. To convert movement into an electrical quantity is

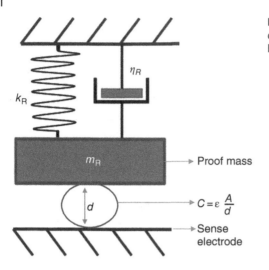

Figure 7.13 Spring, mass, and damping system whose motion will be sensed electrostatically.

crucial for the development of a sensing system. Electrostatic and piezoelectric sensing methods are the most commonly used ones in gravimetric MEMS sensors due to readout electronics compatibility.

Electrostatic sensing is based on the sensing capacitance change between two ports due to displacement. This can be modeled as a spring, mass, and damping system as shown in Figure 7.13.

The proof mass is biased with constant DC voltage and the sense electrode is grounded. The change of the stored charge on electrodes results in a current flow between two electrodes of the capacitor. Equation 7.37 expresses the current caused by charge change:

$$i = \frac{dQ}{dt} = \frac{d(CV_{DC})}{dt} = V_{DC}\frac{dC}{dt} = V_{DC}\frac{dC}{dx} \times \frac{dx}{dt} \tag{7.37}$$

Since the DC voltage is constant, the sense current is related only with the displacement.

Piezoresistive sensing, another popular sensing mechanism, is based on the change in the resistance of a resistor at a certain region of a mechanical structure. This resistance change may be caused by the change of the geometry or the resistivity of the resistor [26]. In metals, geometric effects are dominant on the resistance change. However, in semiconductors, stress and strain change the resistivity of the material, and this phenomenon is called piezoresistivity. Equation 7.38 defines the ratio of resistance change and total resistance:

$$\frac{\Delta R}{R} = (1+2v)\varepsilon + \frac{\Delta \rho}{\rho} \tag{7.38}$$

where v is the Poisson ratio, ε is the strain, and ρ is the resistivity of the material.

Resistance changes are assumed to be linear with a proportionality factor, π, for small stresses. In most applications, long and narrow resistors are used to confine electric field and current [26]. Under these conditions,

$$\frac{\Delta R}{R} = \pi_l \sigma_l + \pi_t \sigma_t \qquad (7.39)$$

π_l and π_t are longitudinal and transverse piezoresistive coefficients, respectively, while σ_l and σ_t are longitudinal and transverse stress applied, respectively.

There are many different resonator designs with a variety of spring types and resonance modes for different purposes. In the following section, resonator types that are commonly used in cell detection application will be investigated.

7.4.2.3 Resonator Structure Types Used for Cell Detection Applications
Cantilever-Based Resonance Sensors Cantilever-based resonators are the most commonly used structures in a range of applications including gas sensing [28], virus detection [29], and metabolic activities of single cells [30]. Figure 7.14 illustrates a cantilever-type resonating structure.

Resonance frequency of the cantilever-type resonator is expressed as

$$f_{res} = \frac{1}{2\pi} \sqrt{\frac{k}{nm}} \qquad (7.40)$$

where k is the spring constant, n is the geometric parameter depending on the mode of resonance and the geometry of the cantilever, and m is the mass of cantilever beam.

For the cantilever in Figure 7.14, k and n parameters can be calculated as in Equation 7.41 and Equation 7.42:

$$k = \frac{Ew}{4}\left(\frac{h}{l}\right)^3 \qquad (7.41)$$

$$n = \frac{33}{140} \qquad (7.42)$$

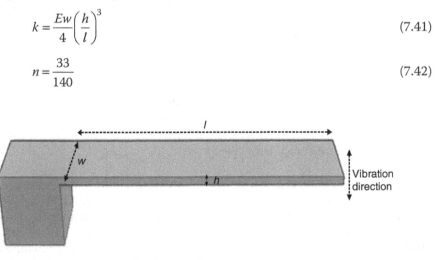

Figure 7.14 Cantilever vibrating in z direction.

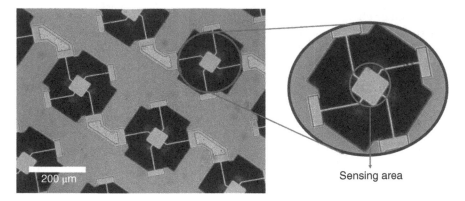

Sensing area

Figure 7.15 A resonant sensor with uniform mass sensitivity. *Source:* Park and Bashir [31]. Reproduced with permission of IEEE.

Since the mass sensitivity of a resonating system depends on resonance frequency, increasing the resonance frequency and decreasing the mass improve the mass sensitivity. Greater sensitivity can be achieved, operating the resonator in higher vibration modes [21].

Equation 7.40 shows that the resonance frequency depends on the mass and the mass addition due to cell capture creates a resonance frequency shift Δf. However, if the added mass is not uniformly distributed, Δf depends on the position of the loaded mass. This problem can be solved by activating certain parts of the cantilever. However, activating certain parts decreases the efficient use of the sensor. To solve the location-based mass sensitivity problem, Park et al. proposed a cantilever-like sensor with a uniform mass sensitivity as shown in Figure 7.15 [31]. Microbeads are located in different regions of the sensing area, and according to experimental results, the difference between maximum and minimum mass is 4.5% of the average mass.

Although this work solves the problem of uniformity in mass detection, the sensor is functional in air, not in liquid (e.g., blood sample). In liquid detection is important for the sake of real-time measurement in blood sample and protection of the cells. However, in-liquid operation introduces viscous damping and decreases quality factor, Q, significantly. The system may be even overdamped and resonance cannot be observed. To solve this issue, suspended microchannel resonators are proposed.

Suspended Microchannel Resonators (SMR)   Suspended microchannel resonators are cantilever structures, in which a microchannel is embedded, Figure 7.16. These cantilevers are operated in air or vacuum. By eliminating the liquid environment around the cantilever, the effect of viscous damping is avoided, and hence, a higher Q is achieved. Using SMRs, fluid density, protein, and

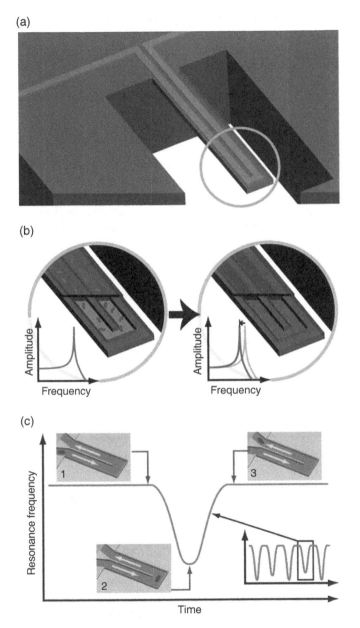

Figure 7.16 (a) SMR system proposed in Ref. [30] is illustrated. (b) Immobilized molecules increase the mass of the channel and create a change in resonance frequency. (c) During particle flow, resonance frequency changes according to position of the particle. Exact mass of the particle can be found by the peak resonance frequency shift. *Source:* Burg et al. [30]. Reproduced with permission of Nature Publishing Group.

single bacteria cell can be detected. Burg et al. proposed an SMR to weigh biomolecules, single cells, and single nanoparticles [30]. Quality factor of the proposed resonator is about 15 000 and resonator mass is 100 ng. In Figure 7.16, the working principle of the system is illustrated.

Fluid flows continuously through the microchannel. There are basically two different measurement techniques. In the first technique, molecules or bacteria cells are bound to the walls of the resonator and the resonance frequency changes permanently. Using immobilized receptors, selective measurement can be achieved. The other technique is based on continuously observing the resonance frequency. When a particle goes through the channel, resonance frequency decreases and reaches its minimum value and then increases again to the initial value. Calculating frequency difference between maximum and minimum resonance frequencies, the mass of the particles can be found. Moreover, counting the number of resonance frequency minima, the number of the particles can be calculated.

Although suspended microchannel resonators have high Q, low mass, and high sensitivity, their channel height is about $2\,\mu m$ due to fabrication limits, which is insufficient for the detection of cells or particles bigger than that. Most biological cells have diameters above $5\,\mu m$, and CTCs are around $15\,\mu m$. For such applications, resonator and microchannel should be separated from each other. Due to requirements of position-independent sensing and high-quality factor in liquid environment for rare cell detection, lateral mode resonators with hydrophobic coatings are developed.

Lateral Mode Resonators Laterally resonating structures are widely used in accelerometer and gyroscope applications. The working principle is similar to cantilever-based resonators except that it has a lateral resonating mode to decrease the viscous damping. A laterally resonating structure for cell sensing has been proposed by our group [32], and a related scheme is given in Figure 7.17.

A PDMS microchannel is placed on top of the resonator. The resonating structure moves laterally to eliminate the squeeze film damper of the liquid flowing through the microchannel. The cell inside the liquid is caught by immobilized antibodies and causes mass change. Mass change results in resonance frequency shift as mentioned in previous sections.

It is shown that the effect of added mass is not position dependent [33]. Only 1.9% standard deviation is observed when a microbead is placed at different locations. The major problem in laterally resonating structures with fingers is fluid leakage through sensing, actuation, and spring gaps while working in fluid environments. This has been solved by our group [34] via air trapping method, whereby the resonator is coated with a thin hydrophobic parylene layer with which fluid leakage between gaps and, hence, corruption of resonator operation are prevented. It is shown that quality factor in liquid decreases only one-third of the quality factor in air. In Figure 7.18, air trapping method is illustrated.

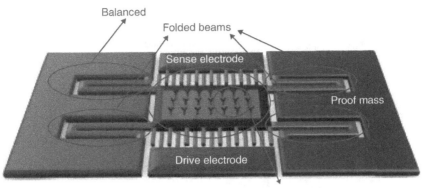

Figure 7.17 Laterally resonating structure for cell sensing.

Parylene-coated lateral resonators for rare cell sensing applications solve chronic problems of both cantilever- and SMR-type structures. As mentioned previously, widely used cantilever-shaped resonators are easy to fabricate, but they suffer from high damping inside the liquid environment. To solve this issue, SMR structures were proposed. Although the sensitivity of the SMRs is very high, these sensors are application limited since the depth of the microchannel is smaller than the radius of most of the rare cells. In parylene-coated laterally resonating structures, the damping problem is alleviated and channel depth is not limited.

7.5 Conclusion and Comments

Accurate quantification and identification of biological cells are crucial in point-of-care diagnosis applications. In this chapter, MEMS-based cell quantification methods based on optical, electrochemical, and gravimetric techniques have been summarized.

The optical sensors are able to do precise quantification. These sensors are especially advantageous in observation. Although there are several successful studies, expenses of optical system are the most important obstacles for the widespread use of such sensors. Optical sensors worth considering for the applications of not only the numbers but also the physical parameters of cells are important.

Electrochemical sensors can also be used for cell quantification. These sensors are favored due to ease of design and fabrication. There are lots of examples for the purpose of both cell quantization and electrical cell characterization. However, drawbacks of the technique allow only a rough quantization.

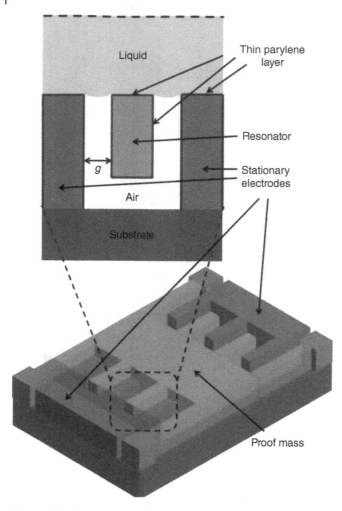

Figure 7.18 Air trapping method.

Gravimetric cell quantification methods using cantilever beam or resonating structures allow single-cell measurements with a high accuracy. This kind of structures can also be used as cell growth sensor to monitor the growth rate of a cell group. In the design stage, the electrical and especially mechanical background theory is very important in the manner of damping, sensitivity, and sensor actuation and sensing.

MEMS technology offers many different techniques of solutions for the cell quantization problem. According to the application requirements, such as sensitivity, selectivity, observability, or ease of fabrication, an appropriate sensor could be selected in the light of the given theory and literature works.

References

1 Yole Developpement, Yole Developpement Press—MEMS and sensors, **2016**. [Online]. Available at http://www.yole.fr/2014-galery-MEMS.aspx#I0002f78f (accessed on June 23, 2017).

2 B. Kuswandi, Nuriman, J. Huskens, and W. Verboom, *Anal. Chim. Acta*, vol. 601, no. 2, pp. 141–155, **2007**.

3 A. Ozcan and E. McLeod, *Annu. Rev. Biomed. Eng.*, vol. 18, pp. 77–102, **2016**.

4 R. Bashir, *Adv. Drug Deliv. Rev.*, vol. 56, no. 11, pp. 1565–1586, **2004**.

5 J. B. Edel, N. P. Beard, O. Hofmann, J. C. deMello, D. D. C. Bradley, and A. J. deMello, *Lab Chip*, vol. 4, no. 2, pp. 136–140, **2004**.

6 B. Yao, G. Luo, L. Wang, Y. Gao, G. Lei, K. Ren, L. Chen, Y. Wang, Y. Hu, Y. Qiu, *Lab Chip*, vol. 5, no. 10, pp. 1041–1047, **2005**.

7 J. A. Chediak, Z. Luo, J. Seo, N. Cheung, L. P. Lee, and T. D. Sands, *Sens. Actuators A Phys.*, vol. 111, no. 1, pp. 1–7, **2004**.

8 M. M. G. Grafton, T. Maleki, M. D. Zordan, L. M. Reece, R. Byrnes, A. Jones, P. Todd, and J. F. Leary, in *Microfluidic MEMS hand-held flow cytometer*, San Francisco, CA, January 22–23, **2011**, vol. 7929, paper no. 79290C–79290C–10.

9 W. Li, T. Knoll, A. Sossalla, H. Bueth, and H. Thielecke, in On-chip integrated lensless fluorescence microscopy/spectroscopy module for cell-based sensors, San Francisco, CA, **2011**, vol. 7894, paper no. 78940Q–78940Q–12.

10 C. Han, S. Pang, D. V. Bower, P. Yiu, and C. Yang, *Anal. Chem.*, vol. 85, no. 4, pp. 2356–2360, **Feb. 2013**.

11 Y. Demircan, S. Orguc, J. Musayev, E. Ozgur, M. Erdem, U. Gunduz, S. Eminoglu, H. Kulah, and T. Akin, Label-free detection of leukemia cells with a lab-on-a-chip system integrating dielectrophoresis and CMOS imaging, in *Transducers—2015 18th International Conference on Solid-State Sensors, Actuators and Microsystems (TRANSDUCERS)*, Anchorage, AK, June 21–25, **2015**, pp. 1589–1592.

12 A. Ozcan and U. Demirci, *Lab Chip*, vol. 8, no. 1, pp. 98–106, **Jan. 2008**.

13 K. Solly, X. Wang, X. Xu, B. Strulovici, and W. Zheng, *Assay Drug Dev. Technol.*, vol. 2, no. 4, pp. 363–372, **2004**.

14 L. Yang, Y. Li, and G. F. Erf, *Anal. Chem.*, vol. 76, no. 4, pp. 1107–1113, **2004**.

15 M. Barreiros dos Santos, J. P. Agusil, B. Prieto-Simón, C. Sporer, V. Teixeira, and J. Samitier, *Biosens. Bioelectron.*, vol. 45, pp. 174–180, **2013**.

16 X. Jiang and M. G. Spencer, *Biosens. Bioelectron.*, vol. 25, no. 7, pp. 1622–1628, **2010**.

17 Y. Wu, J. D. Benson, and M. Almasri, *Biomed. Microdevices*, vol. 14, no. 4, pp. 739–750, **2012**.

18 D. Satake, H. Ebi, N. Oku, K. Matsuda, H. Takao, M. Ashiki, M. Ishida, *Sens. Actuators B Chem.*, vol. 83, no. 1, pp. 77–81, **2002**.

19 M. Moscovici, A. Bhimji, and S. O. Kelley, *Lab Chip*, vol. 13, no. 5, pp. 940–946, **2013**.

20 A. Cagliani and Z. J. Davis, Bulk disk resonator based ultrasensitive mass sensor, in *Proceedings of IEEE Sensors*, Christchurch, New Zealand, October 25–28, **2009**, pp. 1317–1320.

21 F. G. Banica, *Chemical Sensors and Biosensors: Fundamentals and Applications.* John Wiley & Sons, Ltd, Chichester, **2012**.

22 H. Feng, S. B. Sang, W. D. Zhang, G. Li, P. W. Li, J. Hu, S. B. Du, and X. J. Wei, Fundamental Study of the Micro-Cantilever for more Sensitive Surface Stress-Based Biosensor, in *Micro-Nano Technology XIV*, Hangzhou, China, **2013**, vol. 562, pp. 334–338.

23 S. M. Heinrich and I. Dufour, "Fundamental Theory of Resonant MEMS Devices," in O. Brand, I. Dufour, S. M. Heinrich, and F. Josse (Eds.) *Resonant MEMS, Fundamentals, Implementation and Application*, Wiley-VCH, Weinheim, **2015**, pp. 1–28.

24 S. P. Beeby, M. J. Tudor, and N. M. White, *Meas. Sci. Technol.*, vol. 17, no. 12, p. R175, **2006**.

25 G. Piazza, "Piezoelectric Resonant MEMS," in O. Brand, I. Dufour, S. M. Heinrich, and F. Josse (Eds.) *Resonant MEMS, Fundamentals, Implementation and Application*, Wiley-VCH, Weinheim, **2015**, pp. 147–172.

26 O. Brand and S. Pourkamali, "Electrothermal Excitation of Resonant MEMS," in O. Brand, I. Dufour, S. M. Heinrich, and F. Josse (Eds.) *Resonant MEMS, Fundamentals, Implementation and Application*, Wiley-VCH, Weinheim, **2015**, pp. 173–201.

27 A. Rahafrooz and S. Pourkamali, *IEEE Trans. Electron Devices*, vol. 58, no. 4, pp. 1205–1214, **2011**.

28 S. Subhashini and A. V. Juliet, "CO2 Gas Sensor Using Resonant Frequency Changes in Micro-Cantilever," in N. Chaki, N. Meghanathan, and D. Nagamalai, Eds. *Computer Networks & Communications (NetCom): Proceedings of the Fourth International Conference on Networks & Communications*, Springer, New York, **2013**, pp. 75–80.

29 A. Gupta, D. Akin, and R. Bashir, *Appl. Phys. Lett.*, vol. 84, no. 11, pp. 1976–1978, **2004**.

30 T. P. Burg, M. Godin, W. Shen, G. Carlson, J. S. Foster, K. Babcock, and S. R. Manalis, *Nature*, vol. 446, no. 7139, pp. 1066–1069, **2007**.

31 K. Park and R. Bashir, MEMS-based resonant sensor with uniform mass sensitivity, in *TRANSDUCERS 2009—15th International Conference on Solid-State Sensors, Actuators and Microsystems*, Denver, CO, June 21–25, **2009**, pp. 1956–1958.

32 E. Bayraktar, D. Eroglu, A. T. Ciftlik, and H. Kulah, *Proc. IEEE Int. Conf. Micro Electro Mech. Syst.*, pp. 817–820, **2011**.

33 D. Eroglu, E. Bayraktar, and H. Kulah, A laterally resonating gravimetric sensor with uniform mass sensitivity and high linearity, in *2011 16th Int. Solid-State Sensors, Actuators Microsystems Conf. TRANSDUCERS'11*, Beijing, China, **2011**, pp. 2255–2258.

34 D. Eroglu and H. Külah, *J. Microelectromech. Syst.*, vol. 20, no. 5, pp. 1068–1070, **2011**.

8

Lab-on-a-Chip Platforms for Disease Detection and Diagnosis

Ziya Isiksacan[1], Mustafa Tahsin Guler[2], Ali Kalantarifard[1], Mohammad Asghari[1], and Caglar Elbuken[1]

[1] *Institute of Materials Science and Nanotechnology, National Nanotechnology Research Center (UNAM), Bilkent University, Ankara, Turkey*
[2] *Department of Physics, Kirikkale University, Kirikkale, Turkey*

8.1 Introduction

The roots of understanding the liquid behaviors date back to centuries ago. Initial examples cover the studies by Hippocrates (400 BC) and Galen (200 AD) who performed the colorimetric analysis of body fluids like urine to monitor human physiology [1]. With the advancement of microfabrication methods and tools, dexterous handling of liquids in small quantities and development of small fluidic components (valves, pumps, etc.) were witnessed after the 1950s. Seminal studies include the development of miniaturized gas chromatography chip [2] and ink-jet printing technology [3]. In 1990, Andreas Manz proposed the concept of the micro-total analysis systems (μTAS), addressing the possibility that miniaturized integrated systems can be developed to perform biochemical analysis with much lower sample volumes [4]. Then, microfluidics has emerged as a discipline where very low volumes of fluids are manipulated and studied, employing structures in micron length scales. Conventional photo-lithographic techniques used for silicon and glass wafers in microelectronics and MEMS have been inherited for the fabrication of microfluidic chips. Polydimethylsiloxane (PDMS) was utilized by George Whitesides for soft lithography introducing rapid fabrication of microfluidic devices [5].

Years of research have shown that this elastomeric, optically transparent, porous, and biocompatible material is suitable for many microfluidic applications. The utmost goal of μTAS is to integrate laboratory tasks on a single chip, namely, lab-on-a-chip (LOC). The LOC systems, once successfully established,

Biosensors and Nanotechnology: Applications in Health Care Diagnostics, First Edition.
Edited by Zeynep Altintas.

have the benefits of low cost, automation, high throughput, multiplexing, and portability. It is also critical to minimize the number of off-chip components such as power sources and pumps to turn microfluidic systems into *truly* LOC systems. This distinction is especially noteworthy for disease diagnostics at the point of care (POC) where the ASSURED criteria are coined by the World Health Organization (WHO) [6, 7]. The acronym stands for "affordable, sensitive, specific, user-friendly, rapid and robust, equipment free, and deliverable to end-user." Various body fluids such as blood, urine, or saliva can be used to look for specific biomarkers. The most commonly investigated biomarkers include proteins (especially for immunoassays), cells (circulating tumor cells, red blood cells, platelets, etc.), nucleic acids (DNA, RNA, etc.), metabolites (glucose, lactate, etc.), and ionic chemicals (sodium, potassium, etc.) [8, 9]. Depending on the type of biomarker as well as the working mechanism of the LOC platform, a number of detection principles can be proposed. The most preferred ones include optical detection (colorimetric, spectroscopic, plasmonic, fluorescence) and electrical detection (electrochemical, piezoelectric, impedimetric, potentiometric). This chapter discusses different LOC platforms that are employed for the detection and diagnosis of diseases and are categorized into six groups: continuous flow, paper-based, microdroplet, digital microfluidic (DMF), compact disc-based, and wearable platforms. Each platform is explained in the following subsections, and seminal studies as well as the state of the art are discussed. Finally, a future perspective and outlook are provided.

8.2 Continuous Flow Platforms

Continuous flow microfluidic platforms are the conventional microfluidic diagnostic systems where a single-phase fluid flow is controlled inside microfluidic channels. The initial examples of continuous flow systems were fabricated on glass using conventional microfabrication techniques. The patterned glass layer is bonded with another glass piece, forming micrometer-scale delicate structures. After the adaptation of polymers, mostly PDMS, by the microfluidics community, the field witnessed an explosion of research output all across the world. The integrated platforms require on-chip pumps, valves, mixers, and sensors. There has been considerable effort on developing these individual components as summarized in several review articles [10–14].

One of the earliest examples of continuous flow LOC detection systems is the DNA analysis system [15]. It is designed for microfluidic DNA gel electrophoresis and contains sample preparation units such as sample metering, mixing, and heating, as well as integrated fluorescence optical detection unit. The system has a hybrid silicon/glass structure that is defined by photolithographic techniques. The system can meter a specific volume of DNA fragment solution, run it through an electrophoretic separation channel, and perform fluorescent detection of the separated DNA segments.

With the wide use of PDMS in microfluidics community, the number of continuous flow diagnostic microfluidic systems has inflated. A cornerstone for PDMS-related devices was the development of elastomeric valve. Later, these microvalves have been integrated on a single device at a highly scalable level that led to very-large-scale integrated valve-based PDMS microfluidic devices [16]. The fundamental technology behind these systems is the development of two-layer microfluidic systems that are fabricated using soft lithography. One of the layers is utilized as the fluidic layer whereas the second layer is designed as the control layer. The two layers were separated using thin PDMS elastomeric valves that can be actuated using increased pressures to seal the crossing fluid layer.

These two-layer devices allowed exquisite control on the fluid motion along a microchannel network that opened up a plethora of applications. An exemplary system is shown in Figure 8.1, which performs on-chip DNA isolation from whole bacteria [17]. The system is composed of 54 elastomeric valves and can perform parallel processing in three lines. First, a single cell is isolated from a cell solution. Then, cell lysis is performed using a rotary mixer, which is composed of a ring-shaped channel with three sequential valves. The operation of these three valves in a certain sequence results in peristaltic pumping of the fluid, which is utilized as a rotary mixer, enhancing diffusion-limited processes for such biochemical applications. After cell lysis, the mixture is transferred to a DNA affinity column for purification and then transferred to the system outlet. As seen in Figure 8.1, in an area of 2 cm × 2 cm, the system integrates a complicated network of two-layer PDMS channels and achieves sophisticated fluidic operations that can be programmed and scaled up in parallel. Such high-level integration of fluid control at microfluidic scale was a revolution in the continuous flow systems and triggered several other advanced detection and diagnostic devices such as fluorescence-activated cell sorting [18] and single-cell enzyme screening [19].

Nucleic acid detection is a very sensitive and specific strategy that has been successfully implemented on continuous microfluidic systems. Another challenge in the disease diagnosis is the detection of rare cells. Identification of cells in a given solution, for example, blood, is performed using flow cytometers. There have been several remarkable examples of continuous flow cytometry systems [20]. However, detection of rare cells poses another level of complexity since these applications require cell purification and high-throughput analysis systems. Detection of circulating tumor cells (CTCs) is an excellent example of rare cell detection, which is critical for early diagnosis of cancer.

Continuous flow LOC systems provide a very suitable platform for detection of CTCs. An outstanding example of such systems is developed by Prof. Toner's group. They demonstrated a microfluidic system that is composed of

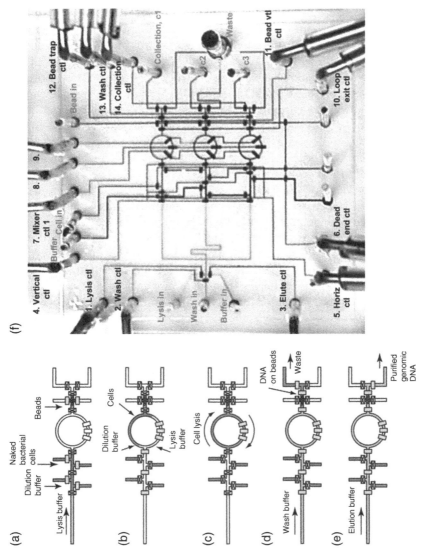

Figure 8.1 Automated on-chip DNA purification. (a–e) The workflow summarizing the on-chip DNA purifier system that has a two-layer structure for fluid flow and valve actuation. Fluid channels and control channels have widths of 100 and 200 μm, respectively. (f) Picture of the parallel DNA isolation protocol. *Source:* Hong et al. [17]. Reproduced with permission of Nature Publishing Group.

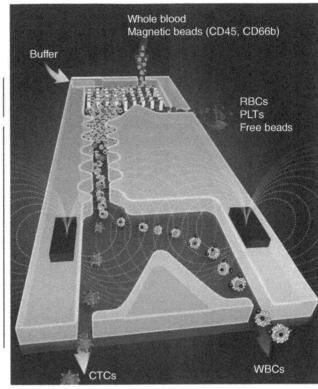

Figure 8.2 CTC isolation chip. Whole blood mixed with antibody-tagged magnetic beads is sent to the channel in parallel with buffer solution. Deterministic lateral displacement compartment of the chip enables the separation of white blood cells (WBCs) and CTCs from the whole blood. Then, WBCs and CTCs are brought on a single line after passing asymmetric focusing elements utilizing inertial microfluidics. Finally, magnetic force is applied over centrally aligned cells that force magnetically labeled CTCs to a different outlet, completing the isolation process. *Source:* Karabacak et al. [22]. Reproduced with permission of Nature Publishing Group. (*See insert for color representation of the figure.*)

the integration of two devices in tandem [21, 22]. The first microfluidic chip is a silicon/glass hybrid that is used to separate white blood cells and CTCs from whole blood sample using size-based deterministic lateral displacement separation technique. Then, these separated cells are transferred to the second PDMS microfluidic chip, which can align the cells using inertial focusing and sort the CTCs by magnetophoresis. A schematic of the system is shown in Figure 8.2. The system was built on a custom rig, and successful isolation of CTCs from a very populated solution was demonstrated. Although not demonstrated in this study, the identification and counting of CTCs can also be integrated using the benefits of continuous flow platforms.

The continuous microfluidic platforms also resulted in several commercial products for disease diagnosis that are remarkable success stories in the field. One of the earliest examples of such systems is Agilent's on-chip gel capillary electrophoresis system, which is composed of a benchtop analyzer unit and single-use cartridges for DNA, RNA, and protein studies [23]. The system works with 1–5 μL of samples and dramatically reduces the sample and reagent consumption. The channels are fabricated in glass and bonded to a plastic housing containing wells for sample, gel solution, and control solutions as seen in Figure 8.3. As the sample is electroosmotically run through a microfluidic channel, separation is achieved, imaging is performed using a laser-induced fluorescence unit, and the results are displayed as electropherograms. The concentration of each sample is calculated by using the calibration control standards. In addition to electrophoretic separation, a cell-based cartridge is also developed for the same platform to monitor transfection efficiency and protein expression and to study apoptosis and gene silencing mechanisms.

Another seminal example of continuous microfluidic platforms is i-STAT whole blood analysis system. This handheld unit provides a platform for measurement of blood gases, electrolytes, cardiac markers, and coagulation factors in the whole blood. The disposable cartridges require only a drop of whole blood and perform amperometric/potentiometric measurement using microfabricated electrodes fabricated on silicon chips and integrated with polymer microfluidic channels. The latest generation of i-STAT cartridges performs

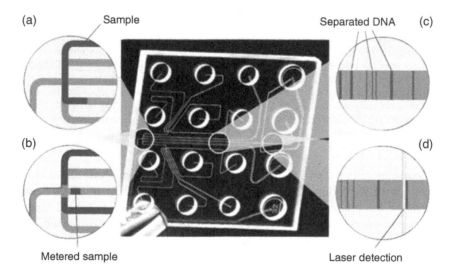

Figure 8.3 Agilent's capillary electrophoresis chip. (a) Sample is delivered to the junction area. (b) Small amount of sample is metered into the detection channel. (c) DNA is electrophoretically separated. (d) Separated DNA is fluorescently detected. *Source:* Mark et al. [23]. Reproduced with permission of Agilent Technologies, Inc.

ELISA assays using on-chip washing/labeling fluids, which proves a very strong platform for the detection of biomarkers for rapid diagnosis of myocardial infarction. These systems were inspiring platforms that fueled the microfluidics research starting from its early years.

A remarkable platform for one of the latest applications of continuous fluidic systems is Ion Torrent's next-generation DNA sequencing system [24]. As opposed to the optical detection-based sequencing technologies, Ion Torrent provides a highly scalable and CMOS-based unit that provides rapid DNA sequencing. The cartridges are composed of microwells with integrated ion-sensitive field-effect transistors (ISFETs) that effectively work as micrometer-size pH meters. Each well contains microbeads with different DNA templates. The nucleotides are flushed sequentially over the wells, and the incorporation of the nucleotide to the DNA strand in a specific well is detected as a pH change as schematically summarized in Figure 8.4. The system has rapidly evolved since 2010 with increased read lengths, higher number of wells, and improved read accuracies.

8.3 Paper-Based LOC Platforms

The intrinsic properties of paper make it a unique choice for microfluidic platforms designed for diagnosis. Paper is composed of entangled cellulose fibers forming a porous structure that makes it hydrophilic. By capillary force, paper can be used as a transport medium without any external power source. Besides, paper is ubiquitous, cheap, biodegradable, and lightweight and can be modified by chemicals; thus, it suits well for biomedical applications and diagnosis [25–30]. Paper-based platforms have existed long before the birth of microfluidics due to the availability and remarkable benefits of paper. It is only after the introduction of several LOC platforms that paper has been re-evaluated as a separate platform that gave rise to Microfluidics 2.0 [31].

In order to control the fluid flow along paper, one needs to define channels on the paper. Development of hydrophobic regions on paper with paraffin and wax to obstruct fluid flow is not a new method and is patented a century ago [32]. It is difficult to spot the earliest functional paper-based diagnostic device due to the lack of records. Yagoda's work in 1937 stands out since it demonstrated the patterning of selected areas on paper using a paraffin stamping method [33] similar to today's paper-based devices. Ring-shaped analysis spots were defined on the paper and employed for the detection of nickel and copper metal ion concentrations.

Conventional paper-based devices are categorized as dipstick and lateral flow devices. The first examples of dipstick paper detection devices were developed by a French chemist, named Jules Maunmené, in 1850 for urine analysis. In order to measure pH of the blood, test strips were introduced in

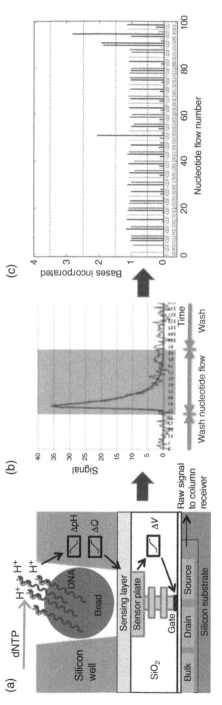

Figure 8.4 (a) Complete single-stranded DNA is cut into small pieces, and every piece is bonded on a single bead for amplification to cover all surfaces of the bead where it is placed on a small pH sensitive well. (b) Nucleotides are flowed through the chip; if there is a conjugate base in the well, then it is incorporated, releasing H^+ ions. Changing pH, resulting from H^+ ions, is detected by CMOS sensor underneath the well. (c) Signals obtained from each well are encoded to obtain the DNA sequence. *Source:* Merriman et al. [24]. Reproduced with permission of John Wiley & Sons.

the 1920s and are still in the market. The widespread use of pH strips is due to their simplicity. By dipping the strip into the solution of interest, the pH of the solution (or for other assays, the analyte concentration) can be determined by a color change on the strip. The pH value can be determined visually using the reference indicator card. However, dipstick tests require manual sample preparation, incubation, and washing steps.

The introduction of lateral flow strips turned the simple dipstick assays into sample-in-result-out paper-based integrated devices that can perform ELISA tests. Lateral flow strips are composed of four main parts. The sample is introduced to the device from the sample inlet port. Then, the sample migrates through the second component, the conjugate pad, which includes reagents and allows for the interaction between the sample and the reagent. Moving through the conjugate pad, the sample reaches the membrane that is generally made of nitrocellulose. Across the membrane, two lines of capture antibodies were immobilized, the first one being an indicator of the presence of analyte and the second one indicating the successful transport of the sample along the membrane. Finally, the sample is collected at the absorbent pad, which is used to soak and store the leftover sample. Both sandwich and competitive assays are implemented as lateral flow assays.

Recently, paper-based platforms have gained a momentum with the advancements in the field of microfluidics [34]. Democratization of diagnostic platforms is an increasing need all across the world, which constrains the complexity of microfluidic components on a single device when cost per test is a major concern. Therefore, researchers moved to paper-based devices, and the field has experienced the resurrection of 100-year-old paper-based systems with advanced functionalities. The developments in microfabrication, printing methods, microelectronics, and microfluidics opened new avenues for paper-based platforms that allow high-scale integration with improved sensitivity and multiplexing [34].

The main enabling factor for the rebirth of paper-based platforms was the development of various low-cost fabrication techniques. In the conventional lateral flow strips, the paper channels were defined simply by cutting. However, in the second-generation paper-based platforms, the channels are defined by creating hydrophobic and hydrophilic regions across the paper (Figure 8.5). In order to pattern the hydrophilic channel region surrounded by hydrophobic barriers, there are fundamentally two approaches: physical and chemical modifications. The paper substrate can be physically modified by the impregnation of hydrophobic materials such as wax, SU-8, PDMS, or polystyrene, which not only turns the fibrous structure to hydrophobic but also blocks the pores. Another approach is to chemically modify the paper using cellulose reactive substances such as alkenyl ketene dimer (AKD). The choice of the fabrication technique is usually dominated by the cost and the minimum feature size required for the channels. Among physical modification techniques, wax

Figure 8.5 Illustration of a single-layer microfluidic paper-based device fabricated by photolithography. *Source:* Martinez et al. [35]. Reproduced with permission of American Chemical Society.

printing is favorable since it is cheap and practical to implement [29]. It only requires a solid wax printer and an oven to first print and then reflow the wax to be absorbed by paper, respectively. This method is mostly used when low-resolution channels suffice for the application. Similarly, PDMS dissolved in hexane or SU-8 negative photoresist [35] is also used for the hydrophobization of paper since these are commonly available materials across most microfluidics laboratories, thanks to the soft lithography process. However, compared to wax printing, SU-8 and PDMS are much costly alternatives. An additional benefit of PDMS-modified paper structures is their flexibility for applications that make use of foldable structures or 3D geometries [26]. The use of photoresist patterning on paper is especially useful when obtaining fine features on paper that are defined through photolithography.

For chemical modification of paper, AKD is the most commonly used agent to create hydrophobic regions that are defined on paper directly using an ink-jet printer [36]. This is a fairly low-cost single-step method. In comparison, some studies demonstrated that the whole paper is turned hydrophobic first, and then the hydrophilic regions are redefined by plasma treatment, employing a two-step method [37]. Among these fabrication techniques, printing-based methods are very advantageous for batch manufacturing of paper-based devices. In addition, the printing technology can also be used for printing of reagents required for the assay using a second head, which allows a fully integrated fabrication scheme that lowers the fabrication cost significantly.

Using the aforementioned techniques, several interesting paper-based devices have been demonstrated, such as detection of glucose and proteins

(Figure 8.5). In addition to these single-layer structures, stacked multilayer paper devices were also demonstrated in more intricate geometries [38]. Similar to multilayer printed circuit boards, fluids can move across different layers through hole-like connections that are similar to vias in printed circuit boards. Such designs achieve complex fluid flow in a smaller footprint.

The most common detection methods used for paper-based platforms are unarguably colorimetric (Figure 8.6a) and electrochemical detection (Figure 8.6b). Colorimetric detection is mostly used for qualitative analysis since it merely depends on a color change based on an enzymatic reaction. One of the earliest demonstrations of colorimetric paper-based detection devices is from Martinez et al., which demonstrates the detection of BSA and glucose using a chromatography paper patterned with SU-8 photoresist. Using an artificial urine solution, detection of two analytes was demonstrated as a semi-quantitative assay as seen in Figure 8.6a [39]. Similarly, well-known enzymatic colorimetric assays were implemented on paper demonstrating detection of uric acid [41], human IgG [42], and bacteria [43].

In an attempt to obtain more quantitative results, electrochemical detection has been developed on paper platforms. Dungchai et al. demonstrated detection of glucose, lactate, and uric acid by integrating electrodes on the paper defined by SU-8 patterning as shown in Figure 8.6b [40]. The low-cost

(a)

[glucose] (mM)		[BSA] (µM)
0		0
2.5		0.38
5.0		0.75
10		1.5
50		7.5
500		75

5 mm

(b)

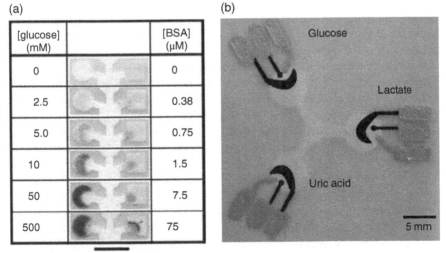

Figure 8.6 Different detection methods used in a paper-based platform. (a) Colorimetric detection of glucose and protein. *Source:* From Martinez et al. [39]. Adapted with permission of John Wiley & Sons, Inc. (b) Electrochemical detection of three analytes implemented on a single device. *Source:* From Dungchai et al. [40]. Reprinted with permission of American Chemical Society.

fabrication of screen-printed carbon or Ag/AgCl electrodes was in line with the benefits of paper-based devices. Therefore, this approach has been widely adapted and opened another dimension for paper-based diagnostic devices.

Overall, paper devices are prone to variability due to environmental conditions such as temperature and humidity. Their clinical performance lags behind other LOC platforms in terms of specificity and sensitivity. In addition, they require larger sample volume due to the retention of the sample in the device during fluid transport. Still, the facts that these systems do not require off-chip components and are fairly low-cost compared to their counterparts indicate that they are attractive alternatives for affordable diagnostic applications.

8.4 Droplet-Based LOC Platforms

Droplet-based platforms are based on the generation and control of minute volumes of microdroplets (dispersed phase) within carrier liquids (continuous phase) inside microchannels. Individual droplets with low volumes allow very precise control over the concentration of the samples of interest as well as the reaction times. In addition, rapid and extremely repeatable generation of monodisperse droplets makes these systems suitable for high-throughput biochemical applications [44, 45]. Basic microfluidic operations that can be carried out using these platforms include transporting, splitting, and mixing of the droplet contents as schematically illustrated in Figure 8.7 [46–50]. In addition to the basic operations, manipulations that change the phase of the droplets can be performed. Some examples include chemical polymerization [51], cell/protein/DNA encapsulation [52, 53], and particle synthesis [54].

Microdroplets are formed using two immiscible fluids using either passive or active techniques [55]. Passive techniques lead to higher throughputs and are more commonly used. The two most preferred geometries used for passive microdroplet generation are T-junction [56] and flow-focusing [57] geometries that can both operate in different flow regimes such as squeezing, dripping, and jetting.

The introduction of the μTAS concept and the investigation of the fundamental physics of microdroplet formation led to the emergence of various droplet-based microfluidic platforms, which are advantageous for reduced processing time due to efficient micromixing inside droplets as well as performing operations (reaction, synthesis, etc.) using pico/nanoliters of liquids isolated inside droplets in a parallel way [58]. For the application of droplet-based platforms for disease detection and diagnosis, optical and electrochemical detection systems are integrated into the LOC platforms. For example, Boedicker et al. developed a system that measures antibiotic susceptibility of bacteria inside microdroplets for drug screening studies [59]. Pekin et al. demonstrated a droplet microfluidic device for highly sensitive quantitative

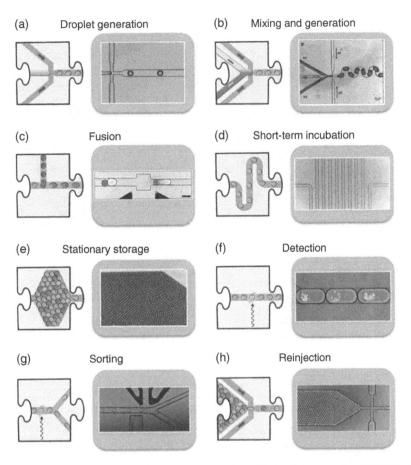

Figure 8.7 A collection of droplet unit operations: (a) droplet generation [46], (b) mixing and generation [47], (c) fusion [48], (d) incubation, (e) storage [49], (f) detection, (g) sorting, and (h) re-injection [46]. *Source:* From Kintses et al. [50]. Adapted with permission from Elsevier. (a and h) *Source:* Schaerli and Hollfelder [46]. Reproduced with permission of Royal Society of Chemistry. (b) *Source:* Huebner et al. [47]. Reproduced with permission of The Royal Society of Chemistry. (e) *Source:* Courtois et al. [49]. Adapted with permission of John Wiley Sons, Inc. (*See insert for color representation of the figure.*)

detection of mutated DNA within complex DNA mixtures as shown in Figure 8.8 [60]. The system performs digital PCR in millions of droplets with picoliter sample volumes obtaining the required sensitivity and specificity to detect rare tumoral DNA in diverse samples. Such an application is very critical for diagnosis of genetic disorders.

Also, droplets can be used for the functional characterization of cell libraries at high throughput by capturing single cells inside microdroplets. Debs et al. developed a platform containing modules for the generation, incubation,

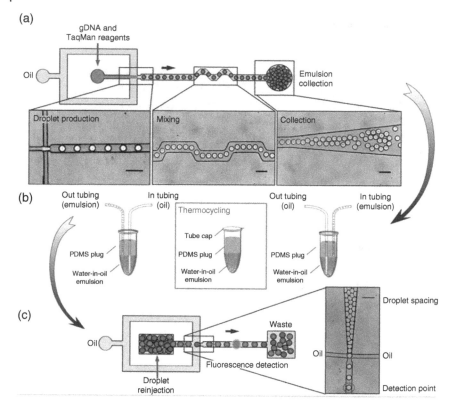

Figure 8.8 Detection of mutated DNA using digital PCR in picoliter microdroplets. (a) An overview of the system showing the production, mixing, and collection of droplets containing DNA, PCR reagents, and TaqMan probes. (b) Thermocycling and amplification of the emulsion. (c) Reinjection of droplets and analysis of the fluorescence signal of individual droplets. *Source:* Pekin et al. [60]. Reproduced with permission of Royal Society of Chemistry.

fusion, and sorting of droplets for hybridoma screening [61]. Since individual cells are isolated inside single droplets, molecules secreted from a single cell can be attained in relatively high concentrations.

Using droplet-based platforms allows detection of diseases at extreme sensitivity and specificity. For example, Juul et al. presented a novel system for the detection of single parasites that cause malaria [62] from a drop of whole blood and saliva samples using in-droplet rolling circle DNA amplification, which does not require thermal cycling. Marcali et al. developed a droplet-based impedimetric hemagglutination detection system that can potentially be used for any agglutination-based study inside microdroplets. The system requires no sample preparation and achieves side injection of blood sample into serum droplets containing agglutinins that are proceeded by electrical detection of agglutination of erythrocytes inside a nanoliter-sized droplet [63].

Despite the high sensitivity obtained using droplet platforms, their commercialization stayed limited mainly due to the relatively complicated system design and the need for several off-chip components. Bio-Rad and RainDance Technologies are two leading companies that produce droplet-based digital PCR devices. More recently, Sphere Fluidics has commercialized a droplet microfluidic system for the analysis and detection of cellular genomic screening.

8.5 Digital Microfluidic-Based LOC Platforms

DMF-based platforms are based on electrowetting-on-dielectric (EWOD) that allows precise and effective control of droplets (for transporting, sorting, merging, etc.) without any need for microchannels [64]. On DMF platforms, droplets are manipulated along virtual channels that are created by sequential activation and deactivation of electrodes that are laid underneath the fluid layer. This is unlike the droplet microfluidic devices explained in the previous section where the liquids are carried inside microchannels and are driven by pumping. This capability frees the DMF systems from all pump and valve units, resulting in compact, electrically controlled LOC platforms as schematically shown in Figure 8.9 [65]. Thanks to the droplet-based operation, DMF systems also benefit from low reagent volume, efficient reagent mixing, and lower reaction times.

EWOD is a phenomenon where one can modify the interfacial tension of polarizable and/or conductive droplets by applying an electric field to hydrophobic, insulated electrodes [66]. This change in the interfacial tension results

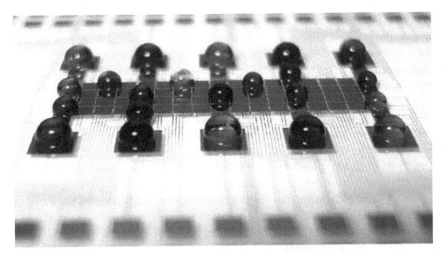

Figure 8.9 A sample digital microfluidic platform system that is capable of working with multiple droplets without requiring any channels, pumps, or valves. *Source:* Jebrail and Wheeler [65]. Reproduced with permission of Elsevier.

in a change in the wettability, therefore contact angle, of the droplet on the surface. This relation is described by the Lippmann-Young equation as follows:

$$\cos\theta(V) = \frac{CV^2}{2\gamma} + \cos\theta_0$$

where θ_0 and θ are the contact angles between the droplet and the surface before and after the voltage application, respectively, C is the capacitance per unit area of the dielectric and the hydrophobic layers, and γ is the interfacial tension between the droplet and the surrounding medium. By turning the voltage "on" and "off," an interfacial tension gradient is formed. This allows the individual droplets to move across the electrodes. Usually, AC actuation is preferred over DC to avoid insulator charging. Based on the application, different media can be used for droplet manipulation. In order to tackle with the evaporation problem, the system can be run in an oil bath.

DMF platforms have two main electrode configurations: single-plane and two-planes (Figure 8.10). In the single-plane configuration, the actuation and ground electrodes are on the bottom surface leaving the fluids of interest open to air (open system). Alternatively, the fluids can be encapsulated by passive covers without compromising system performance. In the two-planes configuration, the ground electrode and actuation electrodes are separated by a spacer and are located on the top and bottom surfaces, respectively (close system). The latter configuration is often used for automated manipulation of droplets in biomedical and chemical fields. The fabrication processes of single-plane and two-plane configurations are similar. Both configurations consist of a substrate, actuation electrodes, and dielectric surface, which are coated with hydrophobic layer. The device can be partitioned into virtual channels in which droplets are manipulated according to their different fluidic functions [67].

One of the earliest diagnostic examples of DMF systems was shown by Richard Fair's group [68]. They showed the colorimetric detection of glucose from physiological samples such as serum, plasma, and urine. The system also demonstrated that multiple solutions can be studied in parallel with a monolithic device. Another critical feature of DMF platforms is their reconfigurability. Sista et al. developed a device for the detection of cardiac troponin-I using a magnetic bead-based immunoassay protocol [69]. On the same system, reconfiguring the electrode actuation, that is, droplet generation/motion sequence, they also implemented a real-time DNA amplification protocol by moving the droplets across temperature zones. The system was used to perform sample DNA amplification for methicillin-resistant staphylococcus aureus (MRSA) detection as an exemplary unit for sample detection of infectious disease [69]. In another example, Barbulovic-Nad et al. introduced a DMF system for complete mammalian cell culture including cell seeding, growth, detachment, and reseeding steps [70]. The system proved that adherent cells can be cultured using this novel platform, which was verified by proliferation

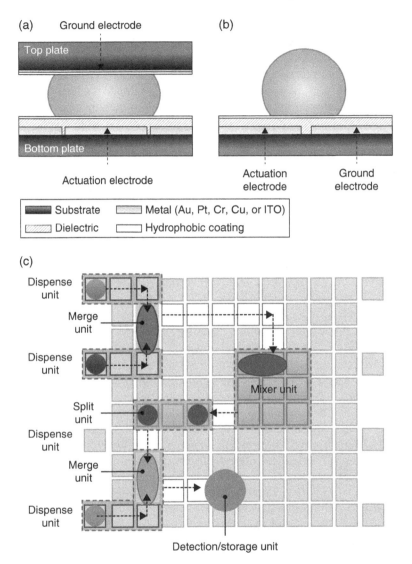

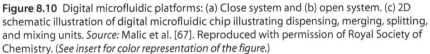

Figure 8.10 Digital microfluidic platforms: (a) Close system and (b) open system. (c) 2D schematic illustration of digital microfluidic chip illustrating dispensing, merging, splitting, and mixing units. *Source:* Malic et al. [67]. Reproduced with permission of Royal Society of Chemistry. (*See insert for color representation of the figure.*)

studies on two cell lines. As a more recent example on cellular studies of DMFs, Ng et al. demonstrated immunocytochemistry in single cells [71]. The assay was performed on adherent cells that are fixed on the top plate. The system was optimized so that shear stress is kept low during medium exchange over the cells while achieving precise and automated fluid delivery. This system can be further extended as a universal platform to study single-cell protein expressions or cell signaling pathways.

A few companies have launched products based on DMFs for biological and chemical detection [72]. It is especially noteworthy that due to the ease of platform construction, automation, and low-cost fabrication, these platforms possess the possibilities to be employed for the detection of a wide variety of biomarkers (DNA, protein, cell, etc.) for disease detection [73]. For example, Advanced Liquid Logic Inc., acquired by Illumina in 2013, developed a benchtop system that can perform DNA, RNA, and protein analyses, proving its high potential as a multi-disease detection platform. Sci-bots is a recent spin-off from Prof. Aaron Wheeler's group aiming for a customizable DMF platform for biochemical diagnostics, cell-based assays, and chemical synthesis.

8.6 CD-Based LOC Platforms

Compact disc (CD)-based LOC systems function based on the rotational motion of CD that creates an outward centrifugal force. Also known as lab-on-a-disc systems, they benefit from the absence of pumps, bubble-free liquid motion, elimination of residual volume, and ease of parallelization [74, 75]. These systems are composed of microfluidic CDs that house the fluids and a CD player-like device that regulates the centrifugal force by controlling the rotation rate (rpm). The channels are mostly constructed using transparent plastics (Figure 8.11) that enables optical detection using a hardware that is similar to the readout units of CD players [76]. The sample is introduced to the center of the CD and carried axially outward using centrifugal force.

Not too surprisingly, optical detection is the most commonly used detection mechanism for CD-based LOC systems. For example, using a colorimetric detection system, Nwankire et al. demonstrated a 5-parameter liver panel test that operates with whole blood. The system integrates plasma separation, which is trivial for CD microfluidic platforms, sample metering, aliquoting, and parallelized detection, which completes the assay in 20 min using less reagent compared to conventional systems [77]. Similarly, fluorescence quantum dots are employed for ultrasensitive detection of botulinum neurotoxin from whole blood, serum, and saliva samples on a CD-based platform [78]. This work presents the implementation of all immunoassay steps on the system, which can readily be adapted for other biomarkers as well. These advanced optical detection systems have their roots in one of the pioneering works in CD microfluidic platforms used for label-free detection of biomolecules [79]. In this study, ligands are immobilized on a CD for immunoassay capturing of molecules. Using the optical setup within the optical disc drive, the error rate of the CD was recorded, and a high correlation was found between the

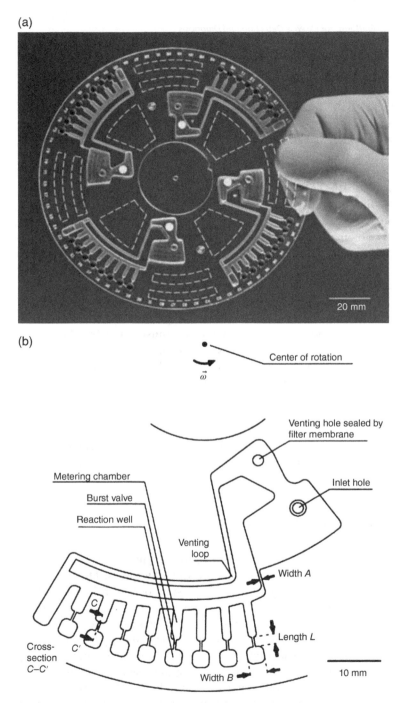

(a)

20 mm

(b)

Center of rotation

$\vec{\omega}$

Venting hole sealed by
filter membrane

Metering chamber

Inlet hole

Burst valve

Reaction well

Venting
loop

Width A

C

Length L

Cross-
section
C–C'

C'

10 mm

Width B

Figure 8.11 (a) Photograph of a CD microfluidic cartridge unit, reaction wells are filled with dye solution. (b) Schematic of the microfluidic design showing some critical components such as inlet, metering chambers, burst valves, and reaction wells. *Source:* Focke et al. [76]. Reproduced with permission of Royal Society of Chemistry. (*See insert for color representation of the figure.*)

biomolecular concentration and the error rate. Centrifugal microfluidic platforms can be used in conjunction with other microfluidic platforms as well. For instance, in a recent study, a CD microfluidic unit is combined with an immunochromatographic strip assay for the very selective detection of influenza A (H1N1) virus detection [80]. The system uses reverse transcriptase loop-mediated isothermal amplification and proves to be a dexterous platform for high specificity pathogen detection.

In an interesting work, electrochemical detection was implemented on a CD microfluidic platform [81]. Electrical slip rings were used to make contact with the spinning disc for amperometric electrochemical detection. The presented system improved the limit of detection for the measurement of C-reactive protein, which is a biomarker for cardiovascular diseases.

Optical disc drive technology that has been around since the 1980s can be easily modified for the formation of CD-based LOC platforms. Optical disc drives contain high-quality and fully integrated laser diodes, photodetectors, lenses, and motors that facilitate the development process for the analyzer. Then, what is left is the development of the CD-type cartridges, which inherit the fabrication methodologies of CDs from the music industry. Therefore, their commercialization was relatively easier compared to the other platforms that require several off-chip components. One of the most well-known centrifugal-based microfluidic companies is Abaxis, which was founded in 1989. Abaxis' Piccolo Xpress system provides a very wide variety of panel assays such as metabolic panel, liver panel, general chemistry, electrolyte panel, and lipid panel. Gyros is another leading company offering CD-based LOC systems solutions targeting several applications ranging from immunogenicity, toxicokinetics, biomarker detection to pharmacodynamics, and impurity testing for therapeutic studies.

8.7 Wearable LOC Platforms

Wearable LOC platforms have been receiving growing interest from academia due to their capacity for reliable health monitoring of wearers. They enable us to obtain information regarding an individual's state of health in a real-time, continuous, and noninvasive manner [82]. Thus, what is aimed is an increase in the quality of life as well as reduction in medical costs associated with diagnostics and therapy [83, 84]. The detection principle is mostly based on electrochemical sensing that provides vital information by measuring physiological analytes from sweat, saliva, or tear. In particular, glucose and lactate are targeted for real-time disease monitoring [85]. Whole blood, serum, or urea, which are mostly used by other LOC platforms, provide higher concentrations and more precise measurements. On the flip side, it is very challenging to make continuous analysis using these fluids. Sweat, saliva, and tear are relatively easy to continuously sample and therefore can be potentially used for continuous, noninvasive monitoring of analytes using wearable platforms. For example, concentrations of the electrolytes (pH, sodium, calcium) and metabolites (glucose, lactate) in sweat provide vital health information [84, 86]. Moreover, thanks to their printed embodiments, these platforms do not have to employ optical or microelectromechanical components, leading to inexpensive and robust sensing systems [87–89].

Saliva known as a very complex biofluid can be employed to monitor hormonal, nutritional, and metabolic parameters [90]. The first LOC platforms aimed at monitoring plaque pH (for studying plaque acidogenicity) and fluoride (for studying fluoride dentifrice efficiency) in saliva [91]. Recently, amperometric saliva LOC platform with wireless transmission feature is fabricated, which measures lactate concentration continuously and noninvasively [92]. In another study, antimicrobial peptides are transferred on a graphene-modified silk substrate for continuous bacterial monitoring [93].

Human sweat can also be used to obtain very important information. For example, sodium for electrolyte imbalance, lactate for cystic fibrosis, ammonium for stress, and calcium for bone mineral loss can be monitored [84]. Sweat can also be employed to measure toxicant [94] and drug-of-abuse concentrations [84]. Sweat-based wearable LOC platforms can be categorized into two types: fabric/flexible plastic-based and epidermal-based LOC platforms. In fabric-based platforms, there is a stable contact between skin and sensor, which allows real-time and continuous measurement of targeted biomarkers in the sweat. Flexible materials used within the platforms allow selective and simultaneous measurement of glucose, lactate, sodium, and potassium ions, as well as skin temperature [82]. Studies for monitoring sodium [95], potassium [96], and chloride levels [97] are also available in the literature. Epidermal-based platforms are mostly tattoo-based LOC platforms where electrodes are printed on temporary tattoos that can have intimate contact with the body for electrochemical sensing [98]. These tattoos are biocompatible, single use, highly resilient to mechanical deformation, and aesthetic, and they do not interfere with the daily routine of the user [84, 86, 98]. The first tattoo-based LOC platform was reported in 2012 [98]. Here, a Ag/AgCl ink is screen printed on a temporary tattoo as an electrode.

Tear contains peptide, electrolytes, lipids, metabolites, ocular epithelial cells, and blood. As such, it is an attractive body fluid for health monitoring. Also, glucose concentrations in blood and tear are shown to correlate well [99]. However, tear is prone to rapid evaporation *in vitro*, limiting its use for POC platforms. The initial studies fabricated bare gold or platinum electrodes on flexible substrates to form strip sensors using photolithography [100]. These studies showed monitoring of transcutaneous oxygen and glucose. However, they were designed such that they caused reflex tear and did not have integrated electronics [84]. The Parviz group at the University of Washington developed a contact lens-based sensor integrated with wireless electronics [101, 102]. The sensor employs glucose oxidase enzyme for continuous glucose monitoring. Google (Alphabet Incorporation) is further developing this sensor system. GlucoWatch is a commercial product that continuously monitors glucose level from interstitial fluid [103], which has been used to measure metabolic diseases, organ failure, drug efficiency, and glucose concentration. The consumers have reported skin irritation, causing the withdrawal of the product from the market. More studies are required to fully utilize interstitial fluid for wearable platforms.

8.8 Conclusion and Outlook

The idea of forming miniaturized integrated systems for biomedical applications has gained popularity within the past decades. Especially, the efforts to combine numerous laboratory tasks on a single chip, that is, LOC, for more effective medical research are noteworthy. This chapter summarized most common LOC platforms that are of use for disease detection and diagnosis. Seminal studies as well as the state of the art in literature were provided. Additionally, successful commercial examples in the market were discussed.

The LOC platforms are promising alternatives to the bulky, lab-based diagnostic equipment. So far, these systems have been mostly adapted in the developed world for rapid diagnostic needs at emergency settings or for advanced biochemical analysis. At the developing world, the driving force for the use of LOC diagnostic platforms was their low cost and simplicity. In spite of the diversity of LOC platforms and their ability to tap into different markets, the number of commercial technologies that made it to the market is still limited considering the wide variety of applications demonstrated in academic publications mainly due to the fact that the engineering of these platforms requires multidisciplinary teams and considerable budget to allow the technologies to move beyond clinical trials successfully. In order to compete with the status quo, LOC platforms have to demonstrate either advanced functionalities or significant reduction in cost. What is so motivating for the researchers in this field is the potential of these technologies to give birth to niche applications.

The twenty-first century is witnessing a shift in the healthcare services. With the increasing awareness of individuals on personal health and healthcare technologies that allow continuous monitoring, our conception of diagnostics will evolve radically. The increasing interest on personalized medicine and telemedicine are some of the indicators of this upcoming transition. Real-time investigation of the health status of a person without interfering with daily activities holds great promise for early detection of a wide variety of abnormalities. This early detection, and therefore early medical intervention, has the potential to dramatically decrease the burden of life-threatening diseases as aimed by wearable LOC platforms. In the near future, rather than using textile or tattoo as a substrate, these platforms can be implanted into the body for continuous monitoring of physicochemical conditions.

Since microfluidics and LOC systems were first introduced in the 1990s, we have seen studies focusing on investigating individual cells in order to understand the building blocks of tissues, organs, and organisms. More recently, a collective approach has gained traction that aims to study the interaction of cells and mimic the functionalities of organs on microfluidic platforms. These systems are called organ-on-a-chip platforms. The ultimate aim of these studies is to create a human-on-a-chip by constructing each organ on the same platform. The human-on-a-chip will simulate the activities of individual organs as well as their collective behaviors. This will revolutionize the position of LOC platforms against conventional systems and the methods used for disease diagnosis as well as personalized treatment.

References

1 Castillo-León, J.; Svendsen, W. E. *Lab-on-a-Chip Devices and Micro-Total Analysis Systems: A Practical Guide*; Springer, Cham, **2014**.

2 Terry, S. C.; Jerman, J. H.; Angell, J. B. *IEEE Transactions on Electron Devices* **1979**, 26, 1880–1886.

3 Bassous, E.; Taub, H.; Kuhn, L. *Applied Physics Letters* **1977**, 31, 135–137.

4 Manz, A.; Graber, N.; Widmer, H. Á. *Sensors and Actuators B: Chemical* **1990**, 1, 244–248.

5 Duffy, D. C.; McDonald, J. C.; Schueller, O. J.; Whitesides, G. M. *Analytical Chemistry* **1998**, 70, 4974–4984.

6 Peeling, R. W.; Holmes, K. K.; Mabey, D.; Ronald, A. *Sexually Transmitted Infections* **2006**, 82, v1–v6.

7 Peeling, R.; Mabey, D. *Clinical Microbiology and Infection* **2010**, 16, 1062–1069.

8 Nahavandi, S.; Baratchi, S.; Soffe, R.; Tang, S.-Y.; Nahavandi, S.; Mitchell, A.; Khoshmanesh, K. *Lab on a Chip* **2014**, 14, 1496–1514.

9 Isiksacan, Z.; Erel, O.; Elbuken, C. *Lab on a Chip* **2016**, 24, 4682–4690.

10 Laser, D. J.; Santiago, J. G. *Journal of Micromechanics and Microengineering* **2004**, 14, R35–R64.

11 Abhari, F.; Jaafar, H.; Yunus, N. A. M. *International Journal of Electrochemical Science* **2012**, 7, 9765–9780.

12 Au, A. K.; Lai, H.; Utela, B. R.; Folch, A. *Micromachines* **2011**, 2, 179–220.

13 Oh, K. W.; Ahn, C. H. *Journal of Micromechanics and Microengineering* **2006**, 16, R13–R39.

14 Nguyen, N.-T.; Wu, Z. *Journal of Micromechanics and Microengineering* **2004**, 15, R1–R16.

15 Burns, M. A.; Johnson, B. N.; Brahmasandra, S. N.; Handique, K.; Webster, J. R.; Krishnan, M.; Sammarco, T. S.; Man, P. M.; Jones, D.; Heldsinger, D. *Science* **1998**, 282, 484–487.

16 Unger, M. A.; Chou, H.-P.; Thorsen, T.; Scherer, A.; Quake, S. R. *Science* **2000**, 288, 113–116.

17 Hong, J. W.; Studer, V.; Hang, G.; Anderson, W. F.; Quake, S. R. *Nature Biotechnology* **2004**, 22, 435–439.

18 Fu, A. Y.; Chou, H.-P.; Spence, C.; Arnold, F. H.; Quake, S. R. *Analytical Chemistry* **2002**, 74, 2451–2457.

19 Thorsen, T.; Maerkl, S. J.; Quake, S. R. *Science* **2002**, 298, 580–584.

20 Shields, IV, C. W.; Reyes, C. D.; López, G. P. *Lab on a Chip* **2015**, 15, 1230–1249.

21 Ozkumur, E.; Shah, A. M.; Ciciliano, J. C.; Emmink, B. L.; Miyamoto, D. T.; Brachtel, E.; Yu, M.; Chen, P.-I.; Morgan, B.; Trautwein, J. *Science Translational Medicine* **2013**, 5, 179ra47.

22 Karabacak, N. M.; Spuhler, P. S.; Fachin, F.; Lim, E. J.; Pai, V.; Ozkumur, E.; Martel, J. M.; Kojic, N.; Smith, K.; Chen, P.-I. *Nature Protocols* **2014**, 9, 694–710.

23 Mark, D.; Haeberle, S.; Roth, G.; von Stetten, F.; Zengerle, R. *Chemical Society Reviews* **2010**, 39, 1153–1182.

24 Merriman, B.; Torrent, I.; Rothberg, J. M.; Team, D. *Electrophoresis* **2012**, 33, 3397–3417.

25 Cate, D. M.; Adkins, J. A.; Mettakoonpitak, J.; Henry, C. S. *Analytical Chemistry* **2014**, 87, 19–41.

26 Bruzewicz, D. A.; Reches, M.; Whitesides, G. M. *Analytical Chemistry* **2008**, 80, 3387–3392.

27 Carrilho, E.; Martinez, A. W.; Whitesides, G. M. *Analytical Chemistry* **2009**, 81, 7091–7095.

28 Hu, J.; Wang, S.; Wang, L.; Li, F.; Pingguan-Murphy, B.; Lu, T. J.; Xu, F. *Biosensors and Bioelectronics* **2014**, 54, 585–597.

29 Lu, Y.; Shi, W.; Jiang, L.; Qin, J.; Lin, B. *Electrophoresis* **2009**, 30, 1497–1500.

30 Wang, J.; Monton, M. R. N.; Zhang, X.; Filipe, C. D.; Pelton, R.; Brennan, J. D. *Lab on a Chip* **2014**, 14, 691–695.

31 Osborn, J. L.; Lutz, B.; Fu, E.; Kauffman, P.; Stevens, D. Y.; Yager, P. *Lab on a Chip* **2010**, 10, 2659–2665.

32 Dieterich, K. Device for grinding broken paper, US Patent, 730617 A, **1902**.

33 Yagoda, H. *Industrial and Engineering Chemistry, Analytical Edition* **1937**, 9, 79–82.

34 López-Marzo, A. M.; Merkoçi, A. *Lab on a Chip* **2016**, 16, 3150–3176.

35 Martinez, A. W.; Phillips, S. T.; Whitesides, G. M.; Carrilho, E. *Analytical Chemistry* **2010**, 82, 3–10.

36 Li, X.; Tian, J.; Garnier, G.; Shen, W. *Colloids and Surfaces B: Biointerfaces* **2010**, 76, 564–570.

37 Li, X.; Tian, J.; Nguyen, T.; Shen, W. *Analytical Chemistry* **2008**, 80, 9131–9134.
38 Martinez, A. W.; Phillips, S. T.; Whitesides, G. M. *Proceedings of the National Academy of Sciences* **2008**, 105, 19606–19611.
39 Martinez, A. W.; Phillips, S. T.; Butte, M. J.; Whitesides, G. M. *Angewandte Chemie International Edition* **2007**, 46, 1318–1320.
40 Dungchai, W.; Chailapakul, O.; Henry, C. S. *Analytical Chemistry* **2009**, 81, 5821–5826.
41 Li, X.; Tian, J.; Shen, W. *Analytical and Bioanalytical Chemistry* **2010**, 396, 495–501.
42 Abe, K.; Kotera, K.; Suzuki, K.; Citterio, D. *Analytical and Bioanalytical Chemistry* **2010**, 398, 885–893.
43 Li, C.-Z.; Vandenberg, K.; Prabhulkar, S.; Zhu, X.; Schneper, L.; Methee, K.; Rosser, C. J.; Almeide, E. *Biosensors and Bioelectronics* **2011**, 26, 4342–4348.
44 Teh, S.-Y.; Lin, R.; Hung, L.-H.; Lee, A. P. *Lab on a Chip* **2008**, 8, 198–220.
45 Foudeh, A. M.; Didar, T. F.; Veres, T.; Tabrizian, M. *Lab on a Chip* **2012**, 12, 3249–3266.
46 Schaerli, Y.; Hollfelder, F. *Molecular Biosystems* **2009**, 5, 1392–1404.
47 Huebner, A.; Sharma, S.; Srisa-Art, M.; Hollfelder, F.; Edel, J. B. *Lab on a Chip* **2008**, 8, 1244–1254.
48 Brouzes, E.; Medkova, M.; Savenelli, N.; Marran, D.; Twardowski, M.; Hutchison, J. B.; Rothberg, J. M.; Link, D. R.; Perrimon, N.; Samuels, M. L. *Proceedings of the National Academy of Sciences of the United States of America* **2009**, 106, 14195–14200.
49 Courtois, F.; Olguin, L. F.; Whyte, G.; Bratton, D.; Huck, W. T.; Abell, C.; Hollfelder, F. *ChemBioChem* **2008**, 9, 439–446.
50 Kintses, B.; van Vliet, L. D.; Devenish, S. R.; Hollfelder, F. *Current Opinion in Chemical Biology* **2010**, 14, 548–555.
51 De Geest, B. G.; Urbanski, J. P.; Thorsen, T.; Demeester, J.; De Smedt, S. C. *Langmuir* **2005**, 21, 10275–10279.
52 He, M.; Sun, C.; Chiu, D. T. *Analytical Chemistry* **2004**, 76, 1222–1227.
53 He, M.; Edgar, J. S.; Jeffries, G. D.; Lorenz, R. M.; Shelby, J. P.; Chiu, D. T. *Analytical Chemistry* **2005**, 77, 1539–1544.
54 Hung, L.-H.; Choi, K. M.; Tseng, W.-Y.; Tan, Y.-C.; Shea, K. J.; Lee, A. P. *Lab on a Chip* **2006**, 6, 174–178.
55 Zhu, P.; Wang, L. *Lab on a Chip* **2017**, 17, 34–75.
56 Thorsen, T.; Roberts, R. W.; Arnold, F. H.; Quake, S. R. *Physical Review Letters* **2001**, 86, 4163.
57 Ganán-Calvo, A. M.; Gordillo, J. M. *Physical Review Letters* **2001**, 87, 274501.
58 Shestopalov, I.; Tice, J. D.; Ismagilov, R. F. *Lab on a Chip* **2004**, 4, 316–321.
59 Boedicker, J. Q.; Li, L.; Kline, T. R.; Ismagilov, R. F. *Lab on a Chip* **2008**, 8, 1265–1272.

60 Pekin, D.; Skhiri, Y.; Baret, J.-C.; Le Corre, D.; Mazutis, L.; Salem, C. B.; Millot, F.; El Harrak, A.; Hutchison, J. B.; Larson, J. W. *Lab on a Chip* **2011**, 11, 2156–2166.

61 El Debs, B.; Utharala, R.; Balyasnikova, I. V.; Griffiths, A. D.; Merten, C. A. *Proceedings of the National Academy of Sciences of the United States of America* **2012**, 109, 11570–11575.

62 Juul, S.; Nielsen, C. J.; Labouriau, R.; Roy, A.; Tesauro, C.; Jensen, P. W.; Harmsen, C.; Kristoffersen, E. L.; Chiu, Y.-L.; Frøhlich, R. *ACS Nano* **2012**, 6, 10676–10683.

63 Marcali, M.; Elbuken, C. *Lab on a Chip* **2016**, 16, 2494–2503.

64 Lee, J.; Moon, H.; Fowler, J.; Schoellhammer, T.; Kim, C.-J. *Sensors and Actuators A: Physical* **2002**, 95, 259–268.

65 Jebrail, M. J.; Wheeler, A. R. *Current Opinion in Chemical Biology* **2010**, 14, 574–581.

66 Fair, R. B. *Microfluidics and Nanofluidics* **2007**, 3, 245–281.

67 Malic, L.; Brassard, D.; Veres, T.; Tabrizian, M. *Lab on a Chip* **2010**, 10, 418–431.

68 Srinivasan, V.; Pamula, V. K.; Fair, R. B. *Lab on a Chip* **2004**, 4, 310–315.

69 Sista, R.; Hua, Z.; Thwar, P.; Sudarsan, A.; Srinivasan, V.; Eckhardt, A.; Pollack, M.; Pamula, V. *Lab on a Chip* **2008**, 8, 2091–2104.

70 Barbulovic-Nad, I.; Au, S. H.; Wheeler, A. R. *Lab on a Chip* **2010**, 10, 1536–1542.

71 Ng, A. H.; Chamberlain, M. D.; Situ, H.; Lee, V.; Wheeler, A. R. *Nature Communications* **2015**, 6, 1.

72 Volpatti, L. R.; Yetisen, A. K. *Trends in Biotechnology* **2014**, 32, 347–350.

73 Shen, H.-H.; Fan, S.-K.; Kim, C.-J.; Yao, D.-J. *Microfluidics and Nanofluidics* **2014**, 16, 965–987.

74 Burger, R.; Kirby, D.; Glynn, M.; Nwankire, C.; O'Sullivan, M.; Siegrist, J.; Kinahan, D.; Aguirre, G.; Kijanka, G.; Gorkin, R. A. *Current Opinion in Chemical Biology* **2012**, 16, 409–414.

75 Tang, M.; Wang, G.; Kong, S.-K.; Ho, H.-P. *Micromachines* **2016**, 7, 26.

76 Focke, M.; Stumpf, F.; Faltin, B.; Reith, P.; Bamarni, D.; Wadle, S.; Müller, C.; Reinecke, H.; Schrenzel, J.; Francois, P. *Lab on a Chip* **2010**, 10, 2519–2526.

77 Nwankire, C. E.; Czugala, M.; Burger, R.; Fraser, K. J.; Glennon, T.; Onwuliri, B. E.; Nduaguibe, I. E.; Diamond, D.; Ducrée, J. *Biosensors and Bioelectronics* **2014**, 56, 352–358.

78 Koh, C.-Y.; Schaff, U. Y.; Piccini, M. E.; Stanker, L. H.; Cheng, L. W.; Ravichandran, E.; Singh, B.-R.; Sommer, G. J.; Singh, A. K. *Analytical Chemistry* **2015**, 87, 922–928.

79 La Clair, J. J.; Burkart, M. D. *Organic and Biomolecular Chemistry* **2003**, 1, 3244–3249.

80 Jung, J.; Park, B.; Oh, S.; Choi, G.; Seo, T. *Lab on a Chip* **2015**, 15, 718–725.

81 Kim, T.-H.; Abi-Samra, K.; Sunkara, V.; Park, D.-K.; Amasia, M.; Kim, N.; Kim, J.; Kim, H.; Madou, M.; Cho, Y.-K. *Lab on a Chip* **2013**, 13, 3747–3754.

82 Gao, W.; Emaminejad, S.; Nyein, H. Y. Y.; Challa, S.; Chen, K.; Peck, A.; Fahad, H. M.; Ota, H.; Shiraki, H.; Kiriya, D. *Nature* **2016**, 529, 509–514.

83 Appelboom, G.; Camacho, E.; Abraham, M. E.; Bruce, S. S.; Dumont, E. L.; Zacharia, B. E.; D'Amico, R.; Slomian, J.; Reginster, J. Y.; Bruyère, O. *Archives of Public Health* **2014**, 72(1), 28.

84 Bandodkar, A. J.; Wang, J. *Trends in Biotechnology* **2014**, 32, 363–371.

85 Gowers, S. A.; Curto, V. F.; Seneci, C. A.; Wang, C.; Anastasova, S.; Vadgama, P.; Yang, G.-Z.; Boutelle, M. G. *Analytical Chemistry* **2015**, 87, 7763–7770.

86 Bandodkar, A. J.; Jia, W.; Wang, J. *Electroanalysis* **2015**, 27, 562–572.

87 Chan, M.; Estève, D.; Fourniols, J.-Y.; Escriba, C.; Campo, E. *Artificial Intelligence in Medicine* **2012**, 56, 137–156.

88 Rodgers, M. M.; Pai, V. M.; Conroy, R. S. *IEEE Sensors Journal* **2015**, 15, 3119–3126.

89 Stoppa, M.; Chiolerio, A. *Sensors* **2014**, 14, 11957–11992.

90 Aguirre, A.; Testa-Weintraub, L.; Banderas, J.; Haraszthy, G.; Reddy, M.; Levine, M. *Critical Reviews in Oral Biology and Medicine* **1993**, 4, 343–350.

91 Preston, A.; Edgar, W. *Journal of Dentistry* **2005**, 33, 209–222.

92 Kim, J.; Valdés-Ramírez, G.; Bandodkar, A. J.; Jia, W.; Martinez, A. G.; Ramírez, J.; Mercier, P.; Wang, J. *Analyst* **2014**, 139, 1632–1636.

93 Mannoor, M. S.; Tao, H.; Clayton, J. D.; Sengupta, A.; Kaplan, D. L.; Naik, R. R.; Verma, N.; Omenetto, F. G.; McAlpine, M. C. *Nature Communications* **2012**, 3, 763.

94 Gamella, M.; Campuzano, S.; Manso, J.; de Rivera, G. G.; López-Colino, F.; Reviejo, A.; Pingarrón, J. *Analytica Chimica Acta* **2014**, 806, 1–7.

95 Schazmann, B.; Morris, D.; Slater, C.; Beirne, S.; Fay, C.; Reuveny, R.; Moyna, N.; Diamond, D. *Analytical Methods* **2010**, 2, 342–348.

96 Guinovart, T.; Parrilla, M.; Crespo, G. A.; Rius, F. X.; Andrade, F. J. *Analyst* **2013**, 138, 5208–5215.

97 Gonzalo-Ruiz, J.; Mas, R.; de Haro, C.; Cabruja, E.; Camero, R.; Alonso-Lomillo, M. A.; Muñoz, F. J. *Biosensors and Bioelectronics* **2009**, 24, 1788–1791.

98 Windmiller, J. R.; Bandodkar, A. J.; Valdés-Ramírez, G.; Parkhomovsky, S.; Martinez, A. G.; Wang, J. *Chemical Communications* **2012**, 48, 6794–6796.

99 Yan, Q.; Peng, B.; Su, G.; Cohan, B. E.; Major, T. C.; Meyerhoff, M. E. *Analytical Chemistry* **2011**, 83, 8341–8346.

100 Kudo, H.; Sawada, T.; Kazawa, E.; Yoshida, H.; Iwasaki, Y.; Mitsubayashi, K. *Biosensors and Bioelectronics* **2006**, 22, 558–562.

101 Thomas, N.; Lähdesmäki, I.; Parviz, B. A. *Sensors and Actuators B: Chemical* **2012**, 162, 128–134.

102 Yao, H.; Shum, A. J.; Cowan, M.; Lähdesmäki, I.; Parviz, B. A. *Biosensors and Bioelectronics* **2011**, 26, 3290–3296.

103 Tierney, M. J.; Tamada, J. A.; Potts, R. O.; Jovanovic, L.; Garg, S.; Team, C. R. *Biosensors and Bioelectronics* **2001**, 16, 621–629.

Section 3

Nanomaterial's Applications in Biosensors and Medical Diagnosis

9

Applications of Quantum Dots in Biosensors and Diagnostics

Zeynep Altintas[1], Frank Davis[2], and Frieder W. Scheller[3]

[1] Technical University of Berlin, Berlin, Germany
[2] Department of Engineering and Applied Design, University of Chichester, Chichester, UK
[3] Institute of Biochemistry and Biology, University of Potsdam, Potsdam, Germany

9.1 Introduction

The unique optical and fluorescence properties of quantum dots (QDs) were first discovered in the early 1980s, but it was not until about 15–20 years later that they first began to be applied in the field of biosensing. Early work using these describes such applications as detection of metal ions and other small molecule sensing as well as cell staining elements [1]. One major advantage of QDs is that they can be synthesized in a range of sizes and with a range of surface substituents [2]. Surface coating of QDs has a critical importance concerning biocompatibility, bioconjugation, and water solubility. Various surface modification methods have been applied to give them biocompatible nature such as surface cap exchange [3], encapsulation in silica shells [4] or micelles [5], and amphiphilic surface [6]. Since their optical and fluorescence properties depend on their size, it opens up the possibility of using a mix of QDs to simultaneously analyze for a range of targets. While developing a biosensor, it is especially important to avoid using large and charged surfaces of QDs to prevent from nonspecific binding of the target molecules. For this, a well-established strategy is the use of poly(ethylene glycol) (PEG) with at least 12–14 units of PEG as surface coating agent [7]. Hydroxyl-coated QDs are capable of reducing nonspecific binding 10- to 20-fold compared with that of protein- and PEG-coated QDs and 140-fold when compared with carboxylate QDs [8].

The application of QDs to bookmarking is one of the major fields of study for these materials. The commercial availability of QDs and their ease of

Biosensors and Nanotechnology: Applications in Health Care Diagnostics, First Edition.
Edited by Zeynep Altintas.
© 2018 John Wiley & Sons, Inc. Published 2018 by John Wiley & Sons, Inc.

functionalization have made them one of the major bioimaging tools. Their use so far has been restricted to *in vitro* and animal experiments, in part due to potential toxicity issues with cadmium-based QDs, which are still used in the majority of research; however research into encapsulating QDs to make them safer or replacing cadmium with less toxic elements may address this issue. Recent years have also witnessed numerous applications of QDs in biosensors. The following sections review the properties of QDs and their use in fluorescence-based imaging and biomedical sensors and water-soluble QDs and their cytotoxicity.

9.2 Quantum Dots: Optical Properties, Synthesis, and Surface Chemistry

Since their discovery by Ekimov [9], QDs are one of the most investigated nanomaterials for biological and medical applications [10]. They have been considered the most promising class of fluorescent labels compared with conventional fluorophores (organic dyes and proteins) due to their high quantum yields (QYs), narrowband emission peaks with a broad excitation wavelength range [11], resistance to photobleaching, relatively long luminescence lifetime (>10 ns), high surface area-to-volume ratio that allows efficient functionalization with biomolecules [12], and, as depicted in Figure 9.1, a wide spectrum of colors that can be emitted by adjusting the QD size and using a single wavelength excitation.

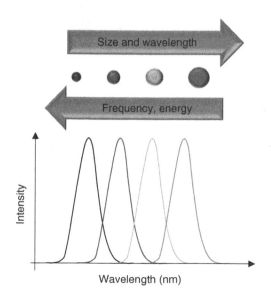

Figure 9.1 Relation of quantum dot size with emission peak using single wavelength excitation.

QDs are generally composed by binary II–VI, III–V, and IV–VI, of the periodic table [1], semiconductor materials (CdTe, CdSe, GaAs, PbSe, InP, etc.) with a diameter range of 2–10 nm [3, 13]. One of the major problems relates to the toxicity of QDs; this is still a controversial issue mostly because of the lack of toxicology studies [10]. The reason QDs are considered toxic is based on two observations: (i) heavy metals are known for their toxicity, and (ii) cadmium-containing QDs can kill cell cultures. Therefore, if they are toxic to cells, they are likely to be toxic to humans [14, 15].

A number of different strategies of surface modification have been used to render QDs biocompatible [1] as shown in Figure 9.2. As an example, there are core–shell structures in which a semiconductor shell, typically zinc sulfide (ZnS), stabilizes the core; furthermore, an additional capping or coating using biocompatible materials or polymeric layers (such as PEG) to the QD core–shell helps to diminish toxicity [16].

QDs have been considered to be more photostable than common organic dyes. However, a significant loss in fluorescence has been recorded upon injection into whole animals and tissues and also in ionic media. The decrease in signal is generally attributed to slow degradation of surface coating or to agents absorbed to the QD surface when exposed to body fluids, which results in fluorescence quenching and surface defects [17–20].

9.3 Biosensor Applications of QDs

During the last decade, QDs have been commonly utilized in biotechnology applications that use fluorescence, such as immunofluorescence assays, cell and animal biology, and DNA array technology. Early applications in the diagnostic area cover immunostaining of membrane proteins, nuclear antigens, actin, and microtubules; immunofluorescence labeling of fixed cells and tissues; and fluorescence-based *in situ* hybridization on chromosomes or combed DNA [21–24]. These smart materials are brighter than dyes due to the compounded effects of extinction coefficients, and they have resistance to bleaching from minutes to hours, which allows the acquisition of well-contrasted and crisp images. Moreover, a very small number of QDs in immunofluorescence assays are enough to produce signal. Even single QDs can be observed in immunocytological circumstances with a substantial sensitivity level of one QD per target compound [25, 26].

Due to their fascinating characteristics, they have also found an enormous attention in the biosensor area. Many biosensing methods have employed QDs to increase the sensitivity, rapidity, and reliability of the biodetection analysis. Li et al. integrated QDs into a lateral flow test strip as a reporter to manufacture a rapid, one-step quantitative detection tool for nitrated ceruloplasmin [27], which is an essential copper-carrying protein in the blood and also plays a key

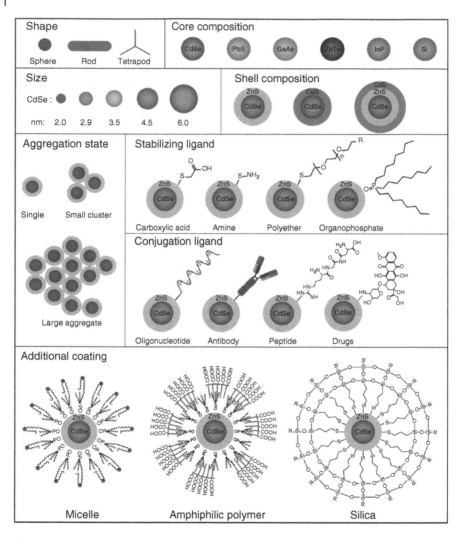

Figure 9.2 Quantum dots (QDs) are a heterogeneous group of materials. Biological fate and toxicity depend on QD physicochemical properties. Shape, core composition, size, and shell composition can be manipulated during QD synthesis. Post-synthesis surface ligands are added to solubilize and target the particles. An additional coating can further protect the QD core from oxidation. Surface chemistry influences the quantum dot's propensity to aggregate, particularly in biological solutions. *Source:* Tsoi et al. [14]. Reproduced with permission of American Chemical Society. (*See insert for color representation of the figure.*)

role in iron metabolism. This enzymatic molecule carries more than 95% of the total copper in healthy individual's plasma. The low levels of ceruloplasmin may indicate several deficiencies including Wilson disease, aceruloplasminemia, Menkes disease, and copper deficiency. The QD-based lateral flow portable biosensor could measure the trace amounts of this biomarker. It was demonstrated that the sensor was capable of detecting $1\,ng\,mL^{-1}$ nitrated ceruloplasmin within 10 min. The presence of non-nitrated biomarker did not show any effect on the biosensor signal that indicated high selectivity. The developed technique has shown a good promise for biomarker detection using the portable fluorescence sensor based on QDs and lateral flow test strip [27].

Similar approach can also be used for the detection of nucleic acid-based biomarkers. Being important potential biomarkers in early cancer diagnosis, the rapid and sensitive quantification of microRNAs (miRNAs) is important with point-of-care diagnosis. Deng et al. developed QD-labeled strip biosensor that relies on target-recycled nonenzymatic amplification strategy for miRNA quantification [28]. In this research, QDs were used as bright photostable labels that supply good detection efficiency for the biosensor. A target-recycled amplification strategy based on sequence-specific hairpin strand displacement process without the assistance of enzymes was additionally introduced to the system to increase sensitivity. The principle of target-recycled nonenzymatic amplification and QD-labeled strip biosensor for amplification product detection is demonstrated in Figure 9.3. The sensing platform could detect miRNAs in the concentration range of 2–200 fmol with a limit of 200 amol. Moreover, miRNA analysis in various tumor cell extracts was comparable with quantitative real-time polymerase chain reaction. The potential practical use of the method was also investigated by testing clinical tumor samples, and a majority of the samples (16 of 20) produced positive signals that demonstrated great promise for simple and early cancer diagnosis [28].

miRNA detection is also possible by combining unique properties of different smart nanomaterials such as graphene QDs. A novel biosensor platform based on this strategy and pyrene-functionalized molecular beacon (p-MB) probes were recently developed. Pyrene was used to trigger specific fluorescence resonance energy transfer (FRET) between graphene QDs and fluorescent dyes that were labeled on p-MBs. A distinctive fluorescent intensity change generated a novel signal for the detection of miRNA targets. The sensor allowed detecting the target biomarkers in the investigation range of 0.1–200 nM. Furthermore, this fluorescence biosensor based on graphene QDs has provided a multiple detection platform for different kinds of miRNAs, which plays a crucial role in precise cancer diagnosis [29].

Implementation of QDs in microfluidics biosensors has been offering novel and effective detection systems for a wide range of target molecules. Very recently, this kind of sensor has been employed for allergen detection using graphene oxide–aptamer–QD complexes. This aptamer-functionalized quencher system was used as probe to undergo conformational change in the case of interaction with food allergens that led to changes in fluorescence because of recovering and fluorescence quenching characteristics of graphene oxide by adsorption and desorption of aptamer-conjugated QDs. Moreover, the developed biosensor allowed one-step homogeneous assay on a microfluidic chip in approximately

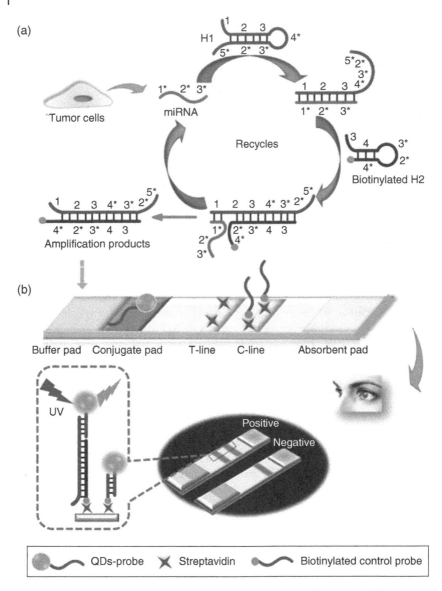

Figure 9.3 (a) Principle of target-recycled nonenzymatic amplification and (b) schematic of QD-labeled strip biosensor for amplification product detection. *Source:* Deng et al. [28]. Reproduced with permission of Elsevier.

10 min to detect food allergens quantitatively. Using a microfluidic system with a miniaturized optical sensor, Ara h1 (a common peanut allergen) could be measured with a detection limit of $56\,ng\,mL^{-1}$, and the outputs of this research display the potential of the method for on-site allergen determination [30].

The detection of disease biomarkers was investigated using dopamine-functionalized QDs based on redox-mediated indirect fluorescence immunoassay approach [31]. The researchers aimed to detect α-fetoprotein (AFP) by utilizing dopamine-functionalized CdSe/ZnS QDs. For this, the detection antibody was conjugated with tyrosinase to act as a bridge connecting the QDs' fluorescence signals with biomarker concentrations. Tyrosinase label used in redox-mediated indirect fluorescence immunoassay catalyzed the dopamine oxidation on the QD-functionalized surface and led to fluorescence quenching in the presence of target biomarker. This method could detect AFP in the concentration range from 10 pM to 100 nM. Upon obtaining a highly sensitive measurement technique, the detection of AFP in real samples was also validated using 40 different AFP-positive samples collected from hepatocellular carcinoma (HCC) patients and 10 different AFP-negative samples as controls. The method clearly differentiated positive and negative samples with 97.5% sensitivity and 100% specificity [31].

The important applications of QDs can also be seen in pathogen detection. A study combined immunomagnetic separation (IMS) and the QD fluorescence to enumerate the bacteria *Escherichia coli* [32]. First the iron oxide core–gold shell ($Fe_3O_4@Au$) magnetic nanoparticles modified with biotinylated antibodies captured the *E. coli* in solution 1, and then chitosan-coated CdTe quantum dots (CdTe QDs) modified with a secondary antibody were added in solution 1. The bacteria were removed from the matrix by employing IMS technique for fluorescence analysis. The selectivity experiments were performed with other bacteria including *Enterobacter aerogenes*, *Enterobacter dissolvens*, *Staphylococcus aureus*, and *Pseudomonas aeruginosa*, and no interference was observed, achieving a limit of detection (LOD) of 30 CFU mL^{-1} with a total analysis time of approximately 120 min.

The same strategy was used by Wu et al. [33] in which a multiplexed system for the detection of three lung cancer biomarkers based on the multicolor QDs was developed by employing micro-magnetic beads as immune carriers. Using this platform, carcinoembryonic antigen (CEA), fragments of cytokeratin 19 (CYRFA 21-1), and neuron-specific enolase (NSE) were concurrently detected in a single sample, as shown in Figure 9.4. LOD of 38 pg mL^{-1} for CEA, 364 pg mL^{-1} for CYRFA 21-1, and 370 pg mL^{-1} for NSE were achieved in this research.

9.4 Other Biological Applications of QDs

Numerous other application areas of QDs have been reported during the last decade. These areas can be listed as gene technology [34, 35], fluorescent labeling of cellular proteins [5], cell tracking [36], *in vivo* animal imaging [37], barriers to use *in vivo* [22], and tumor biology research [38, 39]. Several studies have demonstrated that QD-conjugated oligonucleotide sequences can be

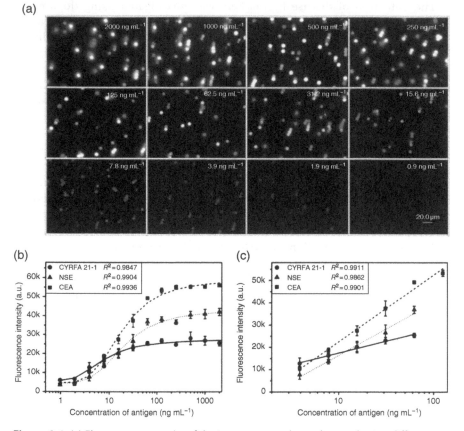

Figure 9.4 (a) Fluorescence graphs of the immune complexes detected using different concentrations of CYRFA 21-1, NSE, and CEA. (b) Standard curves for the detection of CYRFA 21-1, NSE, and CEA in the multiplexed assay (using four-parameter equation fitting). (c) The linear range of the plot in (b). *Source:* Wu et al. [33]. Reproduced with permission of Elsevier.

used in gene technology by attaching these sequences to QDs via surface carboxylic acid groups. Such probes can target to bind with mRNA and DNA [35]. Xu et al. showed that identification and specific labeling of targeted DNA sequences can be accomplished by employing green, blue, and red QDs in different combinations. Peculiar spectral barcodes were obtained with the mixture of these QDs that allowed high degree of multiplexing, which is necessary in complex genetic analyses. In this research, QDs were utilized as microbead system for genotyping of single nucleotide polymorphism by encoding the beads conjugated to allele-specific oligonucleotides. The genomic samples were then amplified to produce biotinylated amplicons prior to the incubation with QD-labeled oligonucleotides and streptavidin Cy5, respectively. Thus, streptavidin Cy5-attached microbeads could interact with biotin on the amplicons. Hybridization reaction was successfully observed due to the combination of QDs and Cy5 signals. The results demonstrated 100% concordance with the widely used TaqMan method for 940 genotypes. Such a technique using QDs offers assays with higher efficiency using much smaller quantities of DNA and also accurate and reliable strategy for multiplexed SNP genotyping [40].

A very recent study on fluorescent labeling of cellular proteins investigated the strategies to label microvesicles (MVs) whose applications are so important since they act as natural carriers that enable to transport biological compounds between cells. These MVs have also shown promising future for therapeutic purposes as delivery vehicles. $Ag_2Se@Mn$ QDs were integrated into magnetic resonance (MR) and near-infrared (NIR) fluorescence imaging for efficient labeling of MVs for *in vivo* dual-mode tracking with high resolution. Very tiny Mn-magnetofunctionalized QDs (around 1.8 nm) can be loaded into MVs through electroporation with high efficiency, and the system provided both MR and NIR fluorescence traceability on MVs. The paired imaging ability of $Ag_2Se@Mn$ QDs confers noninvasive entire-body dual-mode tracking of MVs and also real-time quantitative biodistribution of these vesicles [41].

Synthesis of cadmium-free quantum dots (CFQDs) using ternary I–III–VI semiconducting material has been considered as a potential non-cytotoxic alternative to conventional binary QDs [12, 42]. In this way Liu et al. [12] used a CFQDs system as a probe for tumor-targeted imaging employing CuInSe/ZnS and CuInS/ZnS conjugated with and without tumor-specific vascular target CGKRK (Cys–Gly–Lys–Arg–Lys) peptide through a PEG linker, QD-PEG-P, and QD-PEG, respectively. The control experiments were carried out with phosphate-buffered saline (PBS) solution. As shown in Figure 9.5a, *in vivo* imaging of the QD-PEG-P system achieved a clear identification of breast tumors of mice with a signal 70% higher compared with QD-PEG; the background signal was assigned to residual QDs in the blood. Figure 9.5b displays *ex vivo* continuous-wave (CW) imaging of tissues collected at the end of the *in vivo* imaging; these confirmed the extensive accumulation of QD-PEG-P in tumors and low background in other tissues except the liver. The liver signal was confounded by tissue autofluorescence, as the livers from the PBS-injected mice were also positive. This problem was solved using time-gated (TG) imaging that selectively eliminates short-lived (<10 ns) autofluorescence while leaving long-lived luminescent signals. I–III–VI QDs have a longer photoluminescence lifetime than conventional binary II–VI QDs (100–300 ns vs. 10–20 ns), suggesting that ternary QDs could be superior for TG imaging; this was confirmed with the highest signal detected in the QD–PEG–P tumor, consistent with the *in vivo* imaging results.

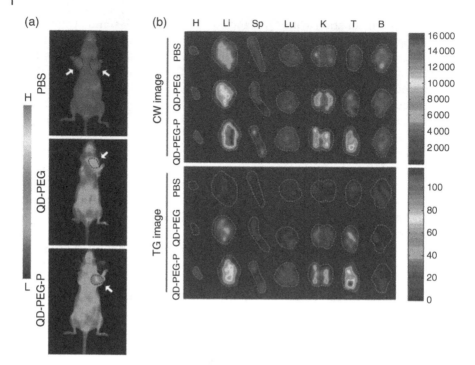

Figure 9.5 Breast tumor targeting with QD-PEG-P. (a) Intravital whole-body fluorescent imaging (ventral view) of MCF10CA1a breast tumor-bearing nude mice at 28 h after intravenous injection of PBS, QD-PEG, or QD-PEG-P. Images were taken under 700 nm channel of LI-COR Pearl Impulse small animal imaging system. White arrows show the position of MCF10CA1a tumors (circled in black). (b) *Ex vivo* CW and TG imaging of the tissues harvested from the mice in (a) after the *in vivo* imaging. B, brain; H, heart; K, kidney; Li, liver; Lu, lung; Sp, spleen; T, tumor. *Source:* Liu et al. [12]. Reproduced with permission of John Wiley & Sons. (*See insert for color representation of the figure.*)

9.5 Water Solubility and Cytotoxicity

Availability of water-soluble QDs has an immense importance in biomedical applications. QDs can be directly prepared in water although they often possess wide size distribution in this case. Three different strategies can be considered to obtain water-soluble QDs: ligand exchange (i), forming micelles through hydrophobic interaction (ii), and silica encapsulation (iii). QDs synthesized in organic solvent generally have hydrophobic surface ligands; however, bifunctional molecules with water-soluble nature can substitute for the hydrophobic ones. For this aim, various water-soluble bifunctional molecules including dithiothreitol [43], mercaptocarbonic acids [HS—$(CH_2)n$—COOH, $n = 1$–15] [25, 44, 45], oligomeric phosphines [25, 46], 2-aminoethanethiol [45], cross-linked dendrons [47], dihydrolipoic acid [48], and peptides [49] can be used while preparing QDs. One end of these molecules connects to the QD surface atom, while the other end has a hydrophilic nature that can react with biomolecules. Despite being a promising technique, ligand exchange alters the physical and chemical states of QD surface atoms; therefore, this quite often dramatically decreases the quantum throughput of QDs.

Phospholipids such as 1,2-dipalmitoyl-*sn*-glycero-3-phosphocholine possess both hydrophilic and hydrophobic ends. They can envelope QDs in the core by establishing oil-in-water micelles via hydrophobic interaction between the surface ligands of the QDs and their hydrophobic ends, thus enabling solubility in water through hydrophilic exterior ends. Using long-chain amphiphilic polymers to create micelle-like structure is a more efficient approach to transfer hydrophobic QDs into the water. For example, poly(acrylic acid) was somewhat grafted with octylamine via EDC coupling to develop an amphiphilic nature, and the micelle-like structures were then created where QDs were encapsulated in the core and −COOH faced outward [21]. This strategy that evolved from the use of amphiphilic polymers generally provides better output than ligand exchange since there is no direct interaction with surface atoms of QDs and thus can protect the original quantum performance to the highest degree. Moreover, having many hydrophobic side chains on the polymer has intensified the hydrophobic interaction to form water-soluble QDs with high stability. These polymers are commercially available in low prices, which make them more favorable materials than the phospholipids, particularly in large-scale production.

Encapsulation of QDs by a silica layer also provides biocompatible and water-soluble agents. Functional organosilicon molecules possessing −SH or −NH$_2$ groups can be integrated into the shell and offer surface functionalities to be used in biomedical applications [50, 51]. The incorporation of organosilicon molecules may lead to a decrease in quantum efficiency, and silica layer should be formed in dilute conditions that are not convenient for large-scale preparation. When silica encapsulation is compared with the other two methods, it results in stable but larger QD particles with a size changing from nm to μm. Ligand exchange and micelle formation strategies do not lead to a significant change in whole particle size, and these methods are also convenient in the case of large quantity production. Additionally, micelle formation keeps the original quantum efficiency at the highest level, and incorporation of PEG molecules into the particles increases the efficiency of targeting and also stability of the QDs. These three strategies have offered water-soluble QDs for many biological applications with a wide range of surface functionalities that can then bind to biomolecule of interest such as antibodies, DNA, RNA, and peptides through commonly used coupling techniques such as EDC and glutaraldehyde chemistries. The last decade has witnessed a great progress in biomarker detection and medical imaging using these water-soluble QDs, and their successful use has been demonstrated both in *in vitro* and *in vivo* investigations.

An important drawback regarding to QDs, particularly for *in vivo* medical applications, is their cytotoxicity. Cadmium-based QDs are the best example for this issue where Cd^{2+} and Pb^{2+} can kill the cells due to their release from QDs [52]. The well-designed surface coating of QDs of high quality is the best solution to avoid this problem as it enables to obtain biologically inert QDs. The coating materials can be selected among inorganic layers such as silica or nontoxic polymers/organic molecules such as PEG. The QDs coated with silica are less toxic than the ones coated with simple molecules, which are commonly used also for biosensor surface development such as 2-aminoethanethiol, mercaptoacetic acid, 11-mercaptoundecanoic acid, and mercaptopropionic acid [52, 53]. Cytotoxicity also depends on several other factors including size, amount of QDs, capping materials, color, processing parameters, and surface coating. In addition to core degradation (Cd^{2+}), formation of free radicals and interaction between QDs and intracellular compartments may result in cytotoxicity. Taking into account all these factors, group III–V QDs can provide better alternatives than that of groups II–IV to prepare low or nontoxic QDs because of a more stable structure arising from covalent bond formation rather than ionic interactions [54, 55].

9.6 Conclusion

Semiconductor QDs exist in a wide number of morphologies and chemistries and this diversity makes them of such interest. They serve as important fluorescent probes for *in vivo* and *in vitro* imaging research as well as for biosensing applications. QDs display superior properties over other fluorescence imaging agents due to their versatile bioconjugation, long-term stability, and adaptable physical properties. The ability to use a number of QDs simultaneously because of the different emissions from different sizes of dot allows simultaneous analysis for a range of conditions. Their use in biosensing technology has led to development of highly sensitive and specific detection tolls for disease biomarkers ranging from proteins to nucleic acids and small molecules. Recent years have recorded numerous successful applications of QDs in various biosensor platforms that demonstrate a great potential for clinical applications and early diagnosis of diseases such as cancer, cardiac problems, and neurodegenerative disorders. Currently, heavy metals and their size have led to extensive discussion about their toxicity, which can range from harmless to very toxic levels based on the used materials. Although the toxicity is not an issue in biosensor-based diagnostic, it requires particular attention in the case of *in vivo* experiments, and this has to be addressed in future developments in the field for bioimaging purposes. The improvements in effective surface modification of QDs could be another direction, which will enhance specificity and sensitivity of the QD-based sensing platforms.

Acknowledgment

Z.A. gratefully acknowledges support from the European Commission, Marie Curie Actions and IPODI as the principle investigator.

References

1 Wegner, K. D.; Hildebrandt, N. *Chem. Soc. Rev.* **2015**, 44, 4792–4834.
2 Petryayeva, E.; Algar, W. R.; Medintz, I. L. *Appl. Spectrosc.* **2013**, 67, 215–252.
3 Bruchez, M.; Moronne, M.; Gin, P.; Weiss, S.; Alivisatos, A. P. *Science* **1998**, 281, 2013–2016.
4 Zhang, Y.; Wang, M.; Zheng, Y.-G., Tan, H.; Hsu, B. Y.-W.; Yang, Z.-C.; Wong, S. Y.; Chang, A. Y.-C.; Choolani, M.; Li, X.; Wang, J. *Chem. Mater.* **2013**, 25, 2976–2985.
5 Dubertret, B.; Skourides, P.; Norris, D. J.; Noireaux, V.; Brivanlou, A. H.; Libchaber, A. *Science* **2002**, 298, 1759–1762.
6 Gao, X.; Cui, Y.; Levenson, R. M.; Chung, L. W. K.; Nie, S. *Nat. Biotechnol.* **2004**, 22, 969–976.
7 Bentzen, E. L.; Tomlinson, I. D.; Mason, J.; Gresch, P.; Warnement, M. R.; Wright, D.; Sanders-Bush, E.; Blakely, R.; Rosenthal, S. J. *Bioconjug. Chem.* **2005**, 16, 1488–1494.
8 Kairdolf, B. A.; Mancini, M. C.; Smith, A. M.; Nie, S. *Anal. Chem.* **2008**, 80, 3029–3034.
9 Ekimov, A. I.; Onushchenko A. A. *JETP Lett.* **1981**, 34, 345–349.
10 Hardman, R. *Environ. Health Perspect.* **2006**, 114, 165–172.
11 Jacak, L.; Hawrylak, P.; Wójs, A. *Quantum Dots.* **2013**. Springer Science & Business Media, Springer-Verlag, Berlin/Heidelberg.
12 Liu, X.; Braun, G. B.; Zhong, H.; Hall, D. J.; Han, W.; Qin, M.; Zhao, C.; Wang, M.; She, Z.-G.; Cao, C.; Sailor, M. J.; Stallcup, W. B.; Ruoslahti, E.; Sugahara, K. N. *Adv. Funct. Mater.* **2016**, 26, 267–276.
13 Dabbousi, B. O.; Rodriguez Viejo, J.; Mikulec, F. V.; Heine, J. R.; Mattoussi, H, Ober, R.; Jensen, K. F.; Bawendi, M. G. *J. Phys. Chem. B* **1997**, 101, 9463–9475.
14 Tsoi, K. M.; Dai, Q.; Alman, B. A.; Chan, W. C. W. *Acc. Chem. Res.* **2013**, 46, 662–671.
15 Chou, L. Y. T.; Chan, W. C. W. *Nat. Nanotechnol.* **2012**, 7, 416–417.
16 Ghaderi, S.; Ramesh, B.; Seifalian, A. M. *J. Drug Target.* **2010**, 19, 475–486.
17 Chen, Y. F.; Rosenzweig, Z. *Anal. Chem.* **2002**, 74, 5132–5138.
18 Li, Y.; Ma, Q.; Wang, X.; Su, X. *Luminescence* **2007**, 22, 60–66.
19 Dyadyusha, L.; Yin, H.; Jaiswal, S.; Brown, T.; Baumberg, J. J.; Booy, F. P.; Melvin, T. *Chem. Commun.* **2005**, 25, 3201–3203.
20 Oh, E.; Liu, R.; Nel, A.; Gemill, K. B.; Bilal, M.; Cohen, Y.; Medintz, I. L. *Nat. Nanotechnol.* **2016**, 11, 479–486.

21 Wu, X.; Liu, H.; Liu, J.; Haley, K. N.; Treadway, J. A.; Larson, J. P.; Ge, N.; Peale, F.; Bruchez, M. P. *Nat. Biotechnol.* **2003**, 21, 41–46.

22 Åkerman, M. E.; Chan, W. C. W.; Laakkonen, P.; Bhatia, S. N.; Ruoslahti, E. *Proc. Natl. Acad. Sci. U. S. A.* **2002**, 99, 12617–12621.

23 Tokumasu F.; Dvorak J. *J. Microsc.* **2003**, 211, 256–261.

24 Lidke, D. S.; Nagy, P.; Heintzmann, R.; Arndt-Jovin, D. J.; Post, J. N.; Grecco, H. E.; Jares-Erijman, E. A.; Jovin, T. M. *Nat. Biotechnol.* **2004**, 22, 198–203.

25 Chan, W. C. W.; Nie, S. *Science* **1998**, 281, 2016–2018.

26 Xiao, Y.; Barker, P. E. *Nucleic Acids Res.* **2004**, 32, 1–5.

27 Li, Z.; Wang, Y.; Wang, J.; Tang, Z.; Pounds, J. G.; Lin, Y. *Anal. Chem.* **2010**, 82, 7008–7014.

28 Deng, H.; Liu, Q.; Wang, X.; Huang, R.; Liu, H.; Lin, Q.; Zhou, X.; Xin, D. *Biosens. Bioelectron.* **2017**, 87, 931–940.

29 Zhang, H.; Wang, Y.; Zhao, D.; Zeng, D.; Xia, J.; Aldalbahi, A.; Wang, C.; San, L.; Fan, C.; Zuo, X.; Mi, X. *ACS Appl. Mater. Interfaces* **2015**, 7, 16152–16156.

30 Weng, X.; Neethirajan, S. *Biosens. Bioelectron.* **2016**, 85, 649–656.

31 Zhang, W.-H.; Ma, W.; Long, Y.-T. *Anal. Chem.* **2016**, 88, 5131–5136.

32 Dogan, Ü.; Kasap, E.; Cetin, D.; Suludere, Z.; Boyaci, I. H.; Türkyılmaz, C.; Ertas, N.; Tamer, U. *Sens. Actuators B* **2016**, 233, 369–378.

33 Wu, S.; Liu, L.; Li, G.; Jing, F.; Mao, H.; Jin, Q.; Zhai, W.; Zhang, H.; Zhao, J.; Jia, C. *Talanta* **2016**, 156–157, 48–54.

34 Xiao, Y.; Barker, P. E. *Nucleic Acids Res.* **2004**, 32(3), e28, 1–5.

35 Han, M. Y.; Gao, X. H.; Su, J. Z.; Nie, S. *Nat. Biotechnol.* **2001**, 19, 631–635.

36 Jaiswal, J. K.; Mattoussi, H.; Mauro, J. M.; Simon, S. M. *Nat. Biotechnol.* **2003**, 21, 47–51.

37 So, M. K.; Xu, C. J.; Loening, A. M.; Gambhir, S. S.; Rao, J. H. *Nat. Biotechnol.* **2006**, 24, 339–343.

38 Zoumi, A.; Yeh, A.; Tromberg, B. J. *Proc. Natl. Acad. Sci. U. S. A.* **2002**, 99, 11014–11019.

39 Brown, E.; Mckee, T.; diTomaso, E.; Pluen, A.; Seed, B.; Boucher, Y.; Jain, R. K. *Nat. Med.* **2003**, 9, 796–800.

40 Xu, H. X.; Sha, M. Y.; Wong, E. Y.; Uphoff, J.; Xu, Y. H.; Treadway, J. A.; Truong, A.; O'Brien, E.; Asquith, S.; Stubbins, M.; Spurr, N. K.; Lai, E. H. *Nucleic Acids Res.* **2003**, 31(8), e43.

41 Zhao, J.-Y.; Chen, G.; Gu, Y.-P.; Cui, R.; Zhang, Z.-L.; Yu, Z.-L.; Tang, B.; Zhao, Y.-F.; Pang, D.-W. *J. Am. Chem. Soc.* **2016**, 138, 1893–1903.

42 Xu, G.; Zeng, S.; Zhang, B.; Swihart, M. T.; Yong, K.-T.; Prasad, P. N. *Chem. Rev.* **2016**, 116, 12234–12327.

43 Pathak, S.; Choi, S. K.; Arnheim, N.; Thompson, M. E. *J. Am. Chem. Soc.* **2001**, 123, 4103–4104.

44 Aldana, J.; Wang, Y. A.; Peng, X. *J. Am. Chem. Soc.* **2001**, 123, 8844–8850.

45 Wuister, S. F.; Swart, I.; van Driel, F.; Hickey, S. G.; de Mello Donega, C. *Nano Lett.* **2003**, 3, 503–507.

46 Kim, S.; Lim, Y. T.; Soltesz, E. G.; De Grand, A. M.; Lee, J.; Nakayama, A.; Parker, J. A.; Mihaljevic, T.; Laurence, R. G.; Dor, D. M.; Cohn, L. H.; Bawendi, M. G.; Frangioni, J. V. *Nat. Biotechnol.* **2004**, 22, 93–97.

47 Kim, S.; Bawendi, M. G. *J. Am. Chem. Soc.* **2003**, 125, 14652–14653.

48 Guo, W.; Li, J. J.; Wang, Y. A.; Peng, X. *Chem. Mater.* **2003**, 15, 3125–3133.

49 Mattoussi, H. M.; Goodman, J. M. E.; Anderson, G. P.; Sundar, V. C.; Mikulec, F. V.; Bawendi, M. G. *J. Am. Chem. Soc.* **2004**, 126, 6115–6123.

50 Gerion, D.; Pinaud, F.; Williams, S. C.; Parak, W. J.; Zanchet, D.; Weiss, S.; Alivisatos, A. P. *Phys. Chem. B* **2001**, 105, 8861–8871.

51 Tan, W.; Wang, K.; He, X.; Zhao, X. J.; Drake, T.; Wang, T.; Bagwe, R. P. *Med. Res. Rev.* **2004**, 24, 621–638.

52 Kirchner, C.; Liedl, T.; Kudera, S.; Pellegrino, T.; Javier, A. M.; Gaub, H. E.; Stölzle, S.; Fertig, N.; Parak, W. J. *Nano Lett.* **2005**, 5, 331–338.

53 Hoshino, A.; Fujioka, K.; Oku, T.; Suga, M.; Sasaki, Y. F.; Ohta, T.; Yasuhara, M.; Suzuki, K.; Yamamoto, K. *Nano Lett.* **2004**, 4, 2163–2169.

54 Lovric, J.; Bazzi, H. S.; Cuie, Y.; Fortin, G. R. A.; Winnik, F. M.; Maysinger, D. *J. Mol. Med.* **2005**, 83, 377–385.

55 Shiohara, A.; Hoshino, A.; Hanaki, K.; Suzuki, K.; Yamamoto, K. *Microbiol. Immunol.* **2004**, 48, 669–675.

10

Applications of Molecularly Imprinted Nanostructures in Biosensors and Medical Diagnosis

Deniz Aktas-Uygun[1], Murat Uygun[1], Zeynep Altintas[2], and Sinan Akgol[3]

[1] Department of Chemistry, Faculty of Science and Arts, Adnan Menderes University, Aydin, Turkey
[2] Technical University of Berlin, Berlin, Germany
[3] Department of Biochemistry, Faculty of Science, Ege University, Izmir, Turkey

10.1 Introduction

Polymers are high molecular weight compounds or macromolecules composed of many repeated units of polymerizable subunits. Physical and chemical properties of these polymeric materials differ from small chemical compounds, and today scientists have synthesized polymeric materials with various shapes, sizes, and functionalities. Polymers are divided into two main subclasses named natural and synthetic polymers, which are both synthesized by polymerization of small molecular weight compounds named monomers and whose main backbone consist of mainly carbon atoms. Proteins, nucleic acids, polysaccharides, and natural rubbers are the most important examples for natural polymers. Proteins are natural polymers and comprised of amino acid units. They have important functions from catalysis to storage and from transportation to protection. Nucleic acids compose the DNA and RNA, which are the main vital elements of life. Synthetic polymers such as nylon, poly(methyl methacrylate), poly(hexamethylene adipamide), poly(ethylene terephthalate), and so on [1–3]consist of organic compounds, and they can be synthesized by using diverse starting monomers and obtained by various chemical processes. These polymers have similar or better characteristics than natural polymers.

Polymers are studied intensively in the fields of polymer chemistry, polymer physics, biophysics, and macromolecular science. They have been substituted for various materials used in the daily life of humans. Today, 90% of chemists are working with polymeric materials in one way or another [4].

Biosensors and Nanotechnology: Applications in Health Care Diagnostics, First Edition.
Edited by Zeynep Altintas.
© 2018 John Wiley & Sons, Inc. Published 2018 by John Wiley & Sons, Inc.

10.2 Molecular Imprinted Polymers

Enzymes, antibodies, and proteins are natural receptors, and they display great affinity for their substrates. Because of this unique selectivity, these natural molecules have been widely used in various chemo/biosensor [5–7], bioassay [8–11], and biomedical diagnostic studies [12–14]. Recognition in all living organisms depends on the intermolecular interaction. Complex structures such as membranes, double-stranded DNA, and cells arise from the combination of these molecular interactions, which are generally weak compared with covalent interaction. These weak interactions are a driving force for vital cellular activity like DNA replication, enzymatic catalysis, protein synthesis, hormonal response, cell adhesion, and so on. The enzymatic "key and lock" theory also depends on the recognition of molecules by "weak interactions." There have been numerous works for mimicking the natural interactions shown by enzymes and antibodies, and a new field in chemistry named biomimetic chemistry has been established. The term "biomimetic" expresses any process that imitates a biochemical reaction for a chemical operation. For example, synthetic enzymes have been synthesized without any macromolecular peptide skeleton by carrying active groups of amino acids in the order of active center of the enzyme [15].

Although natural receptors are highly specific and display good affinity to their substrates, they have important limitations for their practical usage. Their main disadvantage is their poor mechanic and physical stability. Also these biological molecules are very expensive because of the multistep separation techniques. Moreover, they require to be utilized under moderate conditions in order to protect their structure and affinity toward its substrates [16].

Molecularly imprinted polymers (MIPs) are synthetic materials synthesized by polymerizing functional monomers and cross-linkers in the presence of a template molecule. By removing the template molecule, a cavity is created, which is compatible with the size, shape, and functional groups of the template molecule. The shape and the size of this cavity allow recognition of the template molecule or its derivatives, while orientation of the functional groups allows specific binding of the template molecule only [17]. By this way, these MIPs have similar selectivity with their functional biological counterparts, and they provide an increased mechanical and physical stability. These biomimetic MIPs also can resist harsh environmental conditions. Their preparation and functionalization is also so easy, making them much cheaper than their biological counterparts [18–20].

General pathway of MIP synthesis is schematically presented in Figure 10.1. In the first step, functional monomers interact with a template molecule by covalent or non-covalent interactions and a pre-polymerization complex is carried out. Then in the second step, the interacted species are "frozen" by polymerization using cross-linkers. At this point, functional units and template

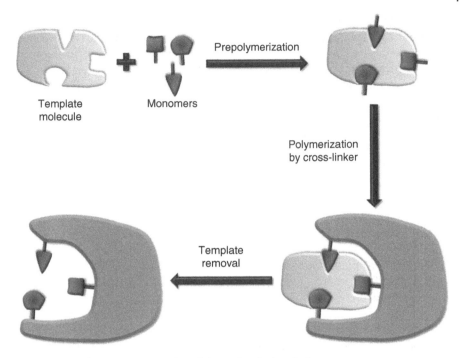

Figure 10.1 Schematic presentation of the molecular imprinting process.

molecule are topologically complementary. Finally in the third step, the template molecule is removed from this polymeric structure. After removing the template, a cavity is created, which is complementary with the template molecule and, under the appropriate conditions, can recognize the template molecule efficiently and selectively by its shape, size, and other physical properties.

The most commonly used functional and cross-linking monomers for MIPs are acrylic-based molecules, which are polymerized by different types of polymerization techniques using various initiators. Forming block polymers are ground for suitable size and then used as column-packing materials [21, 22]. These MIPs can be used as a sensor material by covering the electrode surface [23–26]. Another approach of MIP polymerization for sensor studies is to polymerize the monomers directly on the electrode surface [27–29].

Other polymerization techniques for MIP preparation are precipitation polymerization and emulsion polymerization. Polymers have also been used for various applications such as solid-phase extraction, biosensors, and analytical chemistry [14, 30–34]. Imprinting strategy is a very important parameter for MIP technology and bulk imprinting is widely used wherein the template molecule is added to the pre-polymerization complex and then polymerized. After removing the template, selective cavities are produced all over the polymer structure [35]. For large templates, surface imprinting strategy has been developed, wherein only some part of the template is imprinted on the surface of the polymer structure [36]. In substructure imprinting strategy, characteristic substructures are used instead of a whole molecule [37, 38]. Structural analogs of the template molecule can also be used for the imprinting process. When working with toxic or rare molecules, this strategy can help the imprinting process efficiently [39]. Antibody replica is another imprinting strategy wherein antibodies are used for target molecule as a starting material of MIPs [40]. Also, using a sacrifice layer, a template molecule covered by a thin layer of polymer first produces the binding sites for the template [41, 42].

10.3 Imprinting Approaches

There are two main approaches applied for the preparation of MIPs. One is "covalent" approach developed by Wulff and coworkers [43] where a reversible covalent bond is created between the template and monomers, before the polymerization, and recognition between the template and the polymer depends on the creation and breakdown of this bond. The other approach is "non-covalent" approach, which was proposed by Klaus Mosbach and coworkers [44]. In this approach, pre-organization between template and monomers is created by a non-covalent interaction or (weak) metal coordination interaction, and the following recognition depends on these interactions [15].

Covalent imprinting—One of the key steps in this imprinting approach is to select the type of the covalent bond that will hold monomer and template together. This bond should be stable enough and strong for polymerization reaction. However, this bond should be broken down under mild conditions. Creation and breakdown of this covalent bond should be fast for rapid interaction between template and imprinted polymer. Most preferred bonds for covalent imprinting are boronic acid esters, acetals, ketals, Schiff bases, disulphide bonds, and coordination [45].

Non-covalent imprinting—This approach offers easier imprinting process than that of covalent imprinting. Functional monomer is simply mixed with template molecule and copolymerized by cross-linkers. After the polymerization, template molecule can be removed easily by simple extraction techniques. Because of its simplicity and versatility, this strategy is mostly preferred for imprinting applications. Any non-covalent interaction can be applied for imprinting process. The most appropriate bond for molecular recognition is the hydrogen bond, and for this, various monomers, which carry functional groups (carboxyl, amino, pyridine, hydroxyl, and amide groups), can be chosen [45].

10.4   Molecularly Imprinted Nanostructures

There have been exponential research and application about nanosized materials recently. Various application areas range from clean energy to environmental monitoring and improved materials. The term nanomaterial is defined as a substance whose one or more dimension is varied between 1 and 100 nm [46]. Nanostructured materials have been widely used for biomedical applications such as chemical sensing, biomaterials, and bioanalytical purposes [47, 48].

Nanosized materials exhibit unique physical and chemical properties, and because of this they are highly preferred as a sensor component. They also show improved thermal and electrical properties. One of the important features of nanomaterials is their high surface area per unit of mass that provides a wide functionalization area and more chemical contact with analyte or substance. Their optical properties are also improved in nanosized world. The size, shape, and origin of the nanoparticles are variable: they can be metallic or polymeric and their shape can be spherical or wired [49].

Nanostructured materials can be a component for various types of biosensing devices by employing their magnetic, optical, electrical, and electrochemical behaviors, and they are successfully used for the detection of different types of biomolecules such as nucleic acids, antibodies, proteins, toxins, and so on [50–56]. Nanomaterials can improve the efficiency of biosensors because of their unique properties especially high surface areas. Thus, sensor performance increases, while limit of detection (LOD) and response time reduce [57, 58].

One of the mostly used materials in sensor applications is nanostructured polymers [59]. Application and properties of the polymeric materials can be developed by immobilization of metallic ions and nanomaterials to the polymeric backbone [60, 61].

MIP nanostructure-based sensors have been greatly used for many biosensors applications: A surface imprinting strategy has been applied for lysozyme enzyme. For this, silica nanoparticles were first synthesized, and then lysozymes were immobilized onto the surface using N-(4-vinylbenzyl)iminodiacetic acid (VBIDA) along with N-isopropylacrylamide, methylenebisacrylamide, and acrylamide. VBIDA was used for chelating Cu(II), and this complex was used for the metal chelate affinity recognition for lysozyme. These surface-imprinted nanoparticles showed the improved binding kinetics for lysozyme, and also selectivity of the nanoparticle was enhanced. Moreover, dissociation constant for lysozyme was found to be 4.1×10^{-8} M, which was very comparable with that of the antibody-based conventional techniques [62].

Another surface imprinting study has been carried out by Yang and coworkers [63]. Researchers synthesized magnetic Fe_3O_4/hydroxyapatite polypyrrole nanoparticles for surface imprinting of bilirubin. The bilirubin-imprinted nanoparticles were used for the photoelectrochemical detection of bilirubin. For this, the synthesized bilirubin-imprinted nanoparticles were attached to the surface of the magnetic glassy carbon electrode. It was shown that this biosensor was quite sensitive for bilirubin with a linear range from 1.0 to 17 μM, and its LOD was found to be 0.007 μM. The biosensor showed great selectivity for bilirubin in the medium of interferents, and its affinity was also proven in serum.

Redox-active magnetic MIP nanospheres were synthesized for streptomycin determination in foods. These magnetic MIP nanospheres were prepared using Au(III)-promoted imprinting polymerization in the presence of strepto-mycin. A competitive-type assay format was used for streptomycin determination utilizing glucose oxidase-labeled streptomycin. The magnetic MIP nanospheres showed good dynamic range for streptomycin ($0.05-20 \, \text{ng mL}^{-1}$) with an LOD of $10 \, \text{pg mL}^{-1}$ [64].

Molecularly imprinted folic acid sensors have been synthesized as thin film- and nanoparticle-type sensors by employing two polymer systems (meth-acrylate and acrylate–vinyl pyrrolidone). Folic acid was detected through quartz crystal microbalance (QCM) with low detection limits of 1–30 ppm. Nanoparticle-type sensors showed three times higher sensitivity than the thin film-type sensors for methacrylate-based systems [65].

Molecularly imprinted electrochemical sensors were developed to determine 3,3',5,5'-tetrabromobisphenol A (TBBPA), they were synthesized using nickel nanoparticles with graphene modification on the surface of a carbon electrode. Detection limit for TBBPA was found to be $1.3 \times 10^{-10} \, \text{mol L}^{-1}$, and its linear relationship toward TBBPA ranged from 5.0×10^{-10} to $1.0 \times 10^{-5} \, \text{mol L}^{-1}$. The electrochemical sensor was successfully applied to tap water, rain, and lake water samples for TBBPA determination [66].

CdS quantum dot-doped chitosan matrix was used to develop an urea recognition support film using imprinting technology. The electrochemical sensor could detect urea by differential pulse voltammetry method with the help of enhanced electron transfer behavior and increased surface area of quantum dots. The biosensor has a very wide linear range for urea recognition with a low detection limit ($1.0 \times 10^{-12} \, \text{M}$). Also, it was examined toward some structurally similar compounds, and it showed that the fabricated sensor had excellent recognition ability for urea molecules [67].

Xue and coworkers developed an amperometric sensor for dopamine detection in human serum. The researchers used gold nanoparticle (AuNP)-based molecular imprinting technology for dopamine determination with an LOD of $7.8 \, \text{nmol L}^{-1}$ [68]. They prepared dopamine-imprinted electrode by electropolymerization of p-aminobenzenethiol in the presence of dopamine.

Lovastatin-imprinted nanosensors were synthesized for lovastatin determination in red yeast rice. Lovastatin was detected through a molecularly imprinted QCM sensor that was prepared by self-assembling monolayer formation of allyl mercaptan on the surface of a QCM chip that could detect lovastatin specifically with a LOD of 0.030 nM [69].

AuNP-based electrochemical sensor was constructed using the molecular imprinting technology to detect erythromycin. The MIP film of this sensor was modified by chitosan/platinum nanoparticle/graphene–AuNP double nano-composites. The sensor could detect erythromycin in the linear range of 7.0×10^{-8} to $9.0 \times 10^{-5} \, mol \, L^{-1}$ with a LOD of $2.3 \times 10^{-8} \, mol \, L^{-1}$ [70].

Ascorbic acid was imprinted on a polypyrrole nanoreactor SBA-15, and the imprinting cavity was used for ascorbic acid recognition. For this, SBZ-15 was used as a mesoporous silica matrix, and the imprinting took place onto its hexagonal channels. The ascorbic acid-imprinted nanocomposites demon-strated a significant recognition ability and high adsorption capacity ($83.7 \, mg \, g^{-1}$). The researchers also investigated competitive compounds such as dopamine, paracetamol, and epinephrine, in order to show the selectivity, and the imprinting factor was found to be 3.2, 1.5, and 1.3, respectively [71].

Molecularly imprinted polypyrrole–graphene–AuNP-modified electrodes were used for electrochemical determination of levofloxacin. Electrooxidation of levofloxacin on electrode was developed by the incorporation of graphene–AuNPs onto the sensor structure. The target analyte was detected in the linear concentration range of $1.0–100 \, \mu mol \, L^{-1}$ with an LOD of $0.53 \, \mu mol \, L^{-1}$. The sensors displayed high sensitivity, good selectivity, and reproducibility [72].

Waterborne viruses were detected using high affinity molecularly imprinted nanopolymers with optical sensors. For this study, bacteriophage MS2 was used as a template, and these artificial affinity ligands were synthesized by employing a novel solid-phase polymerization technique. Preparation and application of this MIP-based virus detection platform is schematically pre-sented in Figure 10.2. Prepared MIP-based virus detection assay was success-fully applied in a surface plasmon resonance (SPR) biosensor. The nanosensor showed a high affinity between nano-MIP and bacteriophage MS2 ($\sim 3 \times 10^{-9} \, M$), and LOD was found to be $5 \times 10^6 \, pfu \, mL^{-1}$ [73].

10.5 MIP Biosensors in Medical Diagnosis

MIPs have found diverse applications in biosensor-based investigations. For example, molecularly imprinted polyaniline–polyvinyl sulfonic acid sensor was developed for *para*-nitrophenol (PNP) detection by Roy and coworkers [74]. In this work, electrochemical imprinting of PNP was carried out onto indium tin oxide glass substrate. PNP was detected by using differential pulse voltammetry with an LOD of $1 \times 10^{-3} \, mM$ and a sensitivity of $1.5 \times 10^{-3} \, A \, mM^{-1}$. Bisphenol A (BPA) was determined electrochemically by employing molecu-larly imprinted chitosan–graphene composite film modified with acetylene black paste electrode, which was successfully applied in the determination of BPA in plastic drinking water and canned beverage [75]. Its linear range was found to be $8.0 \, nM$ to $1.0 \, \mu M$ and its LOD was observed at $6.0 \, nM$.

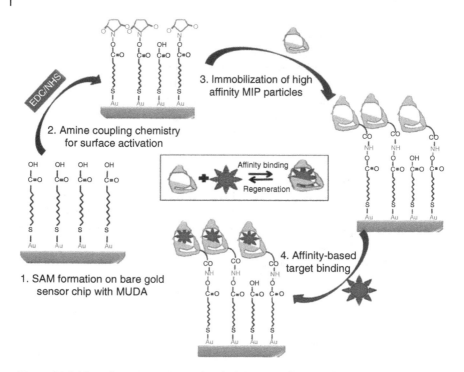

Figure 10.2 Virus detection using molecularly imprinted nanopolymer-functionalized SPR biosensor. *Source:* Altintas et al. [73]. Reproduced with permission of American Chemical Society. (*See insert for color representation of the figure.*)

Quinoxaline-2-carboxylic acid (QCA) was imprinted on a polymer film in order to study a rapid, sensitive, and selective QCA determination using modified carbon electrode by differential pulse voltammetry. For this, sol–gel technology was used in the stepwise modification of multiwalled carbon nanotube–chitosan functional layer on a glassy carbon electrode. Linear current response for QCA was found to be very wide (2.0×10^{-6} to 1.0×10^{-3} mol L^{-1}), and its LOD was 4.4×10^{-7} mol L^{-1} [76]. Imprinted chitosan–acrylamide, graphene, ferrocene composite cryogel biosensor was used to detect microalbumin in urine samples. In this study, human serum albumin was used as a template molecule, and chitosan was modified by the graft copolymerization of acrylamide and N,N'-methylenebisacrylamide. Polymerization was carried out at sub-zero temperature in order to create the cryogel structure. Albumin was detected using differential pulse voltammetry with high sensitivity (investigation range: $1.0 \times 10^{-4} - 1.0 \times 10^{1}$ mg L^{-1}) and a low LOD of 5.0×10^{-5} mg L^{-1}. The biosensor could be stored for 6 weeks without any significant decrease in its selectivity. It also showed good selectivity toward some possible interfering compounds such as ascorbic acid, uric acid, urea, sodium chloride, potassium, and creatinine [77].

Selective and sensitive electrochemical sensor for metronidazole was developed by Chen and coworkers [78]. The sensor was prepared using a core–shell strategy, and the synthesized metronidazole-imprinted magnetic polymers were attached onto the surface of the magnetic glassy carbon electrodes, which showed good stability, high recognition ability and affinity toward metronidazole. The target could be detected in the range from 5.0×10^{-8} to 1.0×10^{-6} M with an LOD of 1.6×10^{-8} M.

An SPR biosensor was developed by imprinting of citrinin onto the gold surface of the SPR chip modified with poly(HEMA-MAGA) film. While linearity range of this sensor was found to be $0.005–1.0 \, ng \, mL^{-1}$, its LOD was $0.0017 \, ng \, mL^{-1}$. The SPR chips were used for the determination of citrinin in red yeast rice samples efficiently [79]. Cardiac troponin T is a specific biomarker for myocardial infarction. Karimian and coworkers prepared molecularly imprinted cardiac troponin T sensors. For this, selective layer was prepared by electropolymerization of *o*-phenylenediamine. The synthesized sensors had good selectivity in the concentration range of $0.009–0.8 \, ng \, mL^{-1}$ and low LOD of $9 \, pg \, mL^{-1}$ [80]. Nitrogen-doped graphene sheet-modified methyl parathion-imprinted sensors were synthesized by Xue and coworkers [81]. The working principle of the sensors relied on cyclic voltammetry and they exhibited a wide determination range $(1.0–10 \, \mu g \, mL^{-1})$ with an LOD of $0.01 \, \mu g \, mL^{-1}$.

Feng and coworkers developed surface-enhanced Raman spectroscopic biosensors for specific detection of α-tocopherol [82]. For this, methacrylic acid was used as a functional monomer, and ethylene glycol dimethacrylate was the cross-linker. These biosensors were used for rapid and selective adsorption and separation of α-tocopherol from oil samples. They also displayed high sensitivity along with a good α-tocopherol recovery in oil samples.

A sequence-specific molecularly imprinted single-stranded oligodeoxyribonucleotide biosensor was prepared by Tiwari et al. for the detection of p53 gene point mutation. Indium tin oxide-coated glass substrate was used for the preparation of the sensor and then characterized by FTIR, SEM, and CV methods. Linear current response of this working electrode was found to be 0.01–300 fM. The biosensor displayed high sensitivity $(0.62 \, \mu A \, fM^{-1})$ with fast response time (14 s) [83]. Molecular imprinting technology was also applied to an impedimetric sensor. For this, polypyrrole-modified pencil graphite electrode was employed for the determination of chlorpyrifos. This organophosphorus pesticide could be quantified up to $4.5 \, \mu g \, L^{-1}$ [84]. Chloramphenicol-selective molecularly imprinted sensor was prepared using carbon paste electrode for milk samples. The MIP sensors exhibited very good recognition properties for chloramphenicol with a lower LOD of 2.0×10^{-9} M [85].

MIP-ionic liquid–graphene composite film-coated electrodes were prepared for electrochemical determination of methyl parathion by Zhao and coworkers [86]. The sensors were prepared by free radical polymerization of methacrylic acid with the template molecule methyl parathion. They demonstrated a very good recognition capability behavior for template molecules with the sensitivity of $12.5\,\mu A\,\mu M^{-1}$. Dopamine was imprinted on SiO_2-coated graphene oxide/ MIP composites and used for recognition of dopamine in human urine samples. The MIP sensor could detect dopamine using differential pulse voltammetry up to $3.0 \times 10^{-8}\,M$ concentration [87]. Oxytetracycline was imprinted on a Prussian blue film-modified sensor. It was detected by its competition with glucose oxidase-labeled oxytetracycline and reduction reaction of H_2O_2. This double amplification strategy increased the selectivity and improved the LOD up to femtomole level [88].

10.6 Diagnostic Applications of MIP Nanostructures

It is very important to detect the chemical and biological agents in medical, agricultural, and environmental fields [89]. For this purpose, selective and sensitive methods have been developed for more effective detection of target compounds. At this point, nanomaterials exhibit their unique properties for improving the effectiveness of the detection unit. They provide opportunity for new recognition and transduction processes. They also improve the signal-to-noise ratio of the detection process with the help of the system component minimization [90] along with the improving the specificity, sensitivity, and stability of the detection system.

Recently, nanomaterial-modified detection systems have been intensively used for the detection of metal ions, small molecules, proteins, and nucleic acids. For this, noble metals, quantum dots, and magnetic materials are used as a nanoparticle component [91–94]. These nanoparticles not only increase the surface area of the detection unit but also exhibit unique optical, electrical, and magnetic properties that improve the detection process. Their surface can also be modified by organic ligands to increase the specificity, selectivity, and practical usage [95]. Another advantage for nanoparticle-modified detection systems in diagnostics is that the required sample volume is very small [96]. These "nanodiagnostics" have been successfully used for various diagnostic assays such as biomarker analysis, cancer diagnosis, diagnostic imaging, and immunoassays.

Selectivity of a sensor is a very important parameter for practical usage, and finding a suitable ligand for target molecule is one of the challenging steps. Selectivity and specificity of the nanoparticulate systems can be improved by using the technology of the molecular imprinting process. The nanosized MIPs have practical advantages such as easy-to-produce properties, scalability, low cost, and flexibility [97]. MIPs act as recognition elements like their biological counterparts such as antibodies and enzymatic receptors, but their mechanical, thermal, and chemical stabilities are much higher. For these reasons, MIPs–nanomaterials have been intensively used for sensing and separation platforms [14, 98].

Wang and coworkers prepared surface-imprinted strategy for determination of the cancer biomarkers. For this, self-assembled monolayers were used by covering the gold-coated silicon chip with hydroxyl-terminated alkanethiol molecules. These sensors were used for potentiometric detection of carcinoembryonic antigen. Researchers could quantify carcinoembryonic antigen in the concentration range of 2.5–250 ng mL^{-1}, and there was no cross-reactivity with hemoglobin molecule [99].

An epitope imprinting technology has been successfully applied to HIV-1-related protein gp41 [100], which was used for the diagnosis and determination of the extent of HIV-1 disease. A QCM chip and a synthetic peptide composed of 35 amino acids, which is analogous to gp41, were used, with the latter as an epitope fragment of the template molecule. The molecularly imprinted QCM chips showed good affinity towards gp41 with a dissociation constant of 3.17 nM. LOD for gp41 was found to be 2 ng mL^{-1}, and this is comparable with the ELISA method. These chips were also used for the detection of gp41 in human urine samples.

Matsui and coworkers combined AuNPs with MIP technology for sensing. Adrenaline was used as a model analyte, and the selectivity and sensitivity of the AuNPs were increased upon imprinting of the adrenaline molecule. The Au-MIP nanoparticles could detect adrenaline selectively up to 5 μM using a UV–Vis spectrophotometer [101]. Magnetic molecularly imprinted poly(ethylene-*co*-vinyl alcohol) nanoparticles were also prepared by Lee and coworkers [102]. These superparamagnetic nanoparticles were successfully applied for disease detection. In this research, albumin, creatinine, lysozyme, and urea-imprinted poly(ethylene-*co*-vinyl alcohol) nanoparticles were synthesized by phase inversion. Rebinding capacities of the nanoparticles varied from 0.76 to 5.97 mg g^{-1}. The nanoparticles were used for separation and sensing studies in human serum and urine samples.

A novel AuNP-modified paper working electrode (Au-PWE) was prepared by electropolymerization of MIPs and used for microfluidic paper-based analytical devices (μ-PADs) [103]. AuNP layer was grown on the cellulose fiber in the PWE. This developed sensor was used for the detection of D-glutamic acid by differential pulse voltammetry, and a linear range of the detection was found to be from 1.2 to 125.0 nM with an LOD of 0.2 nM. Selectivity, reproducibility, and stability profiles of this sensor were also investigated.

Viswanathan et al. prepared protein-imprinted three-dimensional gold nanoelectrodes for the detection of an ovarian cancer biomarker (CA 125) [104]. CA 125-imprinted gold nanoelectrode ensemble was prepared using cyclic voltammetry, differential pulse voltammetry, and electrochemical impedance spectroscopy techniques. The nanosensor demonstrated a good recognition profile with the concentration range of 0.5–400 U mL^{-1} for CA 125, and LOD was found to be 0.5 U mL^{-1}. The sensor was used for the detection of the biomarker in real serum and spiked serum. It was demonstrated that the sensitivity was not significantly affected in the presence of the nonspecific proteins.

Prasad and Pandey prepared D- and L-aspartic acid-imprinted sensors using multiwalled carbon nanotube/pencil graphite electrode. Poly(indole-3-acetic acid) was used as a conductive polymer, and differential pulse anodic stripping voltammetry was used for the determination of aspartic acids. LOD values for D- and L-aspartic acid were found to be 0.025 and 0.016 µM, respectively. The sensors were proposed as a practical sensing platform for the determination of this disease biomarker in clinical settings [105].

Novel molecularly imprinted nanosensors were prepared for cholic acid determination [106]. For this, polymerizable methacryloylamido-cysteine attached AuNPs were synthesized, and methacryloylamido-histidine-Pt(II) monomers were used for metal chelation of cholic acid. Cholic acid binding affinity of these nanosensors was investigated by Langmuir and Scatchard methods, and the affinity constants were found to be $1.48 \times 10^{-4} \, \text{mol L}^{-1}$ and $6.59 \times 10^{-6} \, \text{mol L}^{-1}$, respectively. Also, cholic acid level in blood serum and urine was determined.

Gültekin et al. prepared dipicolinic acid-imprinted nanoshell-modified AuNPs for the recognition of *Bacillus cereus* spores. Methacryloyl iminodiacetic acid Cr(III) was used as chelating agent for dipicolinic acid, which is the main component of *B. cereus* spores. The interaction between the MIP sensors and dipicolinic acid was investigated with fluorescence measurements. Fluorescence of the MIP nanoshell sensor could be selectively quenched by dipicolinic acid [107]. Patra et al. prepared molecularly imprinted ZnO nanostructure-based electrochemical sensor for calcitonin-a clinic marker for thyroid carcinoma. For this, zinc oxide nanostructures were initially modified with tyrosine, and the polymerization took place on the vinyl functionalized electrode surface. Linear working range of the sensor was found to be $9.99 \, \text{ng L}^{-1}$ to $7.919 \, \text{mg L}^{-1}$, and the LOD was obtained as $3.09 \, \text{ng L}^{-1}$ [108].

Tong and coworkers synthesized carbon nanotube-based MIP-modified ceramic carbon electrodes for electrochemical detection of cholesterol, whose quantification was carried out using cyclic voltammetry and linear sweep voltammetry. The cholesterol sensor displayed excellent sensitivity with a linear range of 10–300 nM and a good LOD of 1 nM [109].

10.7 Conclusions

As many technological applications, nanotechnology has opened new opportunities in the area of biomedical diagnosis and biosensors. Nanosized materials have a great surface area to be modified by diverse functional groups with different characteristics, giving intensive application area for them. Nanomaterials also improve the sensitivity and selectivity of the assay platforms. With the help of the molecularly imprinted nanostructures, cheaper biosensors can be prepared easily to be used in disease detection, and these sensors are capable of providing rapid response for the disease biomarkers that can be protein, gene, bacterium, or viruses. However, imprinting large biological molecules such as proteins and viruses has major drawbacks [110, 111]. New approaches such as epitope imprinting, nanoparticle formation of MIPs, and incorporation of computational simulations into the imprinting arena may play an immense role to overcome this problem [111–115]. These approaches can also help avoiding the high level nonspecific interactions between MIPs and their biological targets in biosensor devices, particularly to detect the target molecule from complex media such as blood, urine, and saliva.

References

1 Roiter, Y.; Minko, S. *Journal of the American Chemical Society* **2005**, 127, 15688–15689.
2 Painter, P. C.; Coleman, M. M. *Fundamentals of Polymer Science: An Introductory Text*; CRC Press, Boca Raton: **1997**.
3 McCrum, N. G.; Buckley, C. P.; Bucknall, C. B. *Principles of Polymer Engineering*; Oxford University Press, New York: **1997**.
4 Feizi, T.; Chai, W. *Nature Reviews Molecular Cell Biology* **2004**, 5, 582–588.
5 Lai, K. S.; Ho, N. H.; Cheng, J. D.; Tung, C. H. *Bioconjugate Chemistry* **2007**, 18, 1246–1250.
6 Liu, J. W.; Lu, Y. *Analytical Chemistry* **2003**, 75, 6666–6672.
7 Paniel, N.; Radio, A.; Marty, J. L. *Sensors* **2010**, 10, 9439–9448.
8 Shults, M. D.; Janes, K. A.; Lauffenburger, D. A.; Imperiali, B. *Nature Methods* **2005**, 2, 277–284.
9 Llamas, N. M.; Stewart, L.; Fodey, T.; Higgins, H. C.; Velasco, M. L. R.; Botana, L. M.; Elliott, C. T. *Analytical and Bioanalytical Chemistry* **2007**, 389, 581–587.
10 Stewart, L. D.; Elliott, C. T.; Walker, A. D.; Curran, R. M.; Connolly, L. *Toxicon* **2009**, 54, 491–498.
11 Lee, R.; Tran, M.; Nocerini, M.; Liang, M. *Journal of Biomolecular Screening* **2008**, 13, 210–217.
12 Gubala, V.; Guevel, X. L.; Nooney, R.; Williams, D. E.; MacCraith, B. *Talanta* **2010**, 81, 1833–1839.
13 Arruebo, M.; Valladares, M.; Gonzalez-Fernandez, A. *Journal of Nanomaterials* **2009**, 2009, 1–24.
14 Ding, X.; Heiden, P. A. *Macromolecular Materials and Engineering* **2014**, 299, 268–282.
15 Yan, M.; Ranström, O. *Molecularly Imprinted Materials. Science and Technology*; Mercel Dekker, New York: **2005**.
16 Rathore, N.; Rajan, R. S. *Biotechnology Progress* **2008**, 24, 504–514.
17 Janiak, D. S.; Kofinas, P. *Analytical and Bioanalytical Chemistry* **2007**, 389, 399–404.

18 Ye, L.; Haupt, K. *Analytical and Bioanalytical Chemistry* **2004**, 378, 1887–1897.

19 Turiel, E.; Martin-Esteban, A. *Analytica Chimica Acta* **2010**, 668, 87–99.

20 Guan, G. J.; Liu, B. H.; Wang, Z. Y.; Zhang, Z. P. *Sensors* **2008**, 8, 8291–8320.

21 Cheong, W. J.; Yang, S. H.; Ali, F. *Journal of Separation Science* **2013**, 36, 609–628.

22 Dias, A. C.; Figueiredo, E. C.; Grassi, V.; Zagatto, E. A.; Arruda, M. A. *Talanta* **2008**, 76, 988–996.

23 Kriz, D.; Kempe, M.; Mosbach, K. *Sensors and Actuators B: Chemical* **1996**, 33, 178–181.

24 Kriz, D.; Mosbach, K. *Analytica Chimica Acta* **1995**, 300, 71–75.

25 Kriz, D.; Ramstrum, O.; Svensron, A.; Mosbach, K. *Analytical Chemistry* **1995**, 67, 2142–2144.

26 Hedborg, E.; Winquist, F.; Lundström, I.; Andersson, L. I.; Mosbach, K. *Sensors and Actuators A: Physical* **1993**, 37–38, 796–799.

27 Sharma, P. S.; Pietrzyk-Le, A.; D'Souza, F.; Kutner, W. *Analytical and Bioanalytical Chemistry* **2012**, 402, 3177–3204.

28 Balamurugan, S.; Spivak, D. A. *Journal of Molecular Recognition* **2011**, 24, 915–929.

29 Sharma, P. S.; Iskierko, Z.; Pietrzyk-Le, A.; D'Souza, F.; Kutner, W. *Electrochemistry Communications* **2015**, 50, 81–87.

30 Tarley, C. R. T.; Kubota, L. T. *Analytica Chimica Acta* **2005**, 548, 11–19.

31 Pietrzyk, A.; Suriyanarayanan, S.; Kutner, W.; Maligaspe, E.; Zandler, M. E.; D'Souza, F. *Bioelectrochemistry* **2010**, 80, 62–72.

32 Horemans, F.; Alenus, J.; Bongaers, E.; Weustenraed, A.; Thoelen, R.; Duchateau, J.; Lutsen, L.; Vanderzande, D.; Wagner, P.; Cleij, T. J. *Sensors and Actuators B: Chemical* **2010**, 148, 392–398.

33 Gallago-Gallegos, M.; Garrido, M. L.; Olivas, R. M.; Baravalle, P.; Baggiani, C.; Camara, C. *Journal of Chromatography A* **2010**, 1217, 3400–3407.

34 Xie, C. G.; Zhou, H. K.; Gao, S.; Li, H. F. *Microchimica Acta* **2010**, 171, 355–362.

35 Madhuri, R.; Tiwari, M. P.; Kumar, D.; Mukharji, A.; Prasad, B. B. *Advanced Materials Letters* **2011**, 2, 264–267.

36 Dickert, F. L.; Hayden, O.; Halikias, K. P. *Analyst* **2011**, 126, 766–771.

37 Titirici, M. M.; Sellergren, B. *Analytical and Bioanalytical Chemistry* **2004**, 378, 1913–1921.

38 Tai, D. F.; Lin, C. Y.; Wu, T. Z.; Chen, L. K. *Analytical Chemistry* **2005**, 77, 5140–5143.

39 Kugimiya, A.; Babe, F.; *Polymer Bulletin* **2011**, 67, 2017–2024.

40 Schirhagl, R.; Latif, U.; Podlipna, D.; Blumenstock, H.; Dickert, F. L. *Analytical Chemistry* **2012**, 84, 3908–3913.

41 Nicolescu, T. V.; Sarbu, A.; Ovidiu Dima, S.; Nicolae, C.; Donescu, D. *Journal of Applied Polymer Science* **2013**, 127, 366–374.

42 Schirhagl, R. *Analytical Chemistry* **2014**, 86, 250–261.
43 Wulff, G.; Grobe-Einsler, R.; Sarhan, A. *Die Makromolekulare Chemie* **1977**, 178, 2799–2816.
44 Arshady, R.; Mosbach, K. *Die Makromolekulare Chemie* **1981**, 182, 687–692.
45 Komiyama, M.; Takeuchi, T.; Mukawa, T.; Asanuma, H. *Molecular Imprinting. From Fundamentals to Applications*; Wiley-VCH, Weinheim: **2003**.
46 Lahiff, E.; Lynam, C.; Gilmartin, N.; O'Kennedy, R.; Diamond, D. *Analytical and Bioanalytical Chemistry* **2010**, 398, 1575–1589.
47 Shimizu, K. I.; Chinzei, I.; Nishiyama, H.; Kakimoto, S.; Sugaya, S.; Yokoi, H.; Satsuma, A. *Sensors and Actuators B: Chemical* **2008**, 134, 618–624.
48 Schultes, G.; Schmidt, M.; Truar, M.; Goettel, D.; Freitag-Weber, O.; Werner, U. *Thin Solid Films* **2007**, 515, 7790–7797.
49 Asefa, T.; Duncan, C. T.; Sharma, K. K. *Analyst* **2009**, 134, 1980–1990.
50 Liu, Z. M.; Li, Z. J.; Shen, G. L.; Yu, R. Q. *Analytical Letters* **2009**, 42, 3046–3057.
51 Chen, J.; Winther-Jensen, B.; Lynam, C.; Ngamna, O.; Moulton, S.; Zhang, W.; Wallace, G. G. *Electrochemical and Solid-State Letters* **2006**, 9, H68–H70.
52 Ko, Y. J.; Maeng, J. H.; Ahn, Y.; Hwang, S. Y.; Cho, N. G.; Lee, S. H. *Electrophoresis* **2008**, 29, 3466–3476.
53 Healy, D. A.; Hayes, C. J.; Leonard, P.; McKenna, L.; O'Kennedy, R. *Trends in Biotechnology* **2007**, 25, 125–131.
54 Byrne, B.; Stack, E.; Gilmartin, N.; O'Kennedy, R. *Sensors* **2009**, 9, 4407–4445.
55 Sanvicens, N.; Pastells, C.; Pascual, N.; Marco, M. P. *Trends in Analytical Chemistry* **2009**, 28, 1243–1252.
56 Sepulveda, N.; Gonzalez-Diaz, J. B.; Garcia-Martin, A.; Lechuga, L. M.; Armelles, G. *Physical Review Letters* **2010**, 104, 147401.
57 Hong, C.; Ying, X.; Pin-Gang, H.; Yu-Zhi, F. *Electroanalysis* **2003**, 15, 1864–1870.
58 Sepulveda, B.; Angelome, P. C.; Lechuga, L. M.; Liz-Marzan, L. M. *Nano Today* **2009**, 4, 244–251.
59 Wang, L.; Fine, D.; Sharma, D.; Torsi, L.; Dodabalapur, A. *Analytical and Bioanalytical Chemistry* **2005**, 384, 310–321.
60 Nicolas-Debarnot, D.; Poncin-Epaillard, F. *Analytica Chimica Acta* **2003**, 475, 1–15.
61 Potje-Kamloth, K. *Critical Reviews in Analytical Chemistry* **2002**, 32, 121–140.
62 Chen, H.; Kong, J.; Yuan, D.; Fu, G. *Biosensors and Bioelectronics* **2014**, 53, 5–11.
63 Yang, Z.; Shang, X.; Zhang, C.; Zhu, J. *Sensors and Actuators B: Chemical* **2014**, 201, 167–172.
64 Liu, B.; Tang, D.; Zhang, B.; Que, X.; Yang, H.; Chen, G. *Biosensors and Bioelectronics* **2013**, 41, 551–556.

65 Hussain, M.; Iqbal, N.; Lieberzeit, P. A. *Sensors and Actuators B: Chemical* **2013**, 176, 1090–1095.

66 Chen, H.; Zhang, Z.; Cai, R.; Rao, W.; Long, F. *Electrochimica Acta* **2014**, 117, 385–392.

67 Lian, H.-T.; Liu, B.; Chen, Y.-P.; Sun, X.-Y. *Analytical Biochemistry* **2012**, 426, 40–46.

68 Xue, C.; Han, Q.; Wang, Y.; Wu, J.; Wen, T.; Wang, R.; Hong, J.; Zhou, X.; Jiang, H. *Biosensors and Bioelectronics* **2013**, 49, 199–203.

69 Eren, T.; Atar, N.; Yola, M. L.; Karimi-Maleh, H. *Food Chemistry* **2015**, 185, 430–436.

70 Lian, W.; Liu, S.; Yu, J.; Xing, X.; Li, J.; Cui, M.; Huang, J. *Biosensors and Bioelectronics* **2012**, 38, 163–169.

71 Mehdinia, A.; Aziz-Zanjani, M. O.; Ahmadifar, M.; Jabbari, A. *Biosensors and Bioelectronics* **2013**, 39, 88–93.

72 Wang, F.; Zhu, L.; Zhang, J. *Sensors and Actuators B: Chemical* **2014**, 192, 642–647.

73 Altintas, Z.; Gittens, M.; Guerreiro, A.; Thompson, K.-A.; Walker, J.; Piletsky, S.; Tothill, I. E. *Analytical Chemistry* **2015**, 87, 6801–6807.

74 Roy, A. C.; Nisha, V. S.; Dhand, C.; Ali, Md. A.; Malhotra, B. D. *Analytica Chimica Acta* **2013**, 777, 63–71.

75 Deng, P.; Xu, Z.; Kuang, Y. *Food Chemistry* **2014**, 157, 490–497.

76 Yang, Y.; Fang, G.; Liu, G.; Pan, M.; Wang, X.; Kong, L.; He, X.; Wang, S. *Biosensors and Bioelectronics* **2013**, 47, 475–481.

77 Fatoni, A.; Numnuam, A.; Kanatharana, P.; Limbut, W.; Thavarungkul, P. *Analyst* **2014**, 139, 6160–6167.

78 Chen, D.; Deng, J.; Liang, J.; Xie, J.; Hu, C.; Huang, K. *Sensors and Actuators B: Chemical* **2013**, 183, 594–600.

79 Atar, N.; Eren, T.; Yola, M. L. *Food Chemistry* **2015**, 184, 7–11.

80 Karimian, N.; Vagin, M.; Zavar, M. H. A.; Chamsaz, M.; Turner, A. P. F.; Tiwari, A. *Biosensors and Bioelectronics* **2013**, 50, 492–498.

81 Xue, X.; Wei, Q.; Wu, D.; Li, H.; Zhang, Y.; Feng, R.; Du, B. *Electrochimica Acta* **2014**, 116, 366–371.

82 Feng, S.; Gao, F.; Chen, Z.; Grant, E.; Kitts, D. D.; Wang, S.; Lu, X. *Journal of Agricultural and Food Chemistry* **2013**, 61, 10467–10475.

83 Tiwari, A.; Deshpande, S. R.; Kobayashi, H.; Turner, A. P. F. *Biosensors and Bioelectronics* **2012**, 35, 224–229.

84 Uygun, Z. O.; Dilgin, Y. *Sensors and Actuators B: Chemical* **2013**, 188, 78–84.

85 Alizadeh, T.; Ganjali, M. R.; Zare, M.; Norouzi, P. *Food Chemistry* **2012**, 130, 1108–1114.

86 Zhao, L.; Zhao, F.; Zeng, B. *Sensors and Actuators B: Chemical* **2013**, 176, 818–824.

87 Zeng, Y.; Zhou, Y.; Kong, L.; Zhou, T.; Shi, G. *Biosensors and Bioelectronics* **2013**, 45, 25–33.

88 Li, J.; Li, Y.; Zhang, Y.; Wei, G. *Analytical Chemistry* **2012**, 84, 1888–1893.
89 Diamond, D. *Principles of Chemical and Biological Sensors*; John Wiley & Sons, Inc, New York: **1998**, p. 1–18.
90 Sheehan, P. E.; Whitman, L. J. *Nano Letters* **2005**, 5, 803–807.
91 Alivisatos, P. *Nature Biotechnology* **2004**, 22, 47–52.
92 Niemeyer, C. M. *Angewandte Chemie, International Edition* **2001**, 40, 4128–4158.
93 West, J. L.; Halas, N. J. *Current Opinion in Biotechnology* **2000**, 11, 215–217.
94 Parak, W. J.; Gerion, D.; Pellegrino, T.; Zanchet, D.; Micheel, C.; Williams, S. C.; Boudreau, R.; Le Gros, M. A.; Larabell, C. A.; Alivisatos, A.P. *Nanotechnology* **2003**, 14, R15–R27.
95 Agasti, S. S.; Rana, S.; Park, M.-H.; Kim, C. K.; You, C.-C.; Rotello, V. M. *Advanced Drug Delivery Reviews* **2010**, 62, 316–328.
96 Canfarotta, F.; Whitecombe, M. J.; Piletsky, S. A. *Biotechnology Advances* **2013**, 31, 1585–1599.
97 Tiwari, M. P.; Prasad, A. *Analytica Chimica Acta* **2015**, 853, 1–18.
98 Ma, Y.; Xu, S.; Wang, S.; Wang, L. *Trends in Analytical Chemistry* **2015**, 67, 209–216.
99 Wang, Y.; Zhang, Z.; Jain, V.; Yi, J.; Mueller, S.; Sokolov, J.; Liu, Z.; Levon, K.; Rigas, B.; Rafailovich, M. H. *Sensors and Actuators B: Chemical* **2010**, 146, 381–387.
100 Lu, C.-H.; Zhang, Y.; Tang, S.-F.; Fang, Z.-B.; Yang, H.-H.; Chen, X.; Chen, G.-N. *Biosensors and Bioelectronics* **2012**, 31, 439–444.
101 Matsui, J.; Akamatsu, K.; Nishiguchi, S.; Miyoshi, D.; Nawafune, H.; Tamaki, K.; Sugimoto, N. *Analytical Chemistry* **2004**, 76, 1310–1315.
102 Le, M.-H.; Thomas, J. L.; Ho, M.-H. Yuan, C.; Lin, H.-Y. *Applied Materials and Interfaces* **2010**, 2, 1729–1736.
103 Ge, L.; Wang, S.; Yu, J.; Li, N.; Ge, S.; Yan, M. *Advanced Functional Materials* **2013**, 23, 3115–3123.
104 Viswanathan, S.; Rani, C.; Ribeiro, S.; Delerue-Matos, C. *Biosensors and Bioelectronics* **2012**, 33, 179–183.
105 Prasad, B. B.; Pandey, I. *Electrochimica Acta* **2013**, 88, 24–34.
106 Gültekin, A.; Ersöz, A.; Denizli, A.; Say, R. *Sensors and Actuators B: Chemical* **2012**, 162, 153–158.
107 Gültekin, A.; Ersöz, A.; Hür, D.; Sarıözlü, N. Y.; Denizli, A.; Say, R. *Applied Surface Science* **2009**, 256, 142–148.
108 Patra, S.; Roy, E.; Madhuri, R.; Sharma, P. K. *Analytica Chimica Acta* **2015**, 853, 271–284.
109 Tong, Y.; Li, H.; Guan, H.; Zhao, J.; Majeed, S.; Anjum, S.; Liang, F.; Xu, G. *Biosensors and Bioelectronics* **2013**, 47, 553–558.
110 Altintas, Z. Molecular imprinting technology in advanced biosensors for diagnostics. In *Advances in Biosensors Research*; Nova Science Publishers Inc, New York: **2015**, p. 1–30.

111 Altintas, Z. Advanced imprinted materials for virus monitoring. In *Advanced Molecularly Imprinting Materials*; Wiley-Scrivener Publishing LLC, Beverly: **2016**, p. 389–412.

112 Altintas, Z.; Pocock, J.; Thompson, K.-A.; Tothill, I. E. *Biosensors and Bioelectronics* **2015**, 74, 994–1004.

113 Altintas, Z. Endotoxin monitoring using nanomaterials. In *Advanced Environmental Analysis: Applications of Nanomaterials*; Royal Society of Chemistry, Cambridge: **2016**, p. 91–107.

114 Abdin, M. J.; Altintas, Z.; Tothill, I. E. *Biosensors and Bioelectronics* **2015**, 67, 177–183.

115 Altintas, Z.; Abdin, M. J.; Tothill, A. M.; Karim, K.; Tothill, I. E. *Analytica Chimica Acta* **2016**, 935, 239–248.

11

Smart Nanomaterials

Applications in Biosensors and Diagnostics

Frank Davis[1], Flavio M. Shimizu[2], and Zeynep Altintas[3]

[1] Department of Engineering and Applied Design, University of Chichester, Chichester, UK
[2] São Carlos Institute of Physics (IFSC), University of São Paulo (USP), São Carlos, Brazil
[3] Technical University of Berlin, Berlin, Germany

11.1   Introduction

Recently there has been a great deal of interest in smart nanomaterials. A smart material is defined as "Materials that can significantly change their mechanical properties (such as shape, stiffness, and viscosity), or their thermal, optical, or electromagnetic properties, in a predictable or controllable manner in response to their environment. Materials that perform sensing and actuating functions, including piezoelectrics, electrostrictors, magnetostrictors, and shape-memory alloys" [1]. An example of a smart material is the cyanobiphenyl compounds used in LCD displays, where the molecules form a liquid crystal phase that can be aligned by an electrical field. A commonly used definition of nanomaterial is "material with any external dimension in the nanoscale or having internal structure in the nanoscale, where nanoscale is, in turn, defined as: size range from approximately 1 nm to 100 nm" [2]. These include such materials as metallic, inorganic, or carbon nanoparticles, nanotubes, and nanowires (NWs) as well as planar systems such as graphene or metal chalcogenides.

Biosensors have been a widely studied field of research since Leland Clark developed the first enzyme biosensors in the 1960s. A biosensor can be thought of as a device in which a biological receptor element such as an enzyme or antibody is coupled to a transducer. This allows conversion of a biological interaction into a measurable physical signal. The transducer can be an electrode, a piezoelectric crystal, or an optical detection method such as a surface plasmon resonance (SPR) chip. Figure 11.1 shows a schematic of a simple biosensor. The Clark glucose electrode, for example [3], immobilizes glucose

Biosensors and Nanotechnology: Applications in Health Care Diagnostics, First Edition.
Edited by Zeynep Altintas.
© 2018 John Wiley & Sons, Inc. Published 2018 by John Wiley & Sons, Inc.

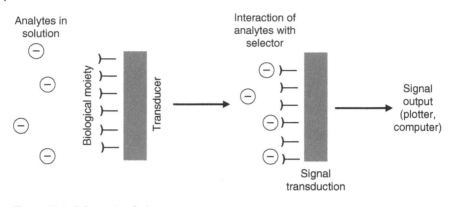

Figure 11.1 Schematic of a biosensor.

oxidase onto a platinum electrode. In the presence of glucose, the enzyme converts it into gluconolactone, consuming oxygen and generating hydrogen peroxide, both of which can be detected electrochemically. Other species that have been used are antibodies, which can be immobilized onto a sensor surface and selectively bind their antigens. This can result in changes of the nature of the surface, for example, binding of the antigen will lead to a mass change (detectable by a piezoelectric transducer), a change in the electrical properties of the surface that can be measured by a number of electrochemical methods, or a thickness change (or refractive index change) measurable by SPR. Biosensors often display advantages in speed and sensitivity over classical methods of measurement, which can be miniaturized, often display lower costs, and do not require skilled laboratory personnel, allowing home use. This is exemplified by the most common of these devices, the electrochemical glucose biosensor, used for blood glucose monitoring for diabetes patients [4, 5]. The use of biosensing technology at home or at a doctor's surgery may allow the rapid measurement of analytes of clinical significance toward various disease markers, without need for waiting for samples to be sent to laboratories, permitting earlier intervention that is frequently of utmost importance. The small size and power requirements of biosensors also potentially allow them to be incorporated into the human body, allowing for continuous monitoring of, for example, glucose. Many reviews have been written on biosensing technology, some of which are given here [4–11].

Nanotechnology has become a major field of research in recent times and the biosensing field is no exception. It is thought that utilizing nanotechnology, especially the use of nanosized materials, could greatly improve the performance of a wide number of biosensor devices. One major advantage will be using nanosized materials and components that allow size reductions in the sensors, making it possible to analyze smaller samples. Invasive methods are required for blood sampling and the smaller the required sample, the less painful these are. Nanomaterials will also increase the surface area of sensing devices such as electrodes, reducing diffusion times and enhancing sensitivities. If reductions in size are great enough, it then becomes possible to measure biological events occurring within single cells.

The purpose of the nanomaterials used in many of these studies is to amplify and transfer a signal resulting from a binding event. Since nanomaterials are of similar sizes to the sensing molecules utilized in biosensors, individual nanomaterials such as metal nanoparticles or nanoelectrodes can be bound to just one or a few sensing recognition elements, which allow for increases in sensitivity. Similarly the small size of the nanomaterial allows radial diffusion, again enhancing sensitivity and lowering detection limits [12].

Biosensors contain recognition elements such as enzymes or antibodies that need to be immobilized onto and interact with a transducer. One issue is that it is difficult to convert the recognition event into a physical change. Nanomaterials can help address this problem; due to their small size, they are comparable in dimensions with the biological molecule. This allows them to interact intimately with the biomolecule, for example, in the case of enzymes, it has proved possible to directly "wire" the enzyme to the electrode surface, enhancing electron transfer and removing the need for mediators. The purpose of a smart nanomaterial could be said to amplify the binding event using a novel property of the materials, for example, a conductive NW could change its conductivity or a quantum dot (QD) could change its fluorescence signature. Also many nanomaterials have novel catalytic effects, enhancing the detection of species such as hydrogen peroxide, produced by oxidase enzymes. There have been a number of reviews and books on the application of nanomaterials and nanomedicine within medical and sensing fields [12–17].

Since the use of nanomaterials has such beneficial effects on biosensor performance, a wide range of different nanomaterials have been studied. Describing all these studies is well beyond the scope of this chapter; therefore we will provide an overview of some of the most commonly utilized such as metal and magnetic nanoparticles (MNPs), carbon nanotubes (CNTs), and other carbon nanostructures, such as graphene and inorganic and conducting polymer NWs.

11.2 Metal Nanoparticles

Metal nanoparticles have been utilized for many years; for example, the red and yellow colors in medieval stained glass could be generated by incorporating gold or silver salts into the glass mix. Medieval glassmakers were unknowingly generating metal nanoparticles within a glass matrix. The chemical, optical, and electronic properties of metal nanoparticles can vary greatly from those of the bulk materials and depend greatly on the size, shape, and substitution of the particle. The biosensor field has utilized a number of metal nanoparticles; however, the field is dominated by gold nanoparticles (AuNPs). AuNPs can be synthesized in a wide variety of sizes and shapes using chemical and electrochemical methods. Gold is one of the most electrically conductive, chemically stable metals and does not easily oxidize or react with water. Also it can be chemically substituted by a number of methods, the foremost being its reaction with sulfur compounds such as thiols due to the formation of a strong Au—SH bond, which allows facile immobilization of a range of biological molecules onto the gold surface under gentle reaction conditions. The optical properties of gold are highly dependent on the excitation of surface plasmons on the gold particle surface and are therefore very sensitive to any transitions occurring at the metal/environment boundary, meaning any recognition events can have large effects on the color. Also, for *in vivo* applications, the low toxicity of gold is much in its favor. There have been a range of reviews describing the use of metal nanoparticles in more detail [18–20].

Nanogold has been widely used within immunosensors, giving enhanced sensitivity. In a simple application, acetylcholinesterase (AChE) could be assayed by a method using AuNPs and acetylthiocholine [21]. The presence of AChE led to the production of thiocholine, which caused aggregation of the AuNPs and a redshift of their plasmon adsorption. The activities of AChE with a concentration as low as $0.6\,mU\,mL^{-1}$ could be assayed. This method could detect AChE inhibitors. The layer-by-layer assembly of AuNPs and an antibody for α-1-fetoprotein gave an immunosensor with a range of $1–250\,ng\,mL^{-1}$ and a detection limit of $0.56\,ng\,mL^{-1}$ [22]. Washing with $4\,M$ urea allowed this sensor to be reused up to eight times. Further work utilized 1,1′-bis-(2-mercapto)-4,4′-bipyridinium dibromide [23] to reduce the detection limit to $0.23\,ng\,mL^{-1}$. An AuNP-modified sol–gel electrode was used to immobilize anti-transferrin to give an immunosensor with a linear range of $1–75\,ng\,mL^{-1}$ and a detection limit of $0.05\,ng\,mL^{-1}$ compared with a detection limit of $1\,ng\,mL^{-1}$ for a system without AuNPs [24]. Citrate-modified AuNPs could be adsorbed onto a gold electrode and used to immobilize anti-IgG [25] to allow potentiometric determination of IgG with a detection limit of $12\,ng\,mL^{-1}$. Similarly AuNP/carbon paste electrodes were used to measure the electrochemistry of a CA19-9 antibody-horseradish peroxidase (HRP) conjugate in solution [26]. Addition of the antigen led to the formation of a complex and reduction of HRP activity, allowing detection of the antigen in human serum. Another anti-interleukin-6–HRP conjugate could be immobilized onto screen-printed graphite electrodes inside a flow cell [27]. Binding of the antigen led to HRP inhibition, allowing determination of interleukin-6 from 5 to $100\,pg\,mL^{-1}$ with a detection limit of $1.0\,pg\,mL^{-1}$. Other systems have also demonstrated increased sensitivity: Gold/nanoparticle/poly(terthiophene)/polymeric dendrimer-modified electrodes could detect $pg\,mL^{-1}$ levels of the lung cancer markers annexin II and MUC5AC [28]. An AuNP-/DNA-modified gold electrode modified with hepatitis B antibody was used in a sandwich-type immunoassay, low antigen levels being determined using a AuNP/HRP/secondary antibody conjugate [29]. Ding et al. used AuNP-tagged antibodies and a nanoporous gold electrode to produce an immunoassay for hepatitis B with a detection limit of 2.3 pg mL^{-1} and a 100-fold better sensitivity than an ELISA process [30].

Many of the sensors utilizing AuNPs are electrochemical in nature; however they can be used with many other methods. One common commercial "biosensor" is the lateral flow assay, where a sample of a fluid such as urine flows along an absorbent material and the presence of a target causes a visible color change, as typified by the common home-use pregnancy test [31]. Variants of these have been made, which include AuNPs [32]; for example, antibodies to pesticides could be immobilized onto AuNPs and then run in a competitive lateral flow assay on nitrocellulose strips where the intensity of the test line is inversely related to pesticide concentration. AuNPs could also be immobilized onto SPR chips and used as a substrate for antibodies to *Salmonella typhimurium* [33]. The chips could detect bacteria with 10-fold increased sensitivity over chips without the AuNP layer. A regenerable electrochemical sensor based on AuNP-modified glassy carbon electrodes (GCEs) substituted with *Salmonella* antibodies also could be constructed and could detect bacteria in the range of 100–10 000 CFU mL^{-1} [34]. Other optical sensing platforms have been designed using AuNPs; for example, 13 nm AuNPs were modified by gelatin and used as a sensor for proteinase activity determination [35]. Proteinase digests the gelatin coating, causing aggregation and a red to blue color shift. Matrix metalloproteinase-2 activity could be determined between 50 and 600 ng mL^{-1}. Extensive reviews of the use of gold and silver nanoparticles (AgNPs) in medical diagnostic applications have been published [17, 36].

AuNPs have also been used in the determination of common blood components such as dopamine, ascorbate, and ureate [37]. One major issue is that the three species have electrochemical peaks that overlap, making accurate determinations difficult. Although not strictly biosensors as they contain no biological recognition element, gold-modified electrodes have been shown to lead to improved peak separations; for example, a layer-by-layer film of AuNPs on a GCE led to enhanced dopamine detection [38]. Enhanced detection of ephedrine and ureate in the presence of excess ascorbate could be obtained using an AuNP-modified polypyrrole-coated GCE [39]; indium tin oxide electrodes when modified by AuNPs have also been shown to be suitable for the detection of ureate [40]. AuNPs could also be incorporated into PEDOT layers and used to detect dopamine and ureate in the presence of ascorbate [41]. A combined AuNP/dendrimer film (Figure 11.2) could be used to immobilize a redox mediator and glucose oxidase [42]. This gave a glucose sensor with a much wider linear range than the corresponding sensor without AuNPs. AuNPs have been shown to stabilize electrochemical glucose biosensors by strongly

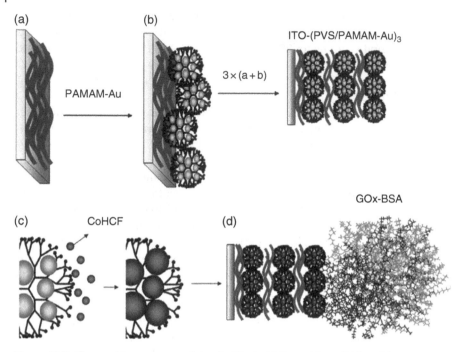

Figure 11.2 Glucose biosensor containing AuNPs. (a, b) Deposition of LbL multilayers containing polyvinyl sulfonate/PAMAM-Au. (c) After deposition of three bilayers, an ITO-(PVS/PAMAM-Au)$_3$@CoHCF electrode was prepared by potential cycling. (d) Immobilizing enzyme using bovine serum albumin (BSA), glutaraldehyde, and glucose oxidase (GOx). *Source:* Crespilho et al. [42]. Reproduced with permission of Elsevier. (*See insert for color representation of the figure.*)

binding the enzyme and preventing leakage from the sensor [43]. Researchers also studied glucose biosensors with and without AuNPs. They showed enhanced linear ranges and sensitivity and demonstrated application of AuNP-mediated electron transfer [44, 45].

AuNPs have been used in the detection of DNA sequences. Oligonucleotides can be detected by hybridization with their corresponding counterstrands. One issue with detection of DNA hybridization is that no electrons or easily detectable chemical products are produced by this, so it is often necessary to amplify the binding event using a smart material. A colorimetric assay was developed using AuNPs modified with thiol-substituted oligonucleotides. Mixing these with as little as 10 fmol of target nucleotides caused hybridization and formation of a network that led to a red to purple color change and a blue product on drying [46]. Heating the polymeric network allowed denaturation and determination of transition temperature, which allowed differentiation of close but not perfect matches. Other scientists modified AuNPs with aptamers to platelet-derived growth factors. Addition of the growth factor led to a red to purple color change [47]. Varying the AuNP size led to different linear ranges of detection (tens to hundreds $nmol\,L^{-1}$) and a lowest detection limit of $3.2\,nmol\,L^{-1}$. The use of metal nanoparticles in genome analysis has been reviewed [36].

Researchers assembled two types of AuNPs substituted with oligonucleotides. These were joined into aggregates by a linking DNA containing an aptamer for adenosine or cocaine [48]. Addition of the aptamers that bind to their target disrupted the linking DNA and rapidly destroyed the aggregates, leading to a purple to red color change. This allowed a dipstick sensor for these materials to be developed [49].

Gold has been a major area of interest in metal nanoparticle research; however other metals have also biosensing applications. AgNPs could be immobilized on electrodes and substituted with antibodies to human serum albumin, microcystin-LR, or penicillin G [50] to give impedimetric sensors with detection limits of $1\,amol\,L^{-1}$, $10\,fmol\,L^{-1}$, and $0.7\,fmol\,L^{-1}$, respectively. Corresponding limits without nanoparticles were $100\,amol\,L^{-1}$, $100\,fmol\,L^{-1}$, and $10\,fmol\,L^{-1}$. Similar results for human serum albumin were obtained with AuNPs. AgNPs have been used as plasmonic sensors. They not only are less costly but also adsorb less in the UV/visible region, thus performing better than gold [51]. Many of the applications of these nanoparticles in the field of biomedicine have been reviewed [51] so only a few will be presented here.

Silver nanotriangles could be deposited onto glass substrate and functionalized with biotin units. These could be used to detect streptavidin in $pmol\,L^{-1}$ to $fmol\,L^{-1}$ levels. Binding caused a redshift in the plasmonic adsorption [52]. Other research group utilized silver nanocubes in the same manner to measure protein adsorption [53], and silver rhomboids were used to measure $100\,nmol\,L^{-1}$ levels of various targets [54]. Toxicity of AgNPs is an issue for *in vivo* applications; however this can be mitigated somewhat by coating the AgNPs with a 2 nm silica coating [55]. These particles were shown to respond to bovine serum albumin (BSA) adsorption. AgNPs were also coated with a DNA aptamer to thrombin. Interaction with the target caused the nanoparticles to aggregate and enhanced their two-photon photoluminescence [56].

Platinum nanoparticle (PtNP)-modified electrodes were used to immobilize anti-IgG in a sandwich-type assay with an alkaline phosphatase-labeled secondary antibody [57]. This antibody then catalyzed copper deposition onto the electrode that inhibited its use in electrochemical hydrogen evolution. IgG could be determined at levels as low as $2\,pg\,mL^{-1}$. Antibodies to prostate-specific antigen were immobilized on a gold electrode, and after the binding of the target treated with PtNP/secondary antibody conjugates, the PtNPs then catalyzed electrochemical hydrogen evolution [58]. Limits of detection of the antigen as low as $1\,fg\,mL^{-1}$ were obtained. PtNP-modified electrodes could also be used in the detection of *Escherichia coli* with a range of 50–$10,000\,CFU\,mL^{-1}$ and a detection limit of $20\,CFU\,mL^{-1}$ [59].

PtNPs could be added to ordered mesoporous carbon to give a composite electrode, which was used as a support for a polypyrrole/glucose oxidase film provided a sensor with a linear range of 0.05–3.70 mmol L^{-1} glucose and was applicable to blood analysis [60]. The combination of PtNPs, silicon nanoparticles, and glucose oxidase resulted in a glucose biosensor. There was catalytic enhancement of H$_2$O$_2$ reduction by the PtNPs, which supplied a linear range as wide as 1×10^{-6} to 2.6×10^{-2} mol L^{-1} glucose [61]. Both platinum and palladium nanoparticles were used along with exfoliated graphite to obtain glucose biosensors [62]. The PtNP-based sensors were shown to be superior and had a linear range up to 2×10^{-2} mol L^{-1} and a detection limit of 1×10^{-6} mol L^{-1}.

Other metals have been used in the direct detection of glucose without the use of enzymes, technically not a biosensor since they contain no biorecognition element but are still worthy of mention. Much of the work on these systems has been reviewed previously [63]. Bimetallic PtRu, PtPd, and PtAu nanoparticles were deposited on a multiwalled CNT (MWCNT) electrode using an ultrasonically assisted electrodeposition process and used for the oxidation of glucose [64]. The best material was the PtRu alloy, which had the largest surface area and highest catalytic ability, allowing linear detection of glucose up to 15 mmol L^{-1} with a detection limit of 0.05 mmol L^{-1} and minimal interference from ascorbic acid, uric acid, acetaminophen, and fructose. A range of other electrochemical sensors based on palladium nanoparticles have been produced [65, 66], as a sensor using palladium nanotubes [67, 68]. Li et al. used a nanoparticulate nickel/copper alloy to produce a nonenzymatic glucose sensor with good performance and minimal response to common interferents [69].

11.3 Magnetic Nanoparticles

In recent years there has been a considerable effort to develop MNPs for biosensing applications. MNPs display several advantages such as low cost of production. They can be synthesized in a wide range of sizes and easily substituted. Most interest appears to be in MNPs with a size range of 10–20 nm since they are superparamagnetic, meaning they respond very quickly to magnetic fields [70]. MNPs can be synthesized by a wide range of methods such as deposition from gas phase and electron beam lithography; by a range of traditional wet chemistry methods such as precipitation or decomposition of soluble organic precursors, oxidation, and synthesis using colloids and other nanoreactors; or microbially. MNPs can be made in a wide variety of formats. They can be metals (iron, cobalt, nickel, or alloys), iron oxides, and related ferrites along with composites of these materials with other inorganics or polymers. Bare MNPs often rapidly agglomerate and thus need to be coated. A number of coatings have been used including surfactants, silica, polymers, and gold. A large number of reviews on the synthesis and uses of these MNPs have been published [70–76]. Because of their size and ability to be easily conjugated to biological molecules, there has been much recent interest for using these in biological and biosensing fields [77–80].

One of the simplest applications of MNPs is to use electromagnets to control their location, for example, to pull them against a sensor surface rather than just relying on diffusion. For example, MNPs labeled with antibodies to a variety of targets can be mixed in various media (serum, saliva, etc.) and bound to their antigen [81]. One electromagnet is used to pull them against a sensing surface (SPR chip) containing a secondary antibody, and then a second electromagnet is used to remove unbound MNPs. Detection of drugs at less than $1\,\text{ng}\,\text{mL}^{-1}$ and troponin below $1\,\text{fg}\,\text{mL}^{-1}$ could be attained in one to a few minutes. The use of magnetic beads in this format has been reviewed elsewhere [82], and detection of a wide variety of species described. This combination of MNPs with a wide variety of optical techniques has been listed elsewhere [80], so a few examples will be given here.

The most common optical technique used with MNPs is SPR. MNPs containing antibodies to ochratoxin could be immobilized onto a gold surface using a magnetic field and were then shown to be capable of detecting the antigen via SPR between 1 and $50\,\text{ng}\,\text{mL}^{-1}$ with a detection limit of $0.94\,\text{ng}\,\text{mL}^{-1}$ [83] or by AC impedance. In another work, an SPR chip was modified by antibodies to β-human chorionic gonadotropin and exposed to the antigen and then MNPs were labeled with a detection antibody [84]; increases in sensitivity by four orders of magnitude compared with direct detection without MNPs were observed (Figure 11.3). An immunosensor for α-fetoprotein could also be constructed using gold-encapsulated Fe_3O_4 MNPs again in a sandwich assay format [85]. SPR measurements gave a linear range of $1.0–200.0\,\text{ng}\,\text{mL}^{-1}$ and a detection limit of $0.65\,\text{ng}\,\text{mL}^{-1}$. Similar assays using aptamers for thrombin [86] gave an immunosensor with a detection limit of $0.017\,\text{nmol}\,\text{L}^{-1}$. Wang et al. used SPR and Fe_3O_4 coated with SiO_2 nanoparticles to detect IgG in the range of $1.25–20\,\mu\text{g}\,\text{mL}^{-1}$. Using $Fe_3O_4/Ag/SiO_2$ nanoparticles lowered the range to $0.30–20\,\mu\text{g}\,\text{mL}^{-1}$ [87]. Nanoparticulate composites of Ag-/Au-coated Fe_3O_4 on a gold electrode could be used to detect IgG. Using the nanocomposites lowered the detection limit by an order of magnitude [88]. Similar results were obtained using Au/Fe_3O_4 nanorods [89]. Agrawal et al. developed a polydimethylsiloxane-based chip containing a magnet, which was combined with MNPs could be used to detect *E. coli* [90].

Combining MNPs with electrochemical sensors has been widely used. For example, Fe_3O_4 nanoparticles coated with gold colloids were used to immobilize AChE [91]. These composites could be magnetically adsorbed onto a suitable carbon-based electrode and utilized to measure the hydrolysis of acetylthiocholine. Exposure to the pesticide dimethoate reduced enzyme activity, allowing the pesticide to be quantified in the range $0.001–10\,\text{ng}\,\text{mL}^{-1}$ with a detection limit of $5.6 \times 10^{-4}\,\text{ng}\,\text{mL}^{-1}$. The electrode could be regenerated

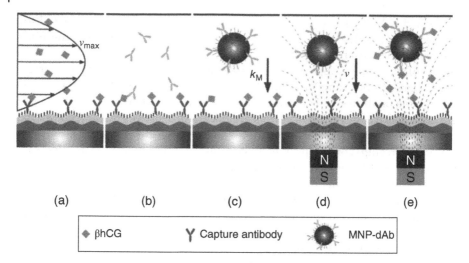

Figure 11.3 Schematics of used detection formats: direct detection (a), sandwich assays with amplification by detection antibody (b), and MNP-dAb without (c) and with (d) applied magnetic field. Detection format consisting of preincubating MNP-dAb with βhCG followed by sandwich assay upon applied magnetic field gradient (e). *Source:* Wang et al. [84]. Reproduced with permission of American Chemical Society. (*See insert for color representation of the figure.*)

by removing used enzyme/MNP composite with a magnetic field. Similar Au/Fe$_3$O$_4$ nanoparticles coated with clenbuterol/BSA conjugates could be adsorbed onto electrodes and used in a competitive electrochemical immunoassay to determine the levels of the drug (range of 0.5–200 ng mL^{-1} and detection limit of 0.22 ng mL^{-1}) [92]. These types of MNPs were coated with antibodies to carcinoembryonic antigen (CEA) [93] and then magnetically immobilized onto a carbon paste electrode to obtain a voltammetric sensor for the antigen (range of 0.005–50 ng mL^{-1} and detection limit of 0.001 ng mL^{-1}), approximately 500 times more sensitive than an ELISA assay. MNPs modified with antibodies to zearalenone could be exposed to HRP antigen and then confined by a magnetic field onto an electrode, and the electrochemical activity of the enzyme was determined; levels of the mycotoxin as low as 7 ng mL^{-1} could be measured [94].

Fe$_3$O$_4$/SiO$_2$ MNPs was confined onto carbon paste electrodes and used to amperometrically determine uric acid (range 0.60–100.0 μmol mL^{-1}, detection limit 0.013 μmol mL^{-1}) [95]. Glucose sensing is also possible with these systems. Fe$_3$O$_4$/SiO$_2$ MNPs were combined with glucose oxidase and HRP and confined on an indium tin oxide electrode to give an amperometric glucose sensor [96] (ranges 0.05–1 and 1–8 mmol mL^{-1}, detection limit 0.01 mmol mL^{-1}) with good stability and with minimal effects from interferents and were suitable for use with human serum. Similar MNPs were combined with MWCNTs and glucose oxidase to allow glucose determination (range $1-30 \, mmol \, mL^{-1}$, detection limit $0.8 \, mmol \, mL^{-1}$) [97], whereas Fe_3O_4 MNPs could be used in a composite with glucose oxidase and polypyrrole [98], magnetically immobilized onto glassy carbon and used to determine glucose potentiometrically in serum (range $0.5-34 \, mmol \, mL^{-1}$, detection limit $0.3 \, mmol \, mL^{-1}$) with high selectivity and good stability (98% performance after 20 days).

Electrochemiluminescence (ECL) wherein light is emitted as a result of electrochemical reactions has been used as a highly sensitive method for the detection of a wide variety of species as recently reviewed [99]. MNPs have been combined with this method; for example, MNPs could be substituted with antibodies to α-fetoprotein and combined in solution first with the antigen and then with a CdS–Au QD substituted with a secondary antibody [100]. A magnetic field was used to immobilize the final composite onto an electrode and ECL generated at a fixed voltage, allowing determination of the antigen with a range from 0.0005 to $5.0 \, ng \, mL^{-1}$ and a detection limit of $0.2 \, pg \, mL^{-1}$. Au/Fe_3O_4 MNPs could be used in a sandwich-type immunosensor along with glucose oxidase and luminol to allow determination of *Bacillus thuringiensis* from 0 to $6 \, ng \, mL^{-1}$ with a detection limit of $0.25 \, pg \, mL^{-1}$ [101].

It has also proved possible to deposit MNPs onto piezoelectric quartz crystal microbalance (QCM) crystals and thereby detect mass changes [80]. A variety of immobilization and amplification methods have been used; for example, the H5N1 virus could be detected using a sandwich-type immunoassay on a QCM crystal [102]. Vancomycin-functionalized MNPs could be attached by a magnetic field to QCM crystal and detect vancomycin-sensitive *Desulfotomaculum* but was unaffected by vancomycin-resistant *Vibrio anguillarum* [103]. MNPs labeled with anti-C reactive protein could be immobilized onto QCM crystals, exposed to antigen, then a HRP-labeled secondary antibody/gold colloid conjugate, and finally the enzyme used to catalyze the oxidation of 3-amino-9-ethylcarbazole, which gave an insoluble product, the mass of which could be determined [104]. The resultant sensor could detect antigen between 0.001 and $100 \, ng \, mL^{-1}$ with a low detection limit of $0.3 \, pg \, mL^{-1}$. A similar method using an $AuFe_3O_4$ MNP/CNT composite and HRP/3-amino-9-ethylcarbazole enhancement was used to determine myoglobin (investigation range: $0.001-5 \, ng \, mL^{-1}$, detection limit: $0.3 \, pg \, mL^{-1}$) [105].

There have been recent studies utilizing the giant magnetoresistive (GMR) effect of layered nonmagnetic and ferromagnetic materials such as MNPs in sensing applications as reviewed here [106]. A competitive assay using a GMR sensor functionalized with antibodies to endoglin was exposed to the antigen along with biotin-labeled endoglin [107]. Further exposure to streptavidin followed by biotin-labeled MNPs led to an assay that could detect the antigen at levels as low as $83 \, fg \, mL^{-1}$ in unprocessed human urine. A similar method used FeCo MNP/antigen conjugates in a competitive assay to determine interleukin-6 in human serum [108].

The Hall effect where voltage differences are generated across an electrical conductor in response to a magnetic field can also be used. This has led to the development of a microfluidic chip that could detect single cells that had been tagged with MNPs, allowing determination of tumor cells in whole blood from cancer patients with higher sensitivity than clinical tests [109]. The same group adapted this technology to determine Gram-positive bacteria [110]. A new field of interest using MNPs is in methods such as diagnostic magnetic resonance (DMR). One of the major advances in medicine has been the development of magnetic resonance imaging (MRI) where it is possible to obtain high-resolution images of internal joints, organs, and so on, without the necessity for high-energy radiation such as X-rays or contrast agents such as barium. This has enabled advances in diagnosis and monitoring of a wide range of conditions. DMR can be thought of as an offshoot of this where nanoparticles can be exposed to magnetic field and their properties measured [111]. These MNPs affect their surroundings, causing changes in the relaxation times of the water molecules surrounding them, which can be measured. Since MNPs such as Fe_3O_4 displayed good biocompatibility and low toxicity, they can potentially be used for *in vivo* measurements as well as on laboratory samples. Again, this is a very wide-ranging field and therefore just a number of examples will be given.

Two types of MNPs could be prepared by substituting the parent MNP with 12-mer oligonucleotides that "recognized" opposing ends of a 24-mer oligonucleotide target. Addition of the target to a mixture of the two MNPs led to aggregation. This caused a decrease in the relaxation time [112], allowing detection of sub-femtomole levels of target. Heating the samples reversed the aggregation and the presence of single mismatches in the target could be determined. In another work, a mixture of MNPs substituted with two different antibodies that bind to different areas of human chorionic gonadotropin was combined. Addition of the antigen caused cross-linking and changes in relaxation times. Levels of target from 0.1 to 1 molecules of analyte per nanoparticle could be determined [113].

Ferrite-coated iron MNPs can also be synthesized and coated with antibodies that bind specifically to pathogenic bacteria. Thousands of MNPs bind to each bacterium, rendering the bacteria superparamagnetic, thereby reducing the relaxation time of billions of surrounding water molecules and allowing their determination by NMR [114]. Classical MRI scanners have to be large enough to accommodate a human; however within this work a miniaturized chip containing microfluids and a microcoil could be used to concentrate and determine the target bacteria (Figure 11.4). In another work by this group, telomerase activity could be assessed. The use of MNPs combined with MRI allowed hundreds of samples to be processed in 384-well plates in tens of minutes and with ultrahigh sensitivity [115]. The same group has used DMR to analyze for a variety of substrates including DNA-cleaving agents [116], protease activity [117], enantiomeric impurities [118], and viruses [119]. A much more detailed review of this work has been published [111].

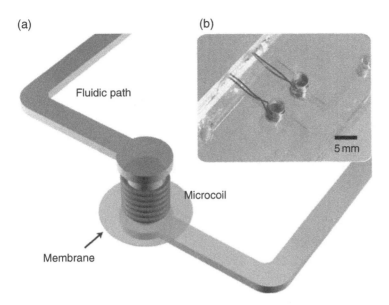

(a)

Fluidic path

Microcoil

Membrane

(b)

5 mm

Figure 11.4 NMR-filter system for bacterial concentration and detection. (a) The system consists of a microcoil and a membrane filter integrated with a microfluidic channel. The microcoil is used for NMR measurements; the membrane filter concentrates bacteria inside the NMR detection chamber to achieve high detection sensitivity. (b) A prototype device with two measurement sites. The NMR detection volume was approximately 1 µL. *Source:* Lee et al. [114]. Reproduced with permission from John Wiley & Sons.

11.4 Carbon Nanotubes

CNTs were discovered in 1991 by Iijima [120], and since then, a variety of CNTs have been developed—from graphene sheets, to single-walled CNTs (SWCNTs) (diameter varying between 0.4 and 2 nm), to double-walled CNTs (DWCNTs), to MWCNTs (outer diameter ranges from 2 to 100 nm [121])—and are now being produced in substantial quantities for various commercial applications [122]. Figure 11.5 depicts the formation of SWCNT and MWCNT structures along with that of graphene.

Depending on the graphene sheet winding direction, SWCNT may have zigzag, armchair, and chiral configurations [123]. Moreover, MWCNTs can exhibit metallic or semiconductive electronic configurations depending on the outer diameter range [124]. Over the last decade CNTs have made new breakthroughs especially in the domain of nanomedicine and biosensing [125], most of which are based on electrochemical detection. CNTs present advantageous properties compared with other nanomaterials for biosensor applications as described by Yang et al. [126]:

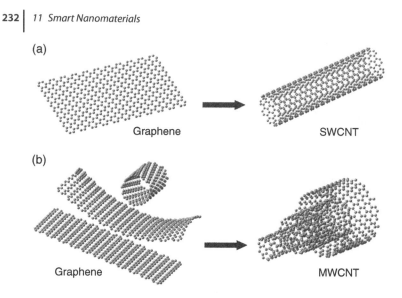

Figure 11.5 Schematic of (a) graphene and single-walled carbon nanotube (SWCNT) and (b) few layer graphene and multiwalled carbon nanotube (MWCNT) structures. *Source:* Vidu et al. [123]. https://www.ncbi.nlm.nih.gov/pmc/articles/PMC4064704/figure/F2/. Used under CC BY.

1) Sensitivity enhancement due to the large surface area. Enzyme immobilization on CNTs has been reported successfully [127], which maintains high biological activity.
2) Fast response time. CNTs have an outstanding ability to mediate fast electron transfer kinetics, hence promoting electron transfer reactions with species such as NADH and hydrogen peroxide [128].
3) Lower redox potential reaction and less surface fouling effects.
4) High stability and longer lifetime.

However their utilization can be problematic because they are insoluble in most solvents. Increasing the solubility of CNTs by functionalization with other nanomaterials, for example, polymers, metal nanoparticles, surfactants, and so on, has been successfully achieved [126].

The detection of small-molecule analytes such as glucose and cholesterol [129] based on nonenzymatic electrochemical biosensors (NEEB) has been widely studied on nanostructured metal oxides (NMOs) [130, 131], graphene [132], and also CNT electrodes [133]. Ye et al. [134] synthesized well-aligned MWCNTs (their scanning electron microscopy (SEM) images are depicted in Figure 11.6a) for nonenzymatic glucose detection using amperometric measurements utilizing GCE and MWCNT electrodes (Figure 11.6b). As expected GCE shows no response to the addition of glucose. In contrast, MWCNT electrode responds rapidly, reaching steady-state signals within 10 s. Experiments demonstrated that a glucose concentration range of $2.0 \, \mu mol \, L^{-1}$

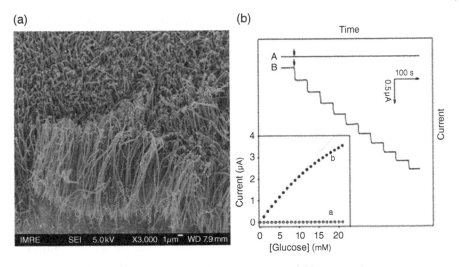

Figure 11.6 (a) SEM image of the well-aligned MWCNTs and (b) current–time response obtained on increasing the glucose concentration from 2.0 μmol L^{-1} to 11 mmol L^{-1} at (A) GCE and (B) MWCNT electrodes. Shown in the inset is the dependence of the current response versus the concentration of glucose at (a) GCE and (b) MWCNT electrodes. *Source:* Ye et al. [134]. Reproduced with permission from Elsevier.

to 11 mmol L^{-1} provided a sensitivity of 4.36 μA cm^{-2} mmol L^{-1} and a detection limit of 1.0 μmol L^{-1}. However interfering compounds (ascorbic and uric acids at 0.1 mmol L^{-1}) produced an increase on current response of 27.0–28.5%.

According to Tian et al. [132], most of the CNTs based on NEEB are functionalized by combining metal (Pt, Au, Cu) or metal oxide (MnO$_2$, NiO, CuO) nanostructured materials [135]. Ensafi et al. [136] developed functionalized MWCNTs decorated with AgNPs on amine chains (AgNPs/F-MWCNTs) NEEB to detect glucose. Hydrodynamic chronoamperometry was used as a detection method, achieving a detection limit of 0.03 μmol L^{-1} and a high sensitivity of 1057.3 mA mmol L^{-1}. In this case interfering compounds such as ascorbic acid (0.07 mmol L^{-1}), dopamine (0.07 mmol L^{-1}), uric acid (0.07 mmol L^{-1}), sucrose (0.70 mmol L^{-1}), and fructose (0.70 mmol L^{-1}) were evaluated. Sucrose and fructose showed insignificant increment on current, but the first three compounds presented an easy increment on oxidation signal; however since signal interference was much lower than the glucose levels, this was solved by diluting the sample to the typical concentrations of the species in human plasma (one thirtieth of those assessed) [135, 137].

Saha et al. [133] report the synthesis of CNTs from coconut oil for cholesterol detection based on NEEB using differential pulse voltammetry. They achieved a sensitivity of approximately 15.31 ± 0.01 μA μmol L^{-1} cm^{-2}, detection limit of 0.017 μmol L^{-1}, and response time of about 6 s. Reproducibility experiments provided a relative standard deviation (RSD) of 3.5% with the sensor retaining approximately 97% of initial current response after 10 days. Interference studies showed small average signal changes of approximately 1.3% using sample solutions of $10^{-5}\,mol\,L^{-1}$ cholesterol, spiked with various amounts of urea, uric acid, glucose, vitamin C, L-alanine, glycine, L-serine, L-phenylalanine, tryptophan, and tyrosine under the same experimental conditions.

Label-free aptamer and immuno-based biosensors for protein assays represent a promising field of biomedical research for noninvasive clinical diagnosis, because biosensor devices simplify the operational procedure, eliminate the labor-intensive labeling steps, avoid handling radioactive compounds, and allow real-time measurements [138, 139]. In this way the label-free detection of salivary α-amylase (SAA) using an aligned carboxylated SWCNT chemiresistor immunosensor utilizing a field-effect transistor (FET) sensor was studied by Tlili et al. [140]. A sensitivity of $3 \times 10^{-4}\,\mu g\,mL^{-1}$ and a detection limit of $6\,\mu g\,mL^{-1}$ were obtained for the concentration range from $10\,\mu g\,mL^{-1}$ to $1\,mg\,mL^{-1}$ of SAA solution measured in phosphate buffer and a detection limit of $7.8\,\mu g\,mL^{-1}$ on artificial salivary solution; both these studies satisfy the clinical work range between 19 and $308\,U\,mL^{-1}$.

As mentioned before, CNTs for stable aggregates present poor solubility; however Zheng et al. [141] observed that SWCNTs are effectively dispersed in water by their sonication in the presence of single-stranded DNA (ssDNA) because of the π-stacking interaction between the aromatic bases and the CNTs. Shahrokhian et al. [138] employed MWCNTs as a conductive transducer for electrochemically monitoring DNA hybridization of hairpin oligonucleotides. This is a signal-off device, that is, the signal is enhanced by the MWCNTs and decreases in the presence of the target. A linear correlation was observed between $10\,pmol\,L^{-1}$ and $0.1\,\mu mol\,L^{-1}$, and the genosensor retained 85% of initial current response after 2 weeks' storage at 4°C.

An electrochemical aptasensor with a thrombin-binding aptamer (TBA) was developed by Park et al. [142] using SWCNTs cast onto a GCE. The TBA was immobilized on SWCNTs through π-stacking interaction, resulting in helical wrapping to the surface as described in Ref. [141]. In the presence of thrombin, the TBA binds with thrombin and the TBA concentration on the SWCNT surface decreases. The remaining amount of TBA can be analyzed by an electrochemical method without any labeling, because the guanine bases of the nucleic acid are measurable by electrochemical methods, as schematically depicted in Figure 11.7. The detection limit of the aptasensor was $10\,nmol\,L^{-1}$ for thrombin detection with a range from $10\,nmol\,L^{-1}$ to $100\,\mu mol\,L^{-1}$.

The detection of BCR/ABL gene (breakpoint cluster region gene and the cellular abl) is of great importance for an early diagnosis, a better prognosis of the disease [143], and an improvement for detecting minimal residual leukemia cells in chronic myelogenous leukemia (CML) patients, especially after undergoing

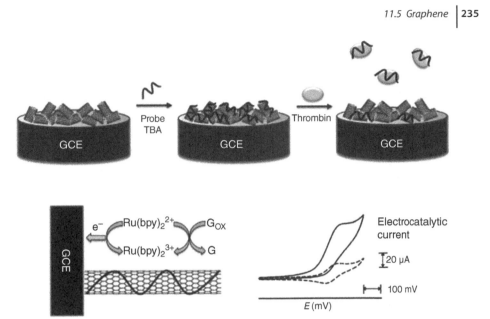

Figure 11.7 Schematic of a label-free electrochemical aptasensor for thrombin detection prepared on a single-walled carbon nanotube (SWCNT)-casted glassy carbon electrode (GCE) and the EC′ reaction mechanism. Inset: the electrocatalytic current of Ru(bpy)$_3^{2+}$ with (solid) and without (dashed) probe TBA that contains guanine. *Source:* Park et al. [142]. Reproduced with permission of John Wiley & Sons.

bone marrow transplantation [144]. To enhance the biosensor performance, Zhang [145] developed a novel label-free detection platform for the BCR/ABL fusion gene from CML based on hybrid Fe$_3$O$_4$ nanoparticle-functionalized CNT electrodes. Significant changes have been observed in the impedance spectra before and after hybridization of the probe ssDNA with the target DNA and with increases in the charge transfer resistance (R_{ct}) from approximately 13 to 27.5 kΩ (Figure 11.8). Under optimal conditions, the dynamic range for detecting the sequence-specific DNA of the BCR/ABL fusion gene was from 1.0×10^{-15} to 1.0×10^{-9} mol L^{-1}, and the detection limit was 2.1×10^{-16} mol L^{-1}. In addition, the DNA electrochemical biosensor was highly selective, being capable of discriminating single-base and double-base mismatched sequences. Stability experiments were carried out; storing the biosensor at 4°C for 7 days led to a decrease in impedance response of approximately 4.6%.

11.5 Graphene

One of the major recent advances within the field of chemo- and biosensors has been the incorporation of novel forms of carbon. A host of new carbon nanostructures such as fullerenes, CNTs, carbon dots, and graphene have been

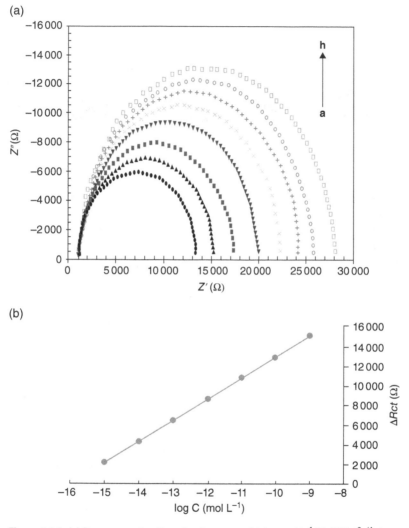

Figure 11.8 (a) Representative Nyquist diagrams of 1.0 mmol L^{-1} [Fe(CN)$_6$]$^{3-/4-}$ in 0.1 mol L^{-1} KCl recorded at a ssDNA/Fe$_3$O$_4$/CNTs/CPE and after the hybridization reaction with different concentrations of the BCR/ABL fusion gene target sequence: (**a**) ssDNA/Fe$_3$O$_4$/CNTs/CPE, (**b**) 1.0×10^{-15}, (**c**) 1.0×10^{-14}, (**d**) 1.0×10^{-13}, (**e**) 1.0×10^{-12}, (**f**) 1.0×10^{-11}, (**g**) 1.0×10^{-10}, and (**h**) 1.0×10^{-9} mol L^{-1}. (b) The plot of ΔR_{ct} versus the logarithm of the BCR/ABL fusion gene target sequence concentrations. *Source:* Zhang [145]. Reproduced with permission of Springer.

investigated. Graphite has a structure in which planes of carbon atoms in a hexagonal arrangement are held together by inter-planar van der Waals forces and within each plane by sp^2-hybridized C—C bonds. In graphene, the number of layers is reduced to just a few layers or even a single layer, making it just a few or one atom thick. It displays a number of useful properties such as high stiffness and optical transparency as well as high surface area and good electrical conductivity. The combination of high surface area and its other physical and electrical properties will potentially enhance the performance of a range of sensors, and a number of reviews have been written on this subject [146–148] as well as a comparison of graphene and CNTs [149].

Initial graphene production was through mechanical exfoliation using adhesive tape [150], but this method is suitable only for small amounts of graphene. High-quality larger films up to 30 inches across can be made by chemical vapor deposition (CVD) onto copper substrates, but for widespread application the production of gram or kilogram quantities of graphene is required. An obvious starting point for graphene production is graphite since it can be thought of as multilayer graphene. Strong oxidizing solutions of nitrate or permanganate in sulfuric acid [151] convert graphite into graphite oxide; this can be dispersed as thin sheets in water and reduced to graphene by agents such as hydrazine [152] or thermally or electrochemically [153]. Often there are some oxygen-containing species present after reduction, meaning reduced graphene oxide has different properties to pure graphene.

Physical exfoliation of graphene using methods such as ultrasonication has been widely studied. One issue is that exfoliated graphene can easily reaggregate; however this can be avoided by the use of certain organic solvents [154] or simple surfactants [155], which keep the materials dispersed. Mechanical exfoliation of graphite has produced high yields of graphene in aqueous or organic solvent systems [156].

Early work modified GCEs with graphene and demonstrated increases in oxidation peak currents and decreases in oxidation overpotentials [157], indicating facilitated reactions at the electrode surface. This was demonstrated for biologically active molecules such as dopamine, uric acid, nucleotide bases, and so on. Since the commercial biosensor market is dominated by glucose monitoring, the use of graphene in glucose biosensors has been widely studied. Glucose oxidase can be utilized along with graphene-based sensors in a wide variety of formats, as reviewed here [147]. Much higher sensitivities were demonstrated for graphene-based sensors compared with similar sensors fabricated with graphene oxide [147]; this is thought to be due to much lower charge transfer resistance for the graphene sensors [158]. For example, glucose oxidase could be immobilized onto electrodes coated with a graphene/PdNP/ chitosan composite and demonstrated a good linear range ($1.0\,\mu mol\,L^{-1}$ to $1.0\,mmol\,L^{-1}$) and low detection limit ($0.2\,\mu mol\,L^{-1}$). This was because of the electrocatalytic activity of the graphene since control electrodes without graphene had lower sensitivity [159]. Graphene could be deposited with a cationic poly(ionic liquid) onto a GCE and used as a substrate to bind glucose oxidase [160]; the resulting sensor demonstrated direct electron transfer between electrode and enzyme and excellent sensitivity and stability.

Direct electron transfer between enzyme and graphene has also been demonstrated for reduced graphene oxide on glassy carbon [161], for graphene–chitosan composites [162], and for graphene/polyethylenimine/ionic liquid composites [163], all of which were suitable for glucose sensing. Graphene has been shown to potentially facilitate superior direct electrochemistry of enzymes compared with, say, CNTs due to a larger surface area for adsorption and faster electron transfer rates [164]. Zhang et al. studied glucose oxidase/reduced graphene oxide biosensors and demonstrated that at low oxygen concentrations, direct electron transfer between graphene and enzyme occurred, whereas at higher oxygen concentrations, the graphene catalyzed the reduction of H_2O_2 [165]. Higher numbers of oxygen functional groups on the reduced graphene oxide were shown to improve enzyme adsorption and sensitivity.

This method is of course not confined to glucose oxidase; a variety of enzymes have been immobilized onto graphene electrodes to develop electrochemical biosensors. Such enzymes include catalase that was immobilized onto a graphene/AuNP composite to give a H_2O_2 sensor; this demonstrated direct electron transfer [166]. Urease could be assembled with multilayer graphene to make a urea sensor [167], and alcohol dehydrogenase could be used along with a graphene/MWCNT composite to construct an electrochemical alcohol biosensor [168]. Graphene, glucose oxidase, and glucoamylase could be assembled along with polyethylenimine using a layer-by-layer method onto glassy carbon, allowing the development of a biosensor for maltose [169]. Surfactant-modified graphene and cholesterol oxidase could be self-assembled into films on electrodes to give an electrochemical cholesterol biosensor (range 0.05–0.35 mmol L^{-1}, detection limit of 0.05 µmol L^{-1}) [170]. Graphene could also be combined with polypyrrole/polystyrene sulfonate to give a stable aqueous dispersion that could be deposited onto a platinum electrode, showed high catalytic activity to the oxidation of uric acid and hydrogen peroxide, and could be combined with xanthine oxidase to develop a hypoxanthine biosensor [171].

Since the nonenzymatic glucose sensors are not strictly biosensors, we will just mention that graphene, due to its catalytic effects and fast electron-transfer kinetics, has been complexed with a wide variety of metals and oxides to successfully detect glucose [146, 148]. One example utilized a graphene foam substituted with cobalt oxide to determine levels of glucose as low as 25 nmol L^{-1} [172]. Fructose could also be detected using an indium tin oxide electrode modified by graphene/Cu/CuO [173], although this was subject to interference by other polysaccharides. Use of capillary electrophoresis followed by detection at a graphene/CuNP electrode allowed separation and determination of mannitol, sucrose, lactose, glucose, and fructose [174].

Graphene has also been examined for its ability to catalyze electrochemical reactions of a variety of small biologically important molecules. One of the most widely investigated is dopamine since its concentration can be directly linked to a number of clinical conditions. Although the detection of dopamine by electrochemical methods in solution is simple, in clinical samples there is a major issue in that the oxidation of dopamine is subject to interference from other compounds with overlapping voltammetric responses such as uric and ascorbic acids. GCEs drop-coated with graphene were shown to allow the determination of dopamine (range 4–100 μmol L^{-1}, detection limit 2.64 μmol L^{-1}) in the presence of 1 mmol L^{-1} ascorbic acid [175]. Other works demonstrated that the sensitivity of commercial screen-printed electrodes to dopamine could be greatly enhanced by drop-coating them with aqueous solutions of ultrasonically exfoliated graphene stabilized by surfactant coatings [176]. A range of other graphene-modified electrodes for dopamine detection are extensively discussed here [146]. Sensitive detection of dopamine in the presence of interferents could be due to a number of factors; for example, dopamine is cationic at physiological pH, whereas ascorbate and ureate are anionic. The flat aromatic surface of graphene complements the aromatic structure of dopamine [146]; for example, graphene nanoflake films on silicon contain a high level of edge defects that are thought to improve the electron-transfer kinetics with dopamine, allowing its determination in the presence of ascorbate [177]. The existence of these interactions has also been suggested to be a factor in improving sensitivity and selectivity; graphene-modified electrodes were shown to be capable of determining dopamine between 5 and 200 μmol L^{-1}, demonstrating superior performance to CNTs due to higher conductivity and the π–π stacking interactions between dopamine and graphene [178].

Other examples of small molecules that can be determined at graphene electrodes include the neurotransmitter serotonin, which can be determined in the presence of ascorbate and dopamine by using GCEs modified with graphene [179]. Alwarappan et al. studied the detection of dopamine and serotonin at graphene electrodes and demonstrated superior sensitivity, signal-to-noise ratio, and stability compared to sensors fabricated using SWCNTs as well as better detection of dopamine in the presence of ascorbic acid [180]. Another neurotransmitter epinephrine could be detected at limits as low as 7 nmol L^{-1} in the presence of ascorbate by using an electrode modified with a graphene/ AuNP composite [181]. Nicotinamide adenine dinucleotide could be detected at graphene-modified GCEs with high electron-transfer rates and low overpotentials [182]; this is important since this molecule is a cofactor for many dehydrogenase enzymes, and kinetics at graphene were shown to be superior to graphite, pyrolytic graphite, and glassy carbon and comparable with CNTs. Amino acids have also been studied at graphene electrodes; for example, tyrosine and tryptophan could be electrochemically detected at graphene-modified electrodes with enhanced sensitivity (limits of detection as low as 0.1 μmol L^{-1}) in the presence of ascorbic and uric acids [183]. Also a graphene/cobalt oxide composite could be used for the electrochemical determination of tryptophan (linear range 0.05–10 μmol L^{-1}, detection limit 0.01 μmol L^{-1}) [184]. A highly sensitive sensor for histidine could be obtained by immobilizing DNA onto a graphene/AuNP composite; this could be cleaved by L-histidine, giving a measurable response with a limit of detection of $0.1 \, pmol \, L^{-1}$ histidine [185].

Graphene has also been utilized in the development of immunosensors; early work coupled antibodies for immunoglobulins to graphene-modified electrodes as previously reviewed [146]. Antibody–antigen reactions do not produce electrons or redox-active species so different methods have been used to detect binding events. The use of graphene as a signal amplifier for producing ultrasensitive biosensors has been extensively reviewed [186]. For example, graphene was deposited onto an FET; aptamers to IgE were immobilized onto this and then binding of IgE could be determined from the draining current [187]. Graphene sheets vertically orientated with respect to the sensor surface could be grown by CVD onto an FET; these were then labeled with AuNP/ antibody conjugates, the resulting sensor being capable of detecting IgG as low as $13 \, pmol \, L^{-1}$ [188]. Disposable electrodes could be modified with graphene and anti-IgG; interrogation by AC impedance gave a label-free immunosensor (range $0.3–7 \, \mu g \, mL^{-1}$ IgG) [189]. Alternatively ferrocene-based labels could be used along with a graphene/AuNP electrode to determine IgG (linear range $1–300 \, ng \, mL^{-1}$, detection limit $0.4 \, ng \, mL^{-1}$) [190].

Other proteins have been studied; for example, aptamers to thrombin were immobilized onto a graphene/porphyrin/GCE [191]. The combination of high conductivity and surface area of the graphene and the electrochemical activity of the porphyrin provided an immunosensor with a range of $5–1500 \, nmol \, L^{-1}$ and detection limit of $0.2 \, nmol \, L^{-1}$. In another work, graphene functionalized with the electroactive dye orange II could be used as the base for aptamers to thrombin and lysozyme, allowing detection at limits as low as 0.35 and $1 \, pmol \, L^{-1}$, respectively [192].

Antibodies to prostate-specific antigen were immobilized onto a graphene/ methylene blue composite electrode and shown to be capable of detecting low levels of the antigen (linear range $0.05–5.00 \, ng \, mL^{-1}$, detection limit $13 \, pg \, mL^{-1}$) [193]. A graphene/AgNP/CNT composite could be used for the detection of chorionic gonadotropin [194]. Antibodies were immobilized onto the electrode and used in a sandwich-type format with an HRP-labeled secondary antibody; the resulting sensor could determine the antigen with a wide linear response from 0.005 to $500 \, mIU \, mL^{-1}$ and detection limit of $0.0026 \, mIU \, mL^{-1}$. Hernandez et al. compared graphene and graphene oxide for immobilizing aptamers with *Staphylococcus aureus* [195]; graphene gave the most reproducible responses and could detect $1 \, CFU \, mL^{-1}$ of the bacterium in 1 min. A multiple target immunosensor was created using an array of screen-printed graphene electrodes; four individual electrodes were coated with different capture antibodies [196]. Treatment with a secondary antibody containing a reactive moiety after exposure to a mixture of the antigens was then used as a starting platform to generate polymerization of an epoxy compound that was then utilized as a scaffold to attach HRP. Detection of four cancer biomarkers—carcinoembryonic antigen (CEA), alpha-fetoprotein (AFP), cancer antigen 125 (CA125), and carbohydrate antigen 153 (CA153) with respective detection limits of 0.01, 0.01, 0.05, and 0.05 ng mL^{-1}—could be attained. A number of other immunosensors based on immobilizing antibodies onto graphene-modified electrodes and their use in electrochemical sensing have been reviewed here [186].

Graphene has been extensively studied for DNA sensing as reviewed previously [146–148, 186] so a brief overview will be given. Initial studies focused on measuring the presence of DNA since all four bases can be electrochemically oxidized, albeit often at high potentials. One of the earliest studies examined both ssDNA and double-stranded DNA (dsDNA) and demonstrated that using linear voltammetry, all four bases in DNA showed separate peaks [157] when a graphene-modified carbon electrode was used; these were not observed for graphite-modified or -unmodified electrodes. It was determined that catalysis at edge defects and the high conductivity led to this behavior [157]. This was confirmed in a later work that also demonstrated detection of DNA bases and other moieties at graphene-modified electrodes and could differentiate between ssDNA and dsDNA [197]. In another work, vertically orientated graphene oxide "walls" could be electrochemically reduced to graphene and then used to detect free nucleotides, ssDNA and dsDNA [198]. dsDNA could be determined in a large linear range from 0.1 fmol L^{-1} to 10 mmol L^{-1}, with the limit of detection estimated to be 9.4 zmol L^{-1}. Single nucleotide polymorphisms could be detected for 20 zmol L^{-1} oligonucleotides (~10 DNA strands mL^{-1}).

Detection of specific DNA strands can be obtained by immobilizing a capture strand of DNA and exposing it to the target strand. Graphene could be solubilized by sonication with pyrenebutyric acid and drop-cast onto a gold electrode and the carboxylic acid groups used to covalently immobilize oligonucleotides [199]. Hybridization with the target in the presence of methylene blue formed dsDNA with intercalated dye molecules; these could be determined electrochemically, allowing detection of the target strand from 1 fmol L^{-1} to 5 pmol L^{-1}, with limit of detection estimated to be 0.38 fmol L^{-1}. Graphene sheets decorated with AuNPs could be used to modify GCEs [200]; thiolated oligonucleotides were attached and then used in a sandwich-type assay with the target strand and a methylene blue-labeled signal strand to determine levels of the target between 1 fmol L^{-1} and 100 nmol L^{-1} with a detection limit of 0.35 fmol L^{-1}. A number of other DNA-based sensors of this type have been developed and are summarized elsewhere [186]; these are capable of deterring DNA levels at sub-pmol L^{-1} levels and can distinguish between fully complementary strands and those with a single mismatch. Graphene/gold composites were used as a basis for a DNA sandwich assay-type sensor where the tracer probe was labeled with HRP and could detect target DNA at a level of 3.4 fmol L^{-1} [201]. Impedimetric sensors for DNA based on adsorbing ssDNA onto a variety of different types of graphene could be constructed [202]; reduced graphene oxide with 3–4 layers was found to be the most effective platform compared with monolayer or multilayer graphenes. Other workers showed that ssDNA could be attached to graphene by π–π stacking or covalent attachment; using impedimetric methods binding of the complementary nucleotide could be determined with a linear range of $50\,fmol\,L^{-1}$ to $1\,\mu mol\,L^{-1}$, three orders of magnitude more sensitive than corresponding graphite-based sensors [203]. Differential pulse voltammetry could be used to distinguish between ssDNA and completely hybridized dsDNA and also detect single nucleotide mismatches due to the different oxidation signals of the four DNA bases.

Other types of graphene biosensors have been constructed; for example, graphene FETs have been used to detect DNA with single-base specificity based on the electronic n-doping effect from the nucleotide bases to graphene [204]. Also a graphene FET has been shown to be capable of detecting both DNA and protein from cellular secretions; binding of DNA causes an increase in conductivity, whereas protein binding has the opposite effect [205]. Graphene also acts as a strain sensor [206], allowing for potential sensing applications, for example, where a biochemical interaction with a functionalized graphene sheet causes a measurable change in strain.

Graphene is often synthesized from graphene oxide; however in some cases the parent materials offer themselves to biosensor construction. A range of graphene oxide biosensors have been reviewed [186]; for example, graphene oxide/methylene blue composites could be used in thrombin aptasensors [207], and in another work a graphene oxide/AuNP/thionine complex [208] could be used as the base to immobilize antibodies to CEA, leading to the development of an extremely sensitive immunosensor for CA125 (range $0.1\,fg\,mL^{-1}$ to $1\,\mu g\,mL^{-1}$, detection limit $0.05\,fg\,mL^{-1}$).

Graphene has shown itself to be a versatile and effective material for improving the performance of a wide range of biosensors. It displays some advantages over the similar CNTs such as lower cost, higher purity, and lower electrical noise and can be more processable [149, 186]. Its flat structure also allows high functionalization and can greatly improve the surface area of sensors, both of which will increase sensitivity. Graphene can also be combined with a range of other nanomaterials such as AuNPs [139], CNTs, and MNPs or be formed into composites with a variety of polymers. It is no surprise that graphene is being so extensively researched.

11.6 Nanostructured Metal Oxides

NMOs play an important part in many technological areas [209]. Moreover, a wide range of possibilities of structural geometries and surface morphology can be achieved as described in the literature: nanorods [210, 211], nanotubes [212, 213], NWs [214–216], nanofibers [217, 218], nanobelts [219, 220], nanoribbons [221], nanoneedles [222, 223], nanospheres [224], nanoflowers [225, 226], nanourchins [227], and others. It is worth mentioning that depending on its electronic structure, the same material can present superconductor, conductor, semiconductor, insulator, or even magnetic character [228]. NMOs have recently aroused much interest as immobilizing matrices for biosensors development because they possess high thermal stability, non-toxicity, large surface area-to-volume ratio, catalytic properties, functional biocompatibility, strong adsorption capability, and enhanced electron transport properties. They provide suitable microenvironments for the immobilization of biomolecules, which, with their other properties, also results in enhanced electron transfer and improved biosensing characteristics [229, 230].

NMOs have been extensively used within biomedical applications, such as biomedical imaging, drug delivery, gene delivery, and biosensing. The development of biosensors has often been focused on small molecule analytes such as glucose, phenol, hydrogen peroxide, cholesterol, urea, and so on [231]. Conventional biosensors are based on oxidase-modified electrode [232], that is, by immobilizing the enzymes glucose oxidase (GOx) [233–235], tyrosinase (Tyr) [236, 237], and cholesterol oxidase (ChOx) [238, 239].

Meanwhile, over the past decade a new generation of glucose sensors has emerged based on nonenzymatic electrochemical biosensors, which have risen at a considerable rate due to their easy fabrication, cost effectiveness, high sensitivity, and fast and accurate measurements [240], and they are one alternative solution for the immobilization and denaturation of enzyme issues found in the fabrication of conventional biosensors. Li et al. [130] developed a nonenzymatic glucose sensor based on copper oxide (CuO) NWs grown on a three-dimensional (3D) porous copper foam (CF) by anodization; this sensor achieved a high sensitivity of $2217.4\,\mu A\,cm^{-2}\,mM^{-1}$ and detection limit of $0.3\,\mu mol\,L^{-1}$ and a linear range encompassing the normal blood glucose range of 4.4–$6.6\,mmol\,L^{-1}$ [241]. One interesting goal achieved in this work was the possibility for noninvasive glucose detection by salivary glucose concentration, estimated to be $0.91 \pm 0.04\,mmol\,L^{-1}$. Moreover, reproducibility tests with six different electrodes presented an RSD of 3.51% and repeatability tests with one electrode gave an RSD = 1.57% (eight measurements), and the sensor was stable for storage up to 15 days (Figure 11.9).

Xu et al. [131] enhanced the sensing properties and developed a flexible nanorod-aggregated flower-like CuO grown carbon fiber fabric (CFF), which could be prepared by a simple, fast, and green hydrothermal method, and they achieved a fast response time and a low detection limit of $1.3\,s$ and $0.27\,mmol\,L^{-1}$, respectively. Reproducibility experiments demonstrated an RSD of 1.53% (eight identically fabricated electrodes) and stability tests showed a sensitivity loss of 9.9% over a period of 1 month.

The mechanism for the oxidation of glucose with nonenzymatic electrochemical biosensors on NMO modified electrode has not been completely

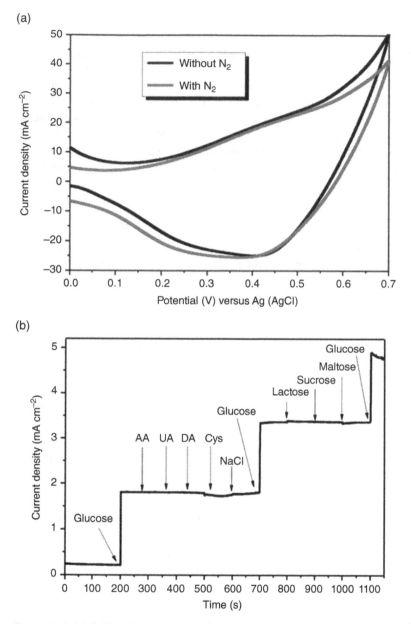

Figure 11.9 (a) Cyclic voltammograms of CuO NWs/CF with or without nitrogen bubbling; (b) anti-interference property of the CuO NWs/CF electrode with initial addition of 1.0 mmol L^{-1} glucose and 0.1 mmol L^{-1} ascorbic acid (AA), uric acid (UA), dopamine (DA), 0.5 mmol L^{-1} cysteine, 0.1 mmol L^{-1} sodium chloride (NaCl), and then again 1.0 mmol L^{-1} glucose, followed by addition of 0.05 mmol L^{-1} lactose, sucrose, and maltose and lastly of 1.0 mmol L^{-1} glucose; (c) reproducibility of six CuO NWs/CF electrodes for detection of 0.5 mmol L^{-1} glucose; the inset shows the repeatability of CuO NWs/CF electrode for detecting 0.5 mmol L^{-1} glucose for eight times; (d) the stability measurement of CuO NWs/ CF electrode for 15 days. *Source:* Li et al. [130]. Reproduced with permission from American Chemical Society.

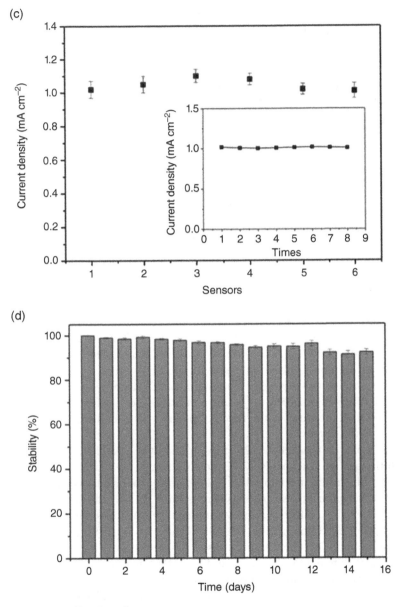

Figure 11.9 (Continued)

established; the most accepted mechanism up to now has been proposed by Marioli and Kuwana [242] for cobalt- or zinc-based materials and is schematically depicted in Figure 11.10. The oxidation is generated by glucose deprotonation and isomerization to its enediol form, followed by adsorption onto the electrode surface and oxidation by M*(II) and M*(III). The M*(III) species are proposed to

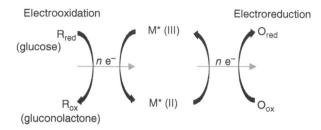

Electrooxidation

Electroreduction

R_{red} (glucose)

R_{ox} (gluconolactone)

M* (III)

M* (II)

$n\,e^-$

O_{red}

O_{ox}

$n\,e^-$

Figure 11.10 Schematic of the mechanism proposed by Marioli and Kuwana, M* = Zn, Co.

act as an electron-transfer medium. During voltammetry measurements, M*(II) on M*O electrode would first be oxidized to M*(III). Then the oxidative M* could catalyze glucose oxidation to generate gluconolactone, and then gluconolactone is further oxidized to gluconic acid [227, 243].

Disease diagnosis can also be achieved using NMOs, as reported by Tak et al. [244]. A nanoflower ZnO matrix fabricated through a hydrothermal method was applied to the DNA-based detection of bacterial meningitis using as probe DNA (p-DNA): 5′-HS-GAT ACG AAT GTG CAG CTG ACA CG-3′, complementary target DNA (t-DNA): 5′-CGT GTC AGC TGC ACA TTC GTA TC-3′ and noncomplementary target DNA (nc-DNA): 5′-GCA CAC ACG TGG TCA AAC GAT TC-3′. p-DNA was immobilized on the nanoflower structure by electrostatic interaction. Differential pulse voltammetry allowed the detection of DNA hybridization with high selectivity, with a sensitivity of $168.64\,\mu A\,ng^{-1}\,\mu L$ over a wide range of $5–240\,ng\,\mu L^{-1}$ and a relatively low detection limit of $5\,ng\,\mu L^{-1}$. The shelf life study of the bioelectrode reveals a high stability over a period more than 16 weeks.

In another work, Perumal et al. [225] developed a hybrid AuNP (30 nm diameter) agglutinated on zinc oxide (ZnO) nanoflowers, creating the "spotted nanoflowers" illustrated in Figure 11.11. They were used as a base to detect DNA from pathogenic leptospirosis-causing strains via hybridization using a thiolated probe DNA. Electrochemical impedance spectroscopy was used as the method of detection, as illustrated in Figure 11.12a; after probe DNA (p-DNA) immobilization the charge transfer resistance (R_{ct}) increased from approximately 0.16 to $0.35\,M\Omega$, which confirms DNA adsorption on NFs. The further hybridization of complementary DNA sequence (t-DNA) process is observed by an increase on R_{ct} to $1.10\,M\Omega$. Figure 11.12b depicts EIS experiments with varying t-DNA concentrations from 10^{-6} to $10^{-13}\,M$, achieving a detection limit of $100\,fmol\,L^{-1}$. Figure 11.12c and d shows the variation at low frequencies (<100 Hz) that is related to the double layer effect [245], which is the most affected region due to the DNA hybridization that occurs on electrode surface. The approach used in this work could ideally be used to achieve multiple diagnoses through array-based technology.

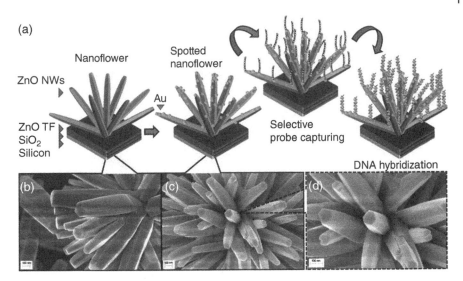

Figure 11.11 (a) Schematic of the steps involved in the synthesis of the spotted nanoflower (NF) DNA bioelectrode. (b) FESEM image of low magnification revealing the flower-like ZnO nanostructure possessing hexagonally shaped tips, which demonstrate the high crystallinity of the prepared ZnO nanowire ends. (c and d) Low- and high-magnification images of spotted NFs indicate that radially oriented NFs have an average length of 2–3 μm and a diameter of approximately 100 nm. *Source:* Perumal et al. [225]. https://www.nature.com/articles/srep12231#comments. Used under CC BY 4.0 http://creativecommons.org/licenses/by/4.0/. (*See insert for color representation of the figure.*)

Early detection of cancer is also a potential application with the use of super-paramagnetic iron oxide nanoparticles (SPIONs), providing it is possible to find a specific cancer biomarker. Bakhtiary et al. [246] reviewed the use of SPIONs as probes for *in vivo* imaging techniques for some tumors such as liver, prostate, brain, breast, lung, colorectal, cervical, and ovarian cancers. In this manner Wang et al. [247] developed a label-free ZnO nanorod-based biosensor for breast cancer detection using the biomarker CA153. Antibodies to CA153 were immobilized onto functionalized ZnO/3-aminopropyltriethoxysilane (APTES) using glutaraldehyde for QCM experiments; these demonstrated a sensitivity of 25.34 ± 0.67 Hz scale^{-1} ($1\,U\,mL^{-1}$) and good linearity (0.99) in the concentration range of 0.5–$26\,U\,mL^{-1}$. A response time of less than 10 s was obtained, and reproducibility tests of 20 biosensors gave an RSD of 2%.

11.7 Nanostructured Hydrogels

Over the last decades, hydrogel (HG) materials have become commonplace in daily life in different forms with a wide range of applications in medical, pharmaceutical [248–250], cosmetic, agricultural, and textile industries [251, 252]

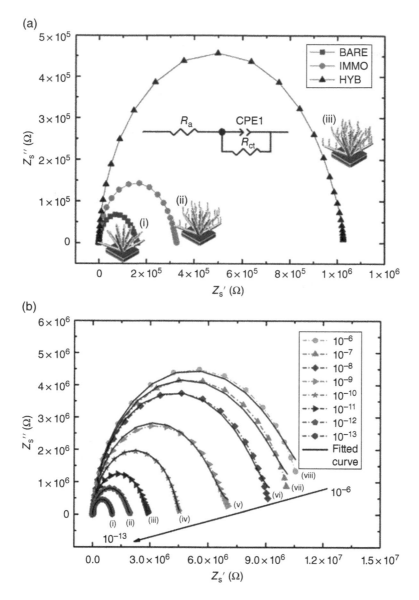

Figure 11.12 (a) Impedance spectra of (i) spotted NF, (ii) spotted NF/p-DNA (probe), and (iii) spotted NF/p-DNA/t-DNA (duplex) bioelectrode; the inset shows the Randles equivalent circuit, where the parameters R_a, R_{ct}, and CPE represent the bulk solution resistance, charge transfer resistance, and constant phase element, respectively. (b) Impedimetric response curve of spotted NF/p-DNA hybridized with different concentrations of complementary target DNA (i–viii), 10 μM to 100 fm.

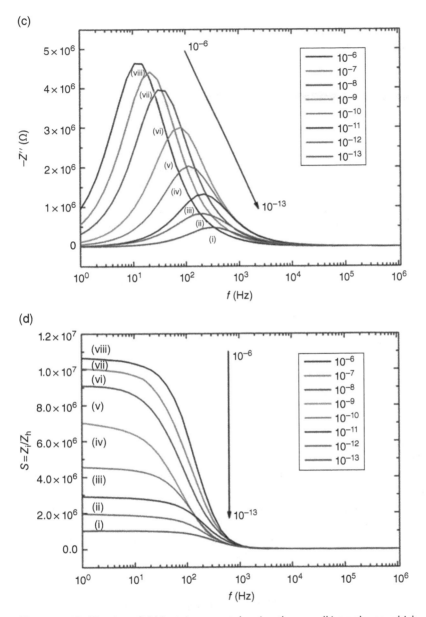

Figure 11.12 (Continued) (c) Imaginary part showing the overall impedance, which decreases, and the peak frequency, which is shifted toward the higher frequencies as the concentration of complementary DNA decreases. (d) The gain curve of spotted NF/p-DNA hybridized at different concentrations. *Source:* Perumal et al. [225]. https://www.nature.com/articles/srep12231#comments. Used under CC BY 4.0 http://creativecommons.org/licenses/by/4.0/. (*See insert for color representation of the figure.*)

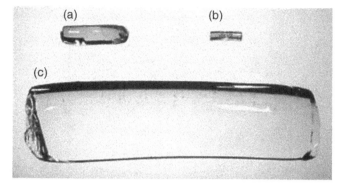

Figure 11.13 Comparison of volumes of a topological gel (polyrotaxane gel) in as-prepared (a), dried (b), and swelling equilibrium (c) states. *Source:* Tanaka et al. [264]. Reproduced with permission of Elsevier.

as well as research fields of stretchable electronics, energy storage, actuators, sensors, bioelectronics, and medical electrodes [253–258].

HGs are 3D polymeric network structures, formed by cross-linking polymer chains [259], which have the ability to change their chemical structure, inducing volume changes in response to external stimuli such as temperature, pH, light, solvent composition, particular chemicals, or electric or magnetic field [260, 261], thereby classifying these materials as stimuli-responsive smart materials. Figure 11.13 shows an example of a polyrotaxane HG in its as-prepared, dried, and fully swollen (equilibrium) states. As observed the gel can swell up to 500 times in volume compared with the as-prepared state. When HGs are used to formulate thin films (<100 nm), this enables a rapid response to external stimuli [263, 264].

According to Laftah [251], the HG's definitions are based mainly on the properties of the polymers and their raw materials. These classifications can be made by considering different approaches, for example, their environmental response (external stimuli), preparation methods (cross-linking or graft polymerization), source of materials (natural, synthetic, or semisynthetic polymers), and morphological characteristics (powder, fiber, membrane, etc.). In a later work, Ullah et al. [262] updated this discussion and classifications as depicted in Figure 11.14.

Although most applications of HGs have been performed within the fields of drug delivery or controlled-release systems [255, 265, 266], the synergistic effect of HGs with organic materials (conducting polymers [267, 268] and ionic liquids [269–271]), inorganic materials (graphene [254, 272] and metal nanoparticles [258, 273]), and biomolecules [274–276] has attracted the attention of many researchers toward the development of sensors and biosensors for the detection of metabolites, neurotransmitters, cells, antibodies, and DNA [130, 277–283].

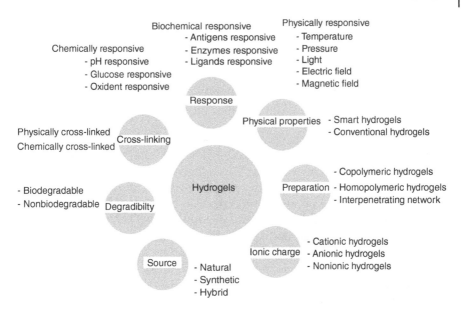

Figure 11.14 Classification of hydrogels based on different properties. *Source:* Ullah et al. [262]. Reproduced with permission of Elsevier.

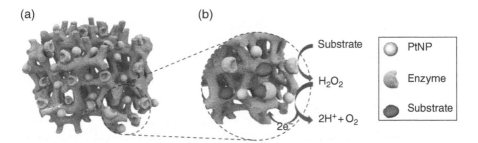

Figure 11.15 Schematic of the general sensing mechanism of a CPH-based electrode platform. (a) PtNPs and enzymes were loaded onto hierarchically 3D porous PAni hydrogel matrices to form PAni hydrogel/PtNP hybrid electrodes. (b) The PtNP-catalyzed sensing process of the biosensor based on PAni/PtNP/enzyme hybrid films. *Source:* Li et al. [279]. Reproduced with permission from American Chemical Society.

Khodagholy et al. [271] used an HG platform combined with room temperature ionic liquids to demonstrate for the first time a solid-state electrolyte on a flexible transistor-based biosensor for the detection of lactic acid, envisioning the wearability of the sensor for real-time health monitoring in a relevant physiological range. Aiming toward advances in healthcare monitoring and clinical diagnostic, workers produced a polyaniline (PAni) HG [278, 279] with PtNPs incorporated in the matrix, as depicted in Figure 11.15a. The PAni HG is hierarchically porous, which favors enzyme immobilization and facilitates the diffusion of H_2O_2 from the enzymatic reaction sites to the PtNPs for further electrochemical oxidation. Figure 11.15b shows the sensing mechanism that occurred for the four systems studied—glucose, uric acid, cholesterol, and triglycerides—achieving a low detection limit of 0.7, 1, 300, and 200 µmol L^{-1}, respectively, with a response time of only 3 s when combined with amperometric measurements.

Nonenzymatic glucose biosensors [131–134, 240] have been extensively studied by using inorganic materials as an alternative solution for the immobilization and denaturation of enzyme issues found upon the fabrication of conventional biosensors [284]. In this manner, HGs containing boronic acid derivatives have attracted much interest as an alternative for nonenzymatic detection of saccharides and saccharide derivatives [285–288]. The strength of boronic acid binding to saccharides is determined by the orientation and relative position of hydroxyl groups; thus boronic acids can differentiate between structurally similar saccharide molecules [285]. Some potential mechanisms are summarized in Figure 11.16.

Gabai et al. [287] employed a number of approaches to characterize the swelling and shrinking that occurs upon glucose binding to a film of boronic acid-containing HG, such as electrochemical impedance spectroscopy, chronopotentiometry, and SPR and QCM measurements, achieving mmol L^{-1} levels of detection.

Usually the saccharide interactions with monoboronic acid groups are weak; this can be improved by utilizing properly positioned multiboronic acids, like phenylboronic acids (PBAs). Zhang et al. [285] evaluated the interaction of glucose with PBA-containing HGs, varying structural parameters such as the position of the boronic acid on the phenyl ring (ortho, meta, and para), the substituent on PBA (fluorine and nitro moieties), and differing linkers to the HG backbone (amino and aminomethyl). They observed a blue- or redshift on diffraction wavelength due to the HG shrinking or swelling depending on the glucose concentration; this is of interest for application as a biosensor or a controlled insulin release platform. The biocompatibility of PBA HGs envisions the development of noninvasive systems similar to contact lenses, in which tear liquid measurements on an eye can be performed. Mesch et al. [288] developed sensors based on localized surface plasmon resonances (LSPR) using PBA HG and glucose; levels of the sugar in the physiological millimolar range were successfully determined.

Wang and Li [289] prepared aptamer–ssDNA cross-linked polymeric HG materials for use within QCM experiments and demonstrated the feasibility of this system for detection of avian influenza viruses (AIV H5N1). Three proportions of aptamer-HG were studied with the ratio of acrylamide, aptamer, and ssDNA of 100:1:1 (HG I), 10:1:1 (HG II), and 1:1:1 (HG III) being formulated and bound to QCM crystals. These were used to measure levels of

(a)   (b)

(c)   (d)

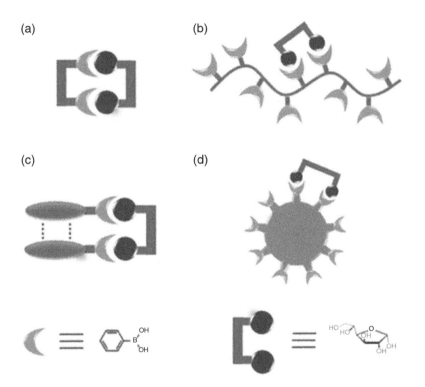

Figure 11.16 Schematic of four strategies for selective saccharide sensing via multivalent boronic acid–saccharide interactions, exemplified by glucose sensing. (a) Synthetic diboronic acids that form 1:1 cyclic boronate esters with glucose, (b) boronic acid-containing polymers that bind glucose with two of the pendant boronic acid moieties, (c) aggregation of simple boronic acids via non-covalent interactions to allow multivalent glucose binding, and (d) boronic acid-conjugated nanomaterials as multivalent scaffolds. Note that the aggregates shown in (c) can be aggregates or saccharide binding altered aggregates of boronic acid or saccharide binding induced aggregates of boronic acid. *Source:* Wu et al. [284]. Reproduced with permission of Royal Society of Chemistry.

the target with detection limits of 1.28, 0.128, and 0.0128 HA units (HAU), respectively. When the HG was exposed to target virus, there was a strong aptamer/virus binding, disrupting the linkage between the aptamer and ssDNA, reducing cross-linking, and thereby causing the HG to abruptly swell. Measurement times were within 30 min with no interference being observed from nontarget AIV subtypes. As the HG III aptasensor was the most efficient, a comparison with an immunosensor where an anti-H5 antibody was directly attached to the QCM crystal was performed to prove the enhanced sensitivity of HG coated system. A limit of detection of 0.128 HAU was achieved with the antibody, thus confirming the efficiency of aptasensor platform. Furthermore it was concluded that the preparation process including the polymerization and pretreatment of HG did not damage or affect the activity/affinity of the aptamer [289].

Continuing the study on the development of aptasensor HG for AIV H5N1 detection, the same research group [290] conjugated ssDNA with QDs and evaluated the fluorescence properties. Binding of the target disrupted the cross-linking interaction and led to swelling of the HG and release of the QDs. Three platforms were interrogated as schematically depicted within Figure 11.17a in the range of 0.4–32 HAU. Increases in fluorescence intensity with the substrate concentration are observed for HGs A and B, while HG C had decreased QD emission. Figure 11.17b helps to explain that, because of the gel structure, with the larger pore sizes of HGs A and B (300–400 nm), virus particles and ssDNA–QD conjugates could penetrate into the matrix, allowing more aptamer/target hybridization, thereby enhancing the fluorescence signal. However HG C has a smaller pore size (100 nm) due to its higher density and degree of cross-linking that results in only a superficial hybridization of virus and aptamer. Although the detection range using QDs was wider (0.4–32 HAU) than the previously studied system [290] (0.0128–0.64 HAU), the detection limit of 0.4 HAU obtained is 30 times higher. However the fluorescence system using QDs has some advantages compared with QCM since QDs can detect multiple targets simultaneously due to their different emission wavelengths, and also the size-dependent property of the HG offers potential specific recognition of targets just by designing a desired pore size and cross-linking density.

11.8 Nanostructured Conducting Polymers

Conducting polymers have been widely used within the fields of chemistry and biosensing [291–294]. Although they tend to be more semiconductor than metallic in nature, they display a wide range of useful properties that have enabled their use within such fields as electronics, optics, composites, sensors, and actuators. There are a wide range of potential polymers, but many tend to be based on either polypyrrole, polyaniline, polythiophene, or poly(3,4-ethylenedioxy thiophene) (PEDOT). Structures of some of these are shown in Figure 11.18.

Each polymer has its own particular set of advantageous properties, but all have good conductivity, flexibility, and environmental stability and can be synthesized both electrochemically and chemically. Synthesis of the materials tends to be straightforward and polymers can be produced as bulk powders, coatings, and thin films. Polyaniline and PEDOT both give optically transparent materials when in the form of thin films; polypyrrole has a low oxidation potential. Polyaniline especially can be easily modified by a variety of chemistries; the other polymers less so but differing monomers can be used and copolymerization carried out. Although conductive polymers tend to be insoluble, the PEDOT/polystyrene sulfonate complex is water soluble.

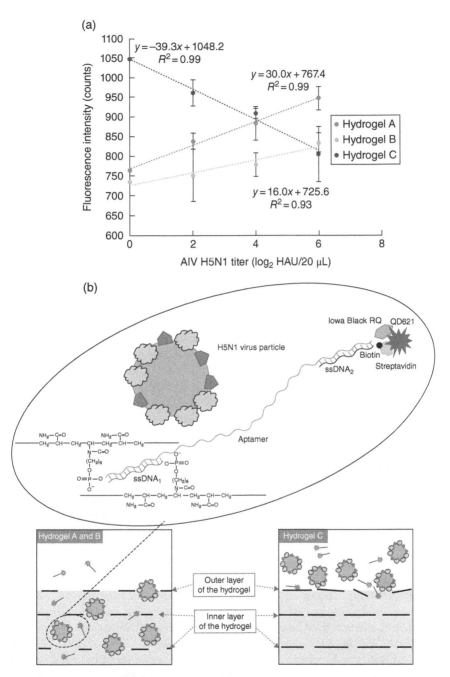

Figure 11.17 (a) Fluorescence intensity changes of three different types of hydrogels, A, B, and C, with acrylamide concentration ratio of 1 : 10 : 100 in response to different titers of AIV H5N1 (2^0, 2^2, 2^4, and 2^6 HAU). The means and error bars (standard deviation) were calculated based on three replicates. (b) Two different reaction mechanisms based on the size-dependent property of the aptamer–hydrogel embedded with $ssDNA_2$-QD conjugates upon target binding for hydrogels A, B, and C. *Source:* Xu et al. [290]. Reproduced with permission of Elsevier.

Polyaniline (PAni)

Polypyrrole (PPy)

Poly(3, 4-ethylenedioxy thiophene) PEDOT

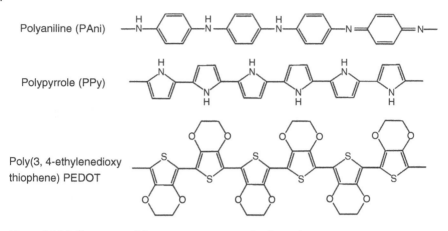

Figure 11.18 Structures of the most common conducting polymers.

The ease of synthesizing these materials has allowed them to be formulated in a variety of nanostructured forms, as reviewed here [294]. As for many other materials, this gives the advantage of high surface area, allowing more intimate interactions with biomolecules of interest, and their small size means a more rapid (usually within seconds) and relatively larger response to these processes [291–294]. However they can be unstable compared with metal or metal oxide-type materials and this has somewhat limited their study.

A number of methods have been used to construct regular-conducting polymer nanostructures, many of which involved the use of a solid or a "soft" template. For example, a range of porous membranes such as alumina or polymeric track-etched membranes containing regular pores can be used as a template. These regular pores through the membrane can be used as template to grow conducting polymer nanotubes or nanofibers by chemical or electrochemical methods [292, 295]. In early work, track-etched membranes were used as a template to grow PEDOT nanotubes, which could be functionalized with glucose oxidase [296]; the resultant membrane was then used directly as the working electrode within a glucose sensor. Small conducting polymer microarrays could be made by a variety of methods, one of which involved using sonochemistry to "punch" holes in an insulating polymer film on an electrode. Polyaniline electrodes of a few μm in diameter could be grown from these holes. Glucose oxidase could be incorporated during the polymerization process to give a glucose sensor [297] or alternatively the microelectrode could be subsequently functionalized with a number of antibodies to develop immunosensors for cancer [298] and stroke marker proteins [299]. Use of these microelectrodes led to a much enhanced sensitivity over planar polymer films as did controlled affinity functionalization over simple entrapment of the antibodies, combining the two processes allowed limits of detection of $1\,pmol\,L^{-1}$ of various targets [298–300].

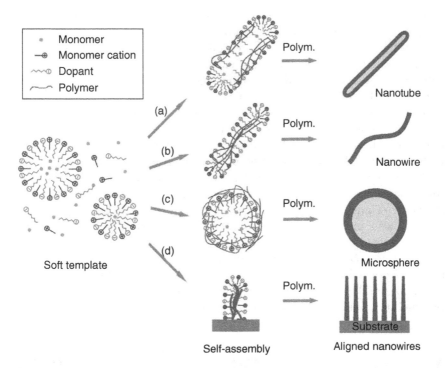

Figure 11.19 Schematic of the mechanism of the soft-template synthesis of different conducting polymer nanostructures: (a) micelles acted as soft templates in the formation of nanotubes. Micelles were formed by the self-assembly of dopants, and polymerization was carried out on the surface of the micelles; (b) nanowires formed by the protection of dopants. The polymerization was carried out inside the micelles; (c) monomer droplets acted as soft templates in the formation of microsphere; and (d) polymerization on the substrate producing aligned nanowire arrays. Nanowires were protected by the dopants, and polymerization was carried out on the tips of nanowires. *Source:* Xia et al. [292]. Reproduced with permission of Elsevier. (*See insert for color representation of the figure.*)

Electrospinning can be utilized to make a wide variety of polymers as nano-sized fibers; for example, polymethyl methacrylate (PMMA) can be spun into fibers, substituted with ferric ions. These ions act as the oxidant to initiate the polymerization of EDOT from the vapor phase to give nanofibers with a PMMA core and a PEDOT shell [301]. The PMMA can then be etched away to leave PEDOT nanotubes.

Ordered nanostructures in solution can also be used to template nanostructured polymer formation as reviewed here [292]. These are often known as "soft template" methods and some are shown in Figure 11.19. Micelles are often used; for example, dodecyl trimethyl ammonium bromide/decan-1-ol micelles could be swollen with pyrrole, which was then polymerized to obtain regular conductive polypyrrole spheres of 60 nm diameter with good conductivity in gram quantities [302]. When polyvinyl alcohol was used as a stabilizer, nanoparticles of 20–60 nm diameter were obtained with narrow size distribution [303]. When sodium (bis 2-ethyl hexyl)sulfosuccinate was used along with ferric ions, rod-shaped reverse micelles could be obtained [304]. Upon reaction with pyrrole, nanotubes of the resultant conducting polymer were formed. A number of works on the synthesis of 1D conducting polymer systems, which have novel properties and also could be used as wires between microelectrodes, have been published [305]. Micellar templates of hexadecyltrimethylammonium bromide and oxalic acid could be used to synthesize polyaniline networks with individual PAni NWs having diameters between 35 and 100 nm [306]. Liu et al. combined persulfate as an oxidant and surfactants to synthesize polypyrrole, polyaniline, or PEDOT in bulk quantities with novel "paper clip-like" structures (Figure 11.20) [307].

It has also proven possible to synthesize anisotropic polymers without templates, for instance, when polyaniline is synthesized in the presence of a stabilizer, poly-*N*-vinyl pyrrolidinone; simply by varying the concentration of the stabilizer, polyaniline nanospheres, nanorods, and nanofibers could all be synthesized [308]. Nanofibers were shown to have the highest levels of oxidation and protonation and superior capacitance and electron-transfer kinetics compared with the other structures. Polyaniline nanotubes could be

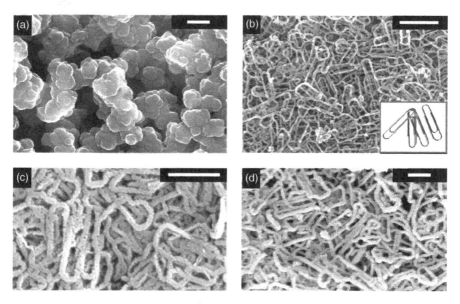

Figure 11.20 SEM images of (a) granular PPy·Cl (scale bar, 200 nm), (b) PPy·Cl nanoclips (scale bar, 1 μm; inset, digital picture of paper clips), (c) PAni HCl nanoclips (scale bar, 1 μm), and (d) PEDOT Cl nanoclips (scale bar, 1 μm). *Source:* Liu et al. [307]. Reproduced with permission from American Chemical Society.

synthesized by a simple method using ammonium persulfate as an oxidant [309]. Chiral polyaniline nanotubes could also be synthesized using camphor sulfonic acid as a dopant; these tubes were electrically conductive and the polymer chains were shown to have a helical structure [310]. It is possible to construct networks of polypyrrole, polyaniline, or PEDOT fibers (50–200 nm diameter), bridging a 2 μm gap between microelectrodes and demonstrate their behavior as transistors [311].

There have been a number of studies on using nanostructured conducting polymers in sensing and biosensing applications, simple applications being the detection of vapors such as ammonia or HCl [291]. Binding of various species was shown to affect the electrical properties of the polymer structures. Polypyrrole nanotubes substituted with carboxylic acid groups could be synthesized within the pores of an alumina membrane, isolated, and then covalently immobilized onto microelectrodes to obtain an FET [312]. The nanotubes could be substituted with glucose oxidase to develop a sensor capable of determining glucose between 0.5 and 20 mmol L^{-1}. Polyaniline nanofibers could be synthesized by interfacial polymerization and then covalently substituted with glucose oxidase; the resultant material was then immobilized on a GCE to give a sensor capable of determining glucose between 0.01 and 1 mmol L^{-1} [313]. Nanoelectrodes with small gaps (20–60 nm) were bridged by electropolymerizing polyaniline/polyacrylate fibers onto which glucose oxidase was electrostatically adsorbed [314]; this allowed construction of a glucose sensor with rapid (<200 ms) response and low oxygen consumption, making it potentially suitable for *in vivo* applications. Wu et al. solubilized boron nitride nanotubes by grafting polyaniline onto their surface; these were then used as a substrate to immobilize glucose oxidase [315]. When immobilized onto carbon electrodes, this resulted in a sensor with rapid response (<3 s), linear range 0.01–5.5 mmol L^{-1}, and LOD of 0.18 μmol L^{-1} glucose. The sensor activity was unaffected by pH between 3 and 7 and maintains high activity even at a relatively high temperature of 60°C.

FETs based on polypyrrole carboxylic acid nanotubes complexed with human olfactory receptors have been constructed, which act as electronic noses [316]. Similar polypyrrole nanotubes were functionalized with human taste receptor protein to develop an electronic tongue [317]. Standards for bitter taste (phenylthiocarbamide and propylthiouracil) could be detected at levels as low as 1 fmol L^{-1}. The same polypyrrole carboxylic acid nanotubes could be immobilized onto amino-modified gold microelectrodes and functionalized with heparin, allowing development of selective and sensitive thrombin sensor [318]. Nanoparticles of carboxylated polypyrrole could be modified with human parathyroid hormone receptor [319] and then incorporated as a close packed array into an FET, giving a rapid method of detecting human parathyroid hormone at levels as low as 48 fmol L^{-1}. The sensor showed good stability and excellent sensitivity in fetal bovine serum.

Mulchandani's group developed a method of growing individual NW electrochemically or as an array; for example, Pd and polypyrrole NWs 500 nm in width, up to 200 nm in thickness, and 7 μm in length could be grown and used as hydrogen and pH sensors, respectively [320, 321]. Electropolymerization of pyrrole together with avidin allowed development of a sensor for the biotin unit; for example, biotinylated DNA caused a rapid change in the NW resistance, even at levels as low as 1 nmol L^{-1} [320]. Similar NWs could be functionalized with an antibody to the cancer marker protein CA125 to give an immunosensor for the protein with a range of 1–1000 U mL^{-1} CA125 and good specific response in blood plasma [321].

Nanovesicles of carboxylated PEDOT containing a human dopamine receptor could be formulated [322] and integrated into an FET system that gave a rapid response (<1 s) to dopamine in human serum with a detection limit of 10 pmol L^{-1}. Polypyrrole NWs (diameter 60–90 nm) were modified by a tripeptide probe (Gly–Gly–His) [323] and then used to sense Cu^{2+} in the concentration range of 20–300 nmol L^{-1}.

Nanostructured films have also been utilized within biosensors; for example, gold nanopillar electrodes were used as a substrate to electrodeposit polypyrrole along with entrapped glucose oxidase to make a glucose sensor, which was five times more sensitive than a similar sensor with planar electrodes [324]. Thin films of polyaniline could be deposited with entrapped uricase, which is removed by hydrolysis, and the resultant porous film then used as a sensor for uricase [325]. Thin films of polyaniline NWs (80–100 nm diameter) could be electrochemically deposited and covalently substituted with oligonucleotides; the resultant films could detect amounts of complementary and noncomplementary strands at levels as low as 1 pmol L^{-1} using differential pulse voltammetry [326].

Polyaniline NWs were electropolymerized with chiral camphorsulfonic acid as the dopant and had been shown to respond differently to different enantiomers of amino acids [327]. Polyaniline nanoparticle/enzyme composites have also been successfully immobilized onto electrodes to provide sensors for hydrogen peroxide [328] and phosphate [329]. The synthesis of molecularly imprinted polypyrrole as a sensor for ascorbic acid [330] and polyaniline boronic acid as a sensor for saccharides has also been proved to be possible [331].

11.9 Conclusions and Future Trends

We have discussed within this work a wide range of nanostructured materials that act as smart materials and the properties of sensors developed from these. Obviously, there are almost infinite possibilities for these materials, not only in isolation but also in combination. These materials have huge potentials to act in synergy, for example, Pt nanoparticles to provide catalytic activity, graphene to provide enhanced surface area, a conductive polymer for conductivity, and an HG for stability and biocompatibility. Current applications of biosensors are often laboratory based although there are home-use applications such as the glucose biosensor. However, as discussed here, the size reductions that could be attained by using nanomaterials allow for the possibility of *in vivo* monitoring, although issues with biocompatibility still need to be addressed. Another field in which these materials could be of use is in wearable chemo- and biosensors for continuous monitoring of the well-being of the wearer or of their environment [332, 333]. Work is already being undertaken in such fields as temporary biosensor tattoos to monitor sweat metabolites or the progress of wound healing. The use of nanomaterials will help address some of the issues required (reductions in size, speed of response, power requirements) to utilize these types of sensors in monitoring the subjects' health and in sport and military applications.

Metal nanostructures are among the oldest of smart materials and will continue to be so because of their ease of synthesis and functionalization as well as their intimate contact with biomolecules, high conductivity, stability, catalytic ability, and relatively low toxicity. There are some issues with reproducibility as sizes and shapes of the particles can be difficult to control and mixed sizes and shapes from the same synthesis may occur. Also in many cases the high optical scrutiny and sensitivity of the plasmon response of these nanomaterials allow the use in optical assays including simple visual assays that produce obvious color changes.

Carbon structures such as graphene and CNTs will also be of interest since they are relatively cheap to produce and display good electrical and catalytic abilities. CNTs especially could be utilized as "nanoneedles" to investigate conditions inside single cells or as tips for AFM probes. However there are issues with both carbon materials. Heterogeneity is a problem as CNTs can often be mixtures of lengths and thicknesses and can be semiconducting or metallic; separation of different types of nanotubes can be problematic and lead to contamination or degradation. Graphene can also be heterogeneous, the flakes can be of different sizes and shapes and be single or a few layers, and there can also be issues with folding or bending. Also due to its high lipophilicity and high surface area, graphene is often prone to contamination, which could be an issue in chemical exfoliation-produced graphene and also may cause it to stack, reverting it back to a more graphite-like structure. However, the high conductivity, catalytic activity, functionality, and other properties of both families of carbon nanomaterials will ensure they remain widely studied within the sensor field of research.

MNPs allow a degree of control, which many other systems lack; for example, a simple magnetic field can draw them to the sensor surface or remove nonspecifically bound species. Also one problem with many forms of sensors is background signal; however there is no magnetic background in almost every biological sample. One field especially of interest is the potential for their combination with MRI technology to perform imaging of conditions rather than just their detection.

The catalytic abilities of NMOs allow them to be used in highly sensitive biodetection protocols. Also of interest is the ability to functionalize NMOs with biological agents and the potential to vary their morphologies, for example, fibers, flowers, and so on, to enhance their properties.

Conducting polymers offer advantages due to their easy synthesis by chemical or electrochemical methods, easy functionalization, often good biocompatibility, and ease of control. However there are potential issues with stability due to the chemical reactions of the polymers, changes in oxidation state, and so on that need to be addressed.

HGs display good biocompatibility and changes in size or shape and electrical or optical properties such as color changes. However improvements in detection range and sensitivity are required for many assays, and there are issues related to robustness and stability. However polymer chemistry is a nature field and there exist many techniques for optimizing the properties offered by HGs.

Overall, we can conclude that various nanostructured materials will all have a large part to play in future biosensing and biomedical applications. The study of these smart nanostructures will allow developments in the field of disease diagnosis and monitoring as well as environmental, agricultural, veterinary, and industrial applications, allowing for improvements in human health and the environment. Deciding which nanomaterials to use in sensors will depend on their stability, selectivity, sensitivity, and reproducibility. In some cases, especially for *in vivo* applications, biocompatibility and toxicity will be vital. The speed and, in the case of imaging applications, the spatial resolution conferred by these materials will be highly beneficial.

One issue with many of the works presented here is that they deal with the response of a sensor system to a single stimulus. However, in biological, physiological, or environmental systems, the "health" of the system is usually a combination of many factors. The potential ability of nanostructured systems to respond to multiple stimuli is therefore of great importance. The possibilities conferred by the small size of these systems allow for the potential to develop chips that could be implanted under the skin to continually monitor the physiological state or to construct smart devices that could, for example, continually monitor glucose levels in a diabetic patient and release insulin when required. However, it is clear that there is still a long way to go before we reach this and a lot of work still to be undertaken.

Acknowledgment

Z.A. gratefully acknowledges support from the European Commission, Marie Curie Actions and IPODI.

References

1 McGraw-Hill; Parker, S. P.. *McGraw-Hill Dictionary of Scientific and Technical Terms*, 6th ed., McGraw-Hill Education, New York, **2003**.
2 ISO/TS 800004-1 *Nanotechnologies: Vocabulary. Part 1: Core Terms*, ISO, Geneva, **2011**.
3 Clark, L.; Lyons, C. *Ann. N. Y. Acad. Sci.* **1962**, 102, 29–45.
4 Wang, J. *Electroanalysis* **2001**, 13, 983–988.
5 Newman, J. D.; Tigwell, L. J.; Turner, A. P. F.; Warner, P. J. Biosensors: A Clearer View. In Turner, A. P. F. (Ed) *Biosensors 2004: The 8th World Congress on Biosensors*, Elsevier, New York, **2004**.
6 Davis, F.; Higson, S. P. J. *Biosens. Bioelectron.* **2005**, 21, 1–20.
7 Cosnier, S. *Electroanalysis* **2005**, 17, 1701–1715.
8 Diaz-Gonzalez, M.; Gonzalez-Garcia, M. B.; Costa-Garci, A. *Electroanalysis* **2005**, 17, 1901–1918.
9 Rodriguez-Mozaz, S.; de Alda, M. J. L.; Barcelo, D. *Anal. Bioanal. Chem.* **2006**, 386, 1025–1041.
10 Centi, S.; Laschi, S.; Mascini, M. *Bioanalysis* **2009**, 1, 1271–1291.
11 Holford, T. R. J.; Davis, F.; Higson, S. P. J. *Biosens. Bioelectron.* **2012**, 34, 12–24.
12 Yun, Y.-H.; Eteshola, E.; Bhattacharya, A.; Dong, Z.; Shim, J.-S.; Conforti, L.; Kim, D.; Schulz, M. J.; Ahn, C. H.; Watts, N. *Sensors* **2009**, 9, 9275–9299.
13 Pickup, J. C.; Zhi, Z.-L.; Khan, F.; Saxl, T.; Birch, D. J. S. *Diabetes Metab. Res. Rev.* **2008**, 24, 604–610.
14 Cash, K. J.; Clark, H. A. *Trends Mol. Med.* **2010**, 16, 584–593.
15 Honeychurch, K. C. Ed. *Nanosensors for Chemical and Biological Applications: Sensing with Nanotubes, Nanowires and Nanoparticles*, Woodhead, Cambridge, **2014**.
16 Li, S.; Ge, Y.; Li, H. *Smart Nanomaterials for Sensor Application*, Bentham, Dubai, **2012**.
17 Choi, S.; Tripathi, A.; Singh, D. *J. Biomed. Nanotech.* **2014**, 10, 3162–3188.
18 Pandey, P.; Datta, M.; Malhotra, B. D. *Anal. Lett.* **2008**, 41, 159–209.
19 Merkoci, A. *Biosens. Bioelectron.* **2010**, 26, 1164–1177.
20 Siangproh, W.; Dungchaib, W.; Rattanaratc, P.; Chailapakul, O. *Anal. Chim. Acta* **2011**, 690, 10–25.
21 Wang, M.; Gu, X.; Zhang, G.; Zhang, D.; Zhu, D. *Langmuir* **2009**, 25, 2504–2507.
22 Zhuo, Y.; Yuan, K.; Chai, Y.; Zhang, Y.; Li, X.; Wang, N.; Zhu, Q. *Sens. Actuators B* **2006**, 114, 631–639.
23 Liang, W.; Yi, W.; Li, S.; Yuan, R.; Chen, A.; Chen, S.; Xiang, G.; Hu, C. *Clin. Biochem.* **2009**, 42, 1524–1530.
24 Yin, T. J.; Wei, W.; Yang, L.; Gao, X.; Gao, Y. *Sens. Actuators B* **2006**, 117, 286–294.

25 Fu, Y. Z.; Yuan, R.; Tang, D.; Chai, Y.; Xu, L. *Colloids Surf. B.* **2005**, 40, 61–66.
26 Du, D.; Xu, X.; Wang, S.; Zhang, A. *Talanta* **2007**, 71, 1257–1262.
27 Liang, K. Z.; Mu, W. J. *Anal. Chim. Acta* **2006**, 580, 128–135.
28 Kim, D. M.; Noh, H. B.; Park, D. S.; Ryu, S. H.; Koo, J. S.; Shim, Y. B. *Biosens. Bioelectron.* **2009**, 25, 456–462.
29 Wu, S.; Zhong, Z.; Wang, D.; Li, M.; Qing, Y.; Dai, N.; Li, Z. *Microchim. Acta* **2009**, 166, 269–275.
30 Ding, C. F.; Li, H.; Hu, K.; Lin, J. M. *Talanta* **2010**, 80, 1385–1391.
31 Posthuma-Trumpie, G. A.; Korf, J.; van Amerongen, A. *Anal. Bioanal. Chem.* **2009**, 393, 569–582.
32 Wang, S.; Zhang, C.; Zhang, Y. *Methods Mol. Biol.* **2009**, 504, 237–252.
33 Ko, S.; Park, T. J.; Kim, H. S.; Kim, J. H.; Cho, Y. J. *Biosens. Bioelectron.* **2009**, 24, 2592–2597.
34 Yang, G. J.; Huang, J. L.; Meng, W. J.; Shen, M.; Jiao, X. A. *Anal. Chim. Acta* **2009**, 11, 159–166.
35 Chuang, Y. C.; Li, J. C.; Chen, S. H.; Liu, T. Y.; Kuo, C. H.; Huang, W. T.; Lin, C. S. *Biomaterials* **2010**, 31, 6087–6095.
36 Larguinho, M.; Baptista, P. V. *J. Proteomics* **2012**, 75, 2811–2823.
37 Lakshmi, D.; Whitcombe, M. J.; Davis, F.; Sharma, P. S.; Prasad, B. B. *Electroanalysis* **2011**, 23, 305–320.
38 Wang, P.; Li, Y.; Huang, X.; Wang, L. *Talanta* **2007**, 73, 431–437.
39 Li, J.; Lin, X. Q. *Anal. Chim. Acta* **2007**, 596, 222–230.
40 Goyal, R. N.; Oyama, M.; Sangal, A.; Singh, S. P. *Indian J. Chem. A Inorg. Bio-Inorg. Phys. Theo. Anal. Chem.* **2005**, 44, 945–949.
41 Mathiyarasu, J.; Senthilkumar, S.; Phani, K. L. N.; Yegnaraman, V. *J. Nanosci. Nanotechnol.* **2007**, 7, 2206–2210.
42 Crespilho, F. N.; Ghica, M. E.; Florescu, M.; Nart, F. C.; Oliveira, O. N.; Brett, C. M. A. *Electrochem. Commun.* **2006**, 8, 1665–1670.
43 Altintas, Z.; Kallempudi, S. S.; Gurbuz, Y. *Talanta* **2014**, 118, 270–276.
44 (a)Luo, X. L.; Xu, J. J.; Du, Y.; Chen, H. Y. *Anal. Biochem.* **2004**, 334, 284–289.; (b)Zhang, S.; Wang, N.; Yu, H.; Niu, Y. *Bioelectrochemistry* **2005**, 67, 15–22.
45 German, N.; Ramanaviciene, A.; Voronovic, J.; Ramanavicius, A. *Microchim. Acta* **2010**, 168, 221–229.
46 Elghanian, R.; Storhoff, J. J.; Storhoff, R. C.; Mucic, R.; Letsinger, R. L.; Mirkin, C. A. *Science* **1997**, 277, 1078–1081.
47 Huang, C. C.; Huang, Y. F.; Cao, Z.; Tan, W.; Chang, H. T. *Anal. Chem.* **2005**, 77, 5735–5741.
48 Liu, J.; Lu, Y. *Angew. Chem. Int. Ed.* **2006**, 45, 90–94.
49 Liu, J.; Mazumdar, D.; Lu, Y. *Angew. Chem. Int. Ed.* **2006**, 45, 7955–7959.
50 Dawan, S.; Kanatharana, P.; Wongkittisuksa, B.; Limbut, W.; Numnuam, A.; Limsakul, C.; Thavarungkul, P. *Anal. Chim. Acta* **2011**, 699, 232–241.
51 Sotiriou, G. S. *WIREs Nanomed. Nanobiotechnol.* **2013**, 5, 19–30.
52 Haes, A. J.; Van Duyne, R. P. *J. Am. Chem. Soc.* **2002**, 124, 10596–10604.

53 Galush, W. J.; Shelby, S. A.; Mulvihill, M. J.; Tao, A.; Yang, P. D.; Groves, J. T. *Nano Lett.* **2009**, 9, 2077–2082.

54 Zhu, S. L.; Li, F.; Du, C. L.; Fu, Y. Q. *Sens. Actuators B* **2008**, 134, 193–198.

55 Sotiriou, G. A.; Sannomiya, T.; Teleki, A.; Krumeich, F.; Voros, J.; Pratsinis, S. E. *Adv. Funct. Mater.* **2010**, 20, 4250–4257.

56 Jiang, C. F.; Zhao, T. T.; Li, S.; Gao, N. Y.; Xu, Q. H. *ACS Appl. Mater. Interfaces* **2013**, 5, 10853–10857.

57 Huang, Y.; Wen, Q.; Jiang, J.-H.; Shen, G.-L.; Yu, R.-Q. *Biosens. Bioelectron.* **2008**, 24, 600–605.

58 Zhang, J.; Ting, B. P.; Khan, M.; Pearce, M. C.; Yang, Y.; Gao, Z.; Ying, J. Y. *Biosens. Bioelectron.* **2010**, 26, 418–423.

59 Cheng, Y.-X.; Liu, Y.-J.; Huang, J.-J.; Feng, Z.; Zian, Y.-Z.; Wu, Z.-R.; Zhang, W.; Jin, L.-T. *Chin. J. Chem.* **2008**, 26, 302–306.

60 Jiang, X.; Wu, Y.; Mao, X.; Cui, X.; Zhu, L. *Sens. Actuators B* **2011**, 153, 158–163.

61 Li, H.; He, J.; Zhao, Y.; Wu, D.; Cai, Y.; Wei, Q.; Yang, M. *Electrochim. Acta* **2011**, 56, 2960–2965.

62 Lu, J.; Do, I.; Drzal, L. T.; Worden, R. M.; Lee, I. *ACS Nano* **2008**, 2, 1825–1832.

63 Toghill, K. E.; Compton, R. G. *Int. J. Electrochem. Sci.* **2010**, 5, 1246–1301.

64 Xiao, F.; Zhao, F. Q.; Mei, D. M.; Mo, Z. R.; Zeng, B. Z. *Biosens. Bioelectron.* **2009**, 24, 3481–3486.

65 Lu, L. M.; Li, H. B.; Qu, F. L.; Zhang, X. B.; Shen, G. L.; Yu, R. Q. *Biosens. Bioelectron.* **2011**, 26, 3500–3504.

66 Safavi, A.; Maleki, N.; Farjami, E. *Biosens. Bioelectron.* **2009**, 24, 1655–1660.

67 Huang, J. S.; Wang, D. W.; Hou, H. Q.; You, T. Y. *Adv. Funct. Mater.* **2008**, 18, 441–448.

68 Wang, Q.; Wang, Q. Y.; Qi, K.; Xue, T. Y.; Liu, C.; Zheng, W. T.; Cui, X. G. *Anal. Methods* **2015**, 7, 8605–8610.

69 Li, X. L.; Yao, J. Y.; Liu, F. L.; He, H. C.; Zhou, M.; Mao, N.; Zhang, Y. *Sens. Actuators B* **2013**, 181, 501–508.

70 Netto, C. G. C. M.; Toma, H. E.; Andrade, L. M. *J. Mol. Catal. B* **2013**, 85–86, 71–92.

71 Aguilar-Arteaga, K.; Rodriguez, J. A.; Barrado, E. *Anal. Chim. Acta* **2010**, 674, 157–165.

72 Philippova, O.; Barabanova, A.; Molchanov, V.; Khokhlov, A. *Eur. Polym. J.* **2011**, 47, 542–559.

73 Akbarzadeh, A.; Samiei, M.; Daravan, S. *Nanoscale Res. Lett.* **2012**, 7, 1–13.

74 Reddy, L. H.; Arias, J. L.; Nicolas, J.; Couvreur, P. *Chem. Rev.* **2012**, 112, 5818–5878.

75 Colombo, M.; Carregal-Romero, S.; Casula, M. F.; Gutiérrez, L.; Morales, M. P.; Bohm, I. B.; Heverhagen, J. T.; Prosperi, D.; Parak, W. J. *Chem. Soc. Rev.* **2012**, 12, 4306–4334.

76 Lu, L.-Y.; Yu, L.-N.; Xu, X.-G.; Jiang, Y. *Rare Met.* **2013**, 32, 323–331.

77 Huang, S.-H.; Juang, R.-S. *J. Nanopart. Res.* **2011**, 13, 4411–4430.
78 Beveridge, J. S.; Stephens, J. R.; Williams, M. E. *Annu. Rev. Anal. Chem.* **2011**, 4, 251–273.
79 Xu, Y.; Wang, E. *Electrochim. Acta* **2012**, 84, 62–73.
80 Rocha-Santos, T. A. P. *Trends Anal. Chem.* **2014**, 62, 28–36.
81 Bruls, D. M.; Evers, T. H.; Kahlman, J. A. H.; van Lankvelt, P. J. W.; Ovsyanko, M.; Pelssers, E. G. M.; Schleipen, J. J. H. B.; de Theije, F. K.; Verschuren, C. A.; van der Wijk, T.; van Zon, J. B. A.; Dittmer, W. U.; Immink, A. H. J.; Nieuwenhuis, J. H.; Prins, M. W. J. *Lab Chip* **2009**, 9, 3504–3510.
82 Palecek, E.; Fojta, M. *Talanta* **2007**, 74, 276–290.
83 Zamfir, L.-G.; Geana, I.; Bourigua, S.; Rotariu, L.; Bala, C.; Errachid, A.; Jaffrezic-Renault, N. *Sens. Actuators B* **2011**, 159, 178–184.
84 Wang, Y.; Dostalek, J.; Knoll, W. *Anal. Chem.* **2011**, 83, 6202–6207.
85 Liang, R.-P.; Yao, G.-H.; Fan, L.-X.; Qiu, J.-D. *Anal. Chim. Acta* **2012**, 737, 22–28.
86 Wang, J.; Zhu, Z.; Munir, A.; Zhou, H. S. *Talanta* **2011**, 84, 783–788.
87 Wang, L.; Sun, Y.; Wang, J.; Wang, J.; Yu, A.; Zhang, H.; Song D. *Colloids Surf. B* **2011**, 84, 484–490.
88 Wang, J.; Song, D.; Zhang, H.; Zhang, J.; Jin, Y.; Zhang, H.; Sun, Y. *Colloids Surf. B* **2013**, 102, 165–170.
89 Zhang, H.; Sun, Y.; Wang, J.; Zhang, J.; Zhang, H.; Zhou, H.; Song, D. *Biosens. Bioelectron.* **2012**, 34, 137–143.
90 Agrawal, S.; Paknikar, K.; Bodas, D. *Microelectron. Eng.* **2014**, 115, 66–69.
91 Gan, N.; Yang, X.; Xie, D.; Wu, T.; Wen, W. *Sensors* **2010**, 10, 625–638.
92 Yang, X.; Wu, F.; Chen, D.-Z.; Lin, H.-W. *Sens. Actuators B* **2014**, 192, 529–535.
93 Li, J.; Gao, H.; Chen, Z.; Wei, X.; Yang, C. F. *Anal. Chim. Acta* **2010**, 665, 98–104.
94 Hervás, M.; López, M. A.; Escarpa, A. *Biosens. Bioelectron.* **2010**, 25, 1755–1760.
95 Arvand, M.; Hassannezhad, M. *Mater. Sci. Eng. C* **2014**, 36, 160–167.
96 Chen, X.; Zhu, J.; Chen, Z.; Xu, C.; Wang, Y.; Yao, C. *Sens. Actuators B* **2011**, 159, 220–228.
97 Baby, T. T.; Ramaprabhu, S. *Talanta* **2010**, 80, 2016–2022.
98 Yang, Z.; Zhang, C.; Zhang, J.; Bai, W. *Biosens. Bioelectron.* **2014**, 51, 268–273.
99 Muzyka, K. *Biosens. Bioelectron.* **2014**, 51, 393–407.
100 Zhou, H.; Gan, N.; Li, T.; Cao, Y.; Zeng, S.; Zheng, L.; Guo, Z. *Anal. Chim. Acta* **2012**, 746, 107–113.
101 Li, J.; Xu, Q.; Wei, X.; Hao, Z. *J. Agric. Food Chem.* **2013**, 61, 1435–1440.
102 Li, D.; Wang, J.; Wang, R.; Li, Y.; Abi-Ghanem, D.; Berghman, L.; Hargis, B.; Lu, H. *Biosens. Bioelectron.* **2011**, 26, 4146–4154.
103 Wan, Y.; Zhang, D.; Hou, B. *Biosens. Bioelectron.* **2010**, 25, 1847–1850.

104 Zhou, J.; Gan, N.; Li, T.; Zhou, H.; Li, X.; Cao, Y.; Wang, L.; Sang, W.; Hu, F. *Sens. Actuators B* **2013**, 178, 494–500.

105 Gan, N.; Wang, L.; Li, T.; Sang, W.; Hu, F.; Cao, Y. *Integr. Ferroelectr.* **2013**, 144, 29–40.

106 Sun, X.; Ho, D.; Lacroix, L.-M.; Xiao, J. Q.; Sun, S. *IEEE Trans. Nanobiosci.* **2012**, 11, 46–53.

107 Srinivasan, B.; Li, Y.; Jing, Y.; Xing, C.; Slaton, J.; Wang, J.-P. *Anal. Chem.* **2011**, 83, 2996–3002.

108 Li, Y. Srinivasan, B.; Jing, Y.; Yao, X.; Hugger, M. A.; Wang, J.-P.; Xing, C. *J. Am. Chem. Soc.* **2010**, 132, 4388–4392.

109 Issadore, D.; Chung, J.; Shao, H.; Liong, M.; Ghazani, A. A.; Castro, C. M.; Weissleder, R.; Lee, H. *Sci. Transl. Med.* **2012**, 141, 1–22.

110 Issadore, D.; Chung, H. J.; Chung, J.; Budin, G.; Weissleder, R.; Lee, H. *Adv. Healthc. Mater.* **2013**, 2, 1224–1228.

111 Haun, J. B.; Yoon, T.-J.; Lee, H.; Weissleder, R. *Wiley Interdiscip. Rev. Nanomed. Nanobiotechnol.* **2010**, 2, 291–304.

112 Josephson, L.; Perez, J. M.; Weissleder, R. *Angew. Chem. Int. Ed. Engl.* **2001**, 4, 3204–3206.

113 Kim, G. Y.; Josephson, L.; Langer, R.; Cima, M. J. *Bioconjug. Chem.* **2007**, 18, 2024–2028.

114 Lee, H.; Yoon, T. J.; Weissleder, R. *Angew. Chem. Int. Ed. Engl.* **2009**, 48, 5657–5660.

115 Grimm, J.; Perez, J. M.; Josephson, L.; Weissleder R. *Cancer Res.* **2004**, 64, 639–643.

116 Perez, J. M.; O'Loughin, T.; Simeone, F. J.; Weissleder, R.; Josephson, L. *J. Am. Chem. Soc.* **2002**, 124, 2856–2857.

117 Zhao, M.; Josephson, L.; Tang, Y.; Weissleder, R. *Angew. Chem. Int. Ed. Engl.* **2003**, 42, 1375–1378.

118 Tsourkas, A.; Hofstetter, O.; Hofstetter, H.; Weissleder, R.; Josephson, L. *Angew. Chem. Int. Ed. Engl.* **2004**, 43, 2395–2399.

119 Perez, J. M.; Simeone, F. J.; Saeki, Y.; Josephson, L.; Weissleder, R. *J. Am. Chem. Soc.* **2003**, 125, 10192–10193.

120 Iijima, S. *Nature* **1991**, 354, 56–58.

121 He, H.; Zuo, P.; Pham-Huy, L. A.; Dramou, P.; Pham-Huy, C.; Xiao, D. *Biomed. Res. Int.* **2013**, 2013, 578290–1/12.

122 Kostarelos, K.; Bianco, A.; Prato M. *Nat. Nanotechnol.* **2009**, 4, 627–633.

123 Vidu, R.; Rahman, M.; Mahmoudi, M.; Enachescu, M.; Poteca, T. D.; Opris, I. *Front. Syst. Neurosci.* **2014**, 8, 91–1/24.

124 Tilmaciu, C.-M.; Morris, M. C. *Front. Chem.* **2015**, 3, 59–121.

125 Faraj, A. A. *Nanomedicine* **2016**, 11, 1431–1445.

126 Yang, N.; Chen, X.; Ren, T.; Zhang, P.; Yang, D. *Sens. Actuators B Chem.* **2015**, 207, 690–715.

127 Jacobs, C. B.; Peairs, M. J.; Venton, B. J. *Anal. Chim. Acta* **2010**, 662, 105–127.

128 Balasubramanian, K.; Burghard, M. *Anal. Bioanal. Chem.* **2006**, 385, 452–468.

129 Saxena, U.; Das, A. B. *Biosens. Bioelectron.* **2016**, 75, 196–205.

130 Li, Z.; Chen, Y.; Xin, Y.; Zhang, Z. *Sci. Rep.* **2015**, 5, 16115–1/8.

131 Xu, W.; Dai, S.; Wang, X.; He, X.; Wang, M.; Hu, C.; Xi, Y. *J. Mater. Chem. B* **2015**, 3, 5777–5785.

132 Tian, K.; Prestgard, M.; Tiwari, A. *Mater. Sci. Eng. C* **2014**, 41, 100–118.

133 Saha, M.; Das, S. *J. Nanostruct. Chem.* **2014**, 4, 94–1/9.

134 Ye, J.-S.; Wen, Y.; Zhang, W. D.; Gan, L. M.; Xu, G. Q.; Sheu, F.-S. *Electrochem. Commun.* **2004**, 6, 66–70.

135 Zhu, Z.; Garcia-Gancedo, L.; Flewitt, A. J.; Xie, H.; Moussy, F.; Milne, W. I. *Sensors* **2012**, 12, 5996–6022.

136 Ensafi, A. A.; Zandi-Atashbar, N.; Rezaei, B.; Ghiaci, M.; Chermahinia, M. E.; Moshiri, P. *RSC Adv.* **2016**, 6, 60926–60932.

137 Ensa, A. A.; Abarghoui, M. M.; Rezaei, B. *Electrochim. Acta* **2014**, 123, 219–226.

138 Shahrokhian, S.; Salimiana, R.; Kalhor, H. R. *RSC Adv.* **2016**, 6, 15592–15598.

139 de-los-Santos-Álvarez, N.; Lobo-Castañón, M.; Miranda-Ordieres, A. J.; Tuñón-Blanco, P. *Trends Anal. Chem.* **2008**, 27, 437–446.

140 Tlili, C.; Cella, L. N.; Myung, N. V.; Shetty, V.; Mulchandani, A. *Analyst* **2010**, 135, 2637–2642.

141 Zheng, M.; Jagota, A.; Semke, E. D.; Diner, B. A.; Mclean, R. S.; Lustig, S. R.; Richardson, R. E.; Tassi, N. G. *Nat. Mater.* **2003**, 2, 338–342.

142 Park, K.; Kwon, S. J.; Kwak, J. *Electroanalysis* **2014**, 26, 513–520.

143 Soares, A. C.; Soares, J. C.; Shimizu, F. M.; Melendez, M. E.; Carvalho, A. L.; Oliveira, O. N. *ACS Appl. Mater. Interfaces* **2015**, 7, 25930–25937.

144 Hershkovitz-Rokah, O.; Modai, S.; Pasmanik-Chor, M.; Toren, A.; Shomron, N.; Raanani, P.; Shpilberg, O.; Granot, G. *Cancer Lett.* **2015**, 356, 597–605.

145 Zhang, W. *J. Appl. Electrochem.* **2016**, 46, 559–566.

146 Fang, Y.; Wang, E. *Chem. Commun.* **2013**, 49, 9526–9539.

147 Wu, S.; He, Q.; Tan, C.; Wang, Y.; Zhang, H. *Small* **2013**, 9, 1160–1172.

148 Wisitsoraat, A.; Tuantranont, A. Graphene-Based Chemical and Biosensors. In Tuantranont, A. (Ed) *Applications of Nanomaterials in Sensors and Diagnostics*, Springer-Heidelberg, New York, **2013**.

149 Yang, W.; Ratinac, K. R.; Ringer, S. P.; Thordarson, P.; Gooding, J. J.; Braet, F. *Angew. Chem. Int. Ed.* **2010**, 49, 2114–2138.

150 Novoselov, K. S.; Geim, A. K.; Morozov, S. V.; Jiang, D.; Zhang, Y.; Dubonos, S. V.; Grigorieva, I. V.; Firsov, A. A. *Science* **2004**, 306, 666–669.

151 Hummers, W. S.; Offeman, R. E. *J. Am. Chem. Soc.* **1958**, 80, 1339–1339.

152 Fowler, J. D.; Allen, M. J.; Tung, V. C.; Yang, Y.; Kaner, R. B.; Weiller, B. H. *ACS Nano* **2008**, 3, 301–306.

153 Huang, D.; Zhang, B.; Zhang, Y.; Zhan, F.; Xu, X.; Shen, Y.; Wang, M. *J. Mater. Chem. A* **2013**, 1, 1415–1420.
154 Coleman, J. N.; Lotya, L.; O'Neill, A.; Bergin, S. D. *Science* **2001**, 331, 568–571.
155 Notley, S. M. *Langmuir* **2012**, 28, 14110–14113.
156 Paton, K. R.; Varrla, E.; Backes, C.; Smith, R. J.; Khan, U.; O'Neill, A.; Boland, C.; Lotya, M.; Istrate, O. M.; King, P.; Higgins, T.; Barwich, S.; May, P.; Puczkarski, P.; Ahmed, I.; Moebius, M.; Pettersson, H.; Long, E.; Coelho, J.; O'Brien, S. E.; McGuire, E. K.; Sanchez, B. M.; Duesberg, G. S.; McEvoy, N.; Pennycook, T. J.; Downing, C.; Crossley, A.; Nicolosi, V.; Coleman, J. N. *Nat. Mater.* **2014**, 13, 624–630.
157 Zhou, M.; Zhai, Y.; Dong, S. *Anal. Chem.* **2009**, 81, 5603–5613.
158 Casero, E.; Parra-Alfambra, A. M.; Petit-Domínguez, M. D.; Pariente, F.; Lorenzo, E.; Alonso, C. *Electrochem. Commun.* **2012**, 20, 63–66.
159. Zeng, Q.; Cheng, J. S.; Liu, X. F.; Bai, H. T.; Jiang, J. H. *Biosens. Bioelectron.* **2011**, 26, 3456–3463.
160 Zhang, Q.; Wu, S.; Zhang, L.; Lu, J.; Verproot, F.; Liu, Y.; Xing, Z.; Li, J.; Song, X. M. *Biosens. Bioelectron.* **2011**, 26, 2632–2637.
161 Wang, Z. J.; Zhou, X. Z.; Zhang, J.; Boey, F.; Zhang, H. *J. Phys. Chem. C* **2009**, 113, 14071–14075.
162 Kang, X. H.; Wang, J.; Wu, H.; Aksay, I. A.; Liu, J.; Lin, Y. H. *Biosens. Bioelectron.* **2009**, 25, 901–905.
163 Shan, C. S.; Yang, H. F.; Song, J. F.; Han, D. X.; Ivaska, A.; Niu, L. *Anal. Chem.* **2009**, 81, 2378–2382.
164 Wang, J.; Yang, S.; Guo, D.; Yu, P.; Li, D.; Ye, J.; Mao, L. *Electrochem. Commun.* **2009**, 11, 1892–1895.
165 Zhang, X.; Liao, Q.; Chu, M.; Liu, S.; Zhang, Y. *Biosens. Bioelectron.* **2014**, 52, 281–287.
166 Huang, K. J.; Niu, D. J.; Liu, X.; Wu, Z. W.; Fan, Y.; Chang, Y. F.; Wu, Y. Y. *Electrochim. Acta* **2011**, 56, 2947–2953.
167 Srivastava, R. K.; Srivastava, S.; Narayanan, T. N.; Mahlotra, B. D.; Vajtai, R.; Ajayan, P. M.; Srivastava, A. *ACS Nano* **2011**, 6, 168–175.
168 Prasannakumar, S.; Manjunatha, R.; Nethravathi, C.; Suresh, G. S.; Rajamathi, M.; Venkatesha, T. V. *J. Solid State Electrochem.* **2012**, 16, 3189–3199.
169 Zeng, G. H.; Xing, Y. B.; Gao, J. A.; Wang, Z. Q.; Zhang, X. *Langmuir* **2010**, 26, 15022–15026.
170 Parlak, O.; Tiwari, A.; Turner, A. P. F.; Tiwari, A. *Biosens. Bioelectron.* **2013**, 49, 53–62.
171 Zhang, J.; Lei, J.; Pan, R.; Xue, Y.; Ju, H. *Biosens. Bioelectron.* **2010**, 26, 371–376.
172 Dong, X.-C.; Xu, H.; Wang, X.-W.; Huang, Y.-X.; Chan-Park, M. B.; Zhang, H.; Wang, L.-H.; Huang, W.; Chen, P. *ACS Nano* **2012**, 6, 3206–3213.
173 Zhou, S.; Wei, D.; Shi, H.; Feng, X.; Xue, K.; Zhang, F.; Song, W. *Talanta* **2013**, 107, 349–355.

174 Chen, Q.; Zhang, L.; Chen, G. *Anal. Chem.* **2012**, 84, 171–178.

175 Kim, Y. R.; Bong, S.; Kang, Y. J.; Yang, Y.; Mahajan, R. K.; Kim, J. S.; Kim, H. *Biosens. Bioelectron.* **2010**, 25, 2366–2369.

176 Walch, N. J.; Davis, F.; Langford, N.; Holmes, J. L.; Collyer, S. D.; Higson, S. P. J. *Anal. Chem.* **2015**, 87, 9273–9279.

177 Shang, N. G.; Papakonstantinou, P.; McMullan, M.; Chu, M.; Stamboulis, A.; Potenza, A.; Dhesi, S. S.; Marchetto, H. *Adv. Funct. Mater.* **2008**, 18, 3506–3514.

178 Wang, Y.; Li, Y.; Tang, L.; Lu, J.; Li, J. *Electrochem. Commun.* **2009**, 11, 889–892.

179 Kim, S. K.; Kim, D.; Jeon, S. *Sens. Actuators B* **2012**, 174, 285–291.

180 Alwarappan, S.; Erdem, A.; Liu, C.; Li, C. Z. *J. Phys. Chem. C* **2009**, 113, 8853–8857.

181 Cui, F.; Zhang, X. *J. Electroanal. Chem.* **2012**, 669, 35–41.

182 Tang, L.; Wang, Y.; Li, Y.; Feng, H.; Lu, J.; Li, J. *Adv. Funct. Mater.* **2009**, 19, 2782–2789.

183 Deng, K. Q.; Zhou, J. H.; Li, X. F. *Colloids Surf. B* **2013**, 101, 183–188.

184 Ye, L.; Luo, D.; Ding, Y.; Liu, B.; Liu, X. *Analyst* **2012**, 137, 2840–2845.

185 Liang, J.; Chen, Z.; Guo, L.; Li, L. *Chem. Commun.* **2011**, 47, 5476–5478.

186 Filip, J.; Kasák, P.; Tkaca, J. *Chem. Pap. Chem. Zvesti* **2015**, 69, 112–133.

187 Ohno, Y.; Maehashi, K.; Matsumoto, K. *J. Am. Chem. Soc.* **2010**, 132, 18012–18013.

188 Mao, S.; Yu, K.; Lu, G.; Chen, J. *Nano Res.* **2011**, 4, 921–930.

189 Loo, A. H.; Bonanni, A.; Ambrosi, A.; Poh, H. L.; Pumera, M. *Nanoscale* **2012**, 4, 921–925.

190 Wang, G.; Gang, X.; Zhou, X.; Zhang, G.; Huang, H.; Zhang, X.; Wang, L. *Talanta* **2013**, 103, 75–80.

191 Zhang, H.; Shuang, S.; Sun, L.; Chen, A.; Qin, Y.; Dong, C. *Microchim. Acta* **2014**, 181, 189–196.

192 Guo, Y.; Han, Y.; Guo, Y.; Dong, C. *Biosens. Bioelectron.* **2013**, 45, 95–101.

193 Mao, K.; Wu, D.; Li, Y.; Ma, H.; Ni, Z.; Yu, H.; Luo, C.; Wei, Q.; Du, B. *Anal. Biochem.* **2012**, 422, 22–27.

194 Lu, J.; Liu, S.; Ge, S.; Yan, M.; Yu, J.; Hu, X. *Biosens. Bioelectron.* **2012**, 33, 29–35.

195 Hernández, R.; Vallés, C.; Benito, A. M.; Maser, W. K.; Xavier Rius, F.; Riu, J. *Biosens. Bioelectron.* **2014**, 54, 553–557.

196 Wu, Y.; Xue, P.; Hui, K. M.; Kang, Y. *Biosens. Bioelectron.* **2014**, 52, 80–187.

197 Lim, C. X.; Hoh, H. Y.; Ang, P. K.; Loh, K. P. *Anal. Chem.* **2010**, 82, 7387–7393.

198 Akhavan, O.; Ghaderi, E.; Rahighi, R. *ACS Nano* **2012**, 6, 2904–2916.

199 Zhang, X.; Gao, F.; Cai, X.; Zheng, M.; Gao, F.; Jiang, S.; Wang, Q. *Mater. Sci. Eng. C Mater. Biol. Appl.* **2013**, 33, 3851–3857.

200 Wang, J.; Shi, A.; Fang, X.; Han, X.; Zhang, Y. *Microchim. Acta* **2014**, 181, 935–940.
201 Liu, A.-L.; Zhong, G.-X.; Chen, J.-Y.; Weng, S.-H.; Huang, H.-N.; Chen, W.; Lin, L.-Q.; Lei, Y.; Fu, F.-H.; Sun, Z.-L.; Lin, X.-H.; Yang, S.-Y. *Anal. Chim. Acta* **2013**, 767, 50–58.
202 Bonanni, A.; Ambrosi, A.; Pumera M. *Chem. Eur. J.* **2012**, 18, 1668–1673.
203 Dubuisson, E.; Yang, Z.; Loh, K. P. *Anal. Chem.* **2011**, 83, 2452–2460.
204 Dong, X.; Shi, Y.; Huang, W. *Adv. Mater.* **2010**, 22, 1649–1653.
205 Mao, S.; Lu, G.; Yu, K.; Bo, Z.; Chen, J. *Adv. Mater.* **2010**, 22, 3521–3526.
206 Sakhaee-Pour, A.; Ahmadian, M. T.; Vafai, A. *Solid State Commun.* **2008**, 147, 336–340.
207 Chen, J. R.; Jiao, X. X.; Luo, H. Q.; Li, N. B. *J. Mater. Chem. B* **2013**, 1, 861–864.
208 Han, J. M.; Ma, J.; Ma, Z. F. *Biosens. Bioelectron.* **2013**, 47, 243–247.
209 Baumer, M.; Freund, H.-J. *Prog. Surf. Sci.* **1999**, 61, 127–198.
210 Liu, J.; Li, Y.; Huang, X.; Zhu, Z. *Nanoscale Res. Lett.* **2010**, 5, 1177–1181.
211 Sarangi, S. N.; Nozaki, S.; Sahu, S. N. *J. Biomed. Nanotechnol.* **2015**, 11, 988–996.
212 Zhao, M.; Li, Z.; Han, Z.; Wang, K.; Zhou, Y.; Huang, J.; Ye, Z. *Biosens. Bioelectron.* **2013**, 49, 318–322.
213 Yang, K.; She, G.-W.; Wang, H.; Ou, X.-M.; Zhang, X.-H.; Lee, C.-S.; Lee, S.-T. *J. Phys. Chem. C* **2009**, 113, 20169–20172.
214 Neveling, D. P.; van den Heever, T. S.; Perold, W. J.; Dicks, L. M. T. *Sens. Actuators B* **2014**, 203, 102–110.
215 Choi, A.; Kim, K.; Jung, H.-I.; Lee, S. Y. *Sens. Actuators B* **2010**, 148, 577–582.
216 Li, X.; Zhao, C.; Liu, X. *Microsyst. Nanoeng.* **2015**, 1, 15014–1/7.
217 Baranowska-Korczyc, A.; Sobczak, K.; Dłużewski, P.; Reszka, A.; Kowalski, B. J. Kłopotowski, Ł.; Elbauma, D.; Fronca, K. *Phys. Chem. Chem. Phys.* **2015**, 17, 24029–24037.
218 Stafiniak, A.; Boratyński, B.; Baranowska-Korczyc, A.; Szyszka, A.; Ramiączek-Krasowska, M.; Prażmowska, J.; Fronc, K.; Elbaum, D.; Paszkiewicz, R.; Tłaczała, M. *Sens. Actuators B* **2011**, 160, 1413–1418.
219 Soejima, T.; Yagyu, H.; Kimizuka, N.; Ito, S. *RSC Adv.* **2011**, 1, 187–190.
220 Zhao, Z.; Tian, J.; Sang, Y.; Cabot, A.; Liu, H. *Adv. Mater.* **2015**, 27, 2557–2582.
221 Devasenathipathy, R.; Karthik, R.; Chen, S.; Ali, M. A.; Mani, V.; Lou, B.-S.; Mohammed, F.; Al-Hemaid, A. *Microchim. Acta* **2015**, 182, 2165–2172.
222 Ye, D.; Liang, G.; Li, H.; Luo, J.; Zhang, S.; Chen, H.; Kong, J. *Talanta* **2013**, 116, 223–230.
223 Cai, B.; Zhao, M.; Wang, Y.; Zhou, Y.; Cai, H.; Ye, Z.; Huang, J. *Ceram. Int.* **2014**, 40, 8111–8116.
224 Reitz, E.; Jia, W. Z.; Gentile, M.; Wang, Y.; Lei, Y. *Electroanalysis* **2008**, 20, 2482–2486.

225 Perumal, V.; Hashim, U.; Gopinath, S. C. B.; Haarindraprasad, R.; Foo, K. L.; Balakrishnan, S. R.; Poopalan, P. *Sci. Rep.* **2015**, 5, 12231–1/12.

226 Sun, S.; Zhang, X.; Sun, Y.; Yang, S.; Songa, X.; Yang, Z. *Phys. Chem. Chem. Phys.* **2013**, 15, 10904–10913.

227 Sun, S.; Zhang, X.; Sun, Y.; Yang, S.; Song, X.; Yang, Z. *ACS Appl. Mater. Interfaces* **2013**, 5, 4429–4437.

228 Reich, E. S. *Nature* **2013**, 495, 17.

229 Malhotra, B. D.; Das, M.; Solanki, P. R. *J. Phys. Conf. Ser.* **2012**, 358, 012007–1/8.

230 Solanki, P. R.; Kaushik, A.; Agrawal, V. V.; Malhotra, B. D. *NPG Asia Mater.* **2011**, 3, 17–24.

231 Zhang, Y.; Nayak, T. R.; Hong, H.; Cai, W. *Curr. Mol. Med.* **2013**, 13, 1633–1645.

232 Chen, C.; Xie, Q.; Yang, D.; Xiao, H.; Fu, Y.; Tan, Y.; Yao, S. *RSC Adv.* **2013**, 3, 4473–4491.

233 Zang, J. F.; Li, C. M.; Cui, X. Q.; Wang, J.; Sun, X.; Dong, H.; Sun, C. Q. *Electroanalysis* **2007**, 19, 1008–1014.

234 Dai, Z. H.; Shao, G. J.; Hong, J. M.; Bao, J. C.; Shen, J. *Biosens. Bioelectron.* **2009**, 24, 1286–1291.

235 Zhou, C.; Xu, L.; Song, J.; Xing, R.; Xu, S.; Liu, D.; Song, H. *Sci. Rep.* **2014**, 4, 7382–1/9.

236 Haddaoui, M.; Raouafi, N. *Sens. Actuators B* **2015**, 219, 171–178.

237 Gu, B. X.; Xu, C. X.; Zhu, G. P.; Liu, S. Q.; Chen, L. Y.; Li, X. S. *J. Phys. Chem. B* **2009**, 113, 377–381.

238 Umar, A.; Rahman, M. M.; Al-Hajry, A.; Hahn, Y. B. *Talanta* **2009**, 78, 284–289.

239 Singh, S. P.; Arya, S. K.; Pandey, P.; Malhotra, B. D.; Saha, S.; Sreenivas, K.; Gupta, V. *Appl. Phys. Lett.* **2007**, 91, 063901.

240 Naik, K. K.; Kumar, S.; Rout, C. S.; *RSC Adv.* **2015**, 5,74585–74591.

241 Khatiba, K. M. E.; Hameed, R. M. A. *Biosens. Bioelectron.* **2011**, 26, 3542–3548.

242 Marioli, J. M.; Kuwana, T. *Electrochim. Acta* **1992**, 37, 1187–1197.

243 Naik, K. K.; Rout, C. S. *RSC Adv.* **2015**, 5, 79397–79404.

244 Tak, M.; Gupta, V.; Tomar, M. *Biosens. Bioelectron.* **2014**, 59, 200–207.

245 Taylor, D. M.; Macdonald, A. G. *J. Phys. D Appl. Phys.* **1987**, 20, 1277–1283.

246 Bakhtiary, Z.; Saei, A. A.; Hajipour, M. J.; Raoufi, M.; Vermesh, O.; Mahmoudi, M. *Nanomed. Nanotechnol. Biol. Med.* **2016**, 12, 287–307.

247 Wang, X.; Yua, H.; Lu, D.; Zhang, J.; Deng, W. *Sens. Actuators B* **2014**, 195, 630–634.

248 Peppas, N. A.; Bures, P.; Leobandung, W.; Ichikawa, H. *Eur. J. Pharm. Biopharm.* **2000**, 50, 27–46.

249 Caldorera-Moore, M.; Peppas, N. A. *Adv. Drug Deliv. Rev.* **2009**, 61, 1391–1401.

250 Li, J.; Mo, L.; Lu, C.-H.; Fu, T.; Yang, H.-H.; Tan, W. *Chem. Soc. Rev.* **2016**, 45, 1410–1431.

251 Laftah, W. A.; Hashim, S.; Ibrahim, A. N. *Polym.-Plast. Technol. Eng.* **2011**, 50, 1475–1486.

252 Qiu, Y.; Park, K. *Adv. Drug Deliv. Rev.* **2001**, 53, 321–339.

253 Zhang, X.; Guan, Y.; Zhang, Y. *Biomacromolecules* **2012**, 13, 92–97.

254 Huang, P.; Chen, W.; Yan, L. *Nanoscale* **2013**, 5, 6034–6039.

255 Hendrickson, G. R.; Lyon, L. A. *Soft Matter* **2009**, 5, 29–35.

256 He, J.; Xu, F.; Cao, Y.; Liu, Y.; Li, D. *J. Phys. D Appl. Phys.* **2016**, 49, 055504/1–6.

257 Xiang, X.; Zhang, H.; Zhang, W.; Guo, H.; Yang, Z.; Zhang, X.; Zhang, Y.; Li, Q.; Zhang, H. *RSC Adv.* **2016**, 6, 24946–24951.

258 Dion, J. R.; Burns, D. H. *Talanta* **2011**, 83, 1364–1370.

259 Pan, L.; Yu, G.; Zhai, D.; Lee, H. R.; Zhao, W.; Liu, N.; Wang, H.; Tee, B. C.-K.; Shi, Y.; Cui, Y.; Bao, Z. *Proc. Natl. Acad. Sci. U. S. A.* **2012**, 109, 9287–9292.

260 Benito-Lopez, F.; Antoñana-Díez, M.; Curto, V. F.; Diamond, D.; Castro-López, V. *Lab Chip.* **2014**, 14, 3530–3538.

261 Kenna, N. M.; Calvert, P.; Morrin, A. *Analyst* **2015**, 140, 3003–3011.

262 Ullah, F.; Othman, M. B. H.; Javed, F.; Ahmad, Z.; Akil, H. M. *Mater. Sci. Eng. C* **2015**, 57, 414–433.

263 Kozlovskaya, V.; Kharlampieva, E.; Erel, I.; Sukhishvili, S. A. *Soft Matter* **2009**, 5, 4077–4087.

264 Tanaka, Y.; Gong, J. P.; Osada, Y. *Prog. Polym. Sci.* **2005**, 30, 1–9.

265 Hoare, T. R.; Kohane, D. S. *Polymer* **2008**, 49, 1993–2007.

266 Brahim, S.; Narinesingh, D.; Guiseppi-Elie, A. *Biosens. Bioelectron.* **2002**, 17, 973–981.

267 Zhao, Y.-X.; Ren, K.-F.; Sun, Y.-X.; Li, Z.-J.; Ji, J. *RSC Adv.* **2014**, 4, 24511–24517.

268 Rong, Q.; Han, H.; Feng, F.; Ma, Z. *Sci. Rep.* **2015**, 5, 11440/1–8.

269 Ziołkowski, B.; Diamond, D. *Chem. Commun.* **2013**, 49, 10308–10310.

270 Benito-Lopez, F.; Byrne, R.; Raduta, A. M.; Vrana, N. E.; McGuinness, G.; Diamond, D. *Lab Chip* **2010**, 10, 195–201.

271 Khodagholy, D.; Curto, V. F.; Fraser, K. J.; Gurfinkel, M.; Byrne, R.; Diamond, D.; Malliaras, G. G.; Benito-Lopez, F.; Owens, R. M. *J. Mater. Chem.* **2012**, 22, 4440–4443.

272 Burrs, S. L.; Yamaguchi, H.; Vanegas, D. C.; Gomes, C.; Bhargava, M.; Mechulan, N.; McLamore, E. S.; Hendershot, P. *Analyst* **2015**, 140, 1466–1476.

273 Kestwal, R. M.; Bagal-Kestwal, D.; Chiang, B. H. *Anal. Chim. Acta* **2015**, 886, 143–150.

274 Sharma, A.; Rawat, K.; Solanki, P. R.; Bohidar, H. B. *Anal. Methods* **2015**, 7, 5876–5885.

275 Zhang, Z.; Tang, Z.; Su, T.; Li, W.; Wang, Q. *RSC Adv.* **2015**, 5, 47244–47247.

276 Ivekovic, D.; Milardovic, S.; Grabaric, B. S. *Biosens. Bioelectron.* **2004**, 20, 872–878.

277 Brahim, S.; Narinesingh, D.; Guiseppi-Elie, A. *Biosens. Bioelectron.* **2002**, 17, 53–59.

278 Zhai, D.; Liu, B.; Shi, Y.; Pan, L.; Wang, Y.; Li, W.; Zhang, R.; Yu, G. *ACS Nano* **2013**, 7, 3540–3546.

279 Li, L.; Wang, Y.; Pan, L.; Shi, Y.; Cheng, W.; Shi, Y.; Yu, G. *Nano Lett.* **2015**, 15, 1146–1151.

280 Xiong, X.; Wu, C.; Zhou, C.; Zhu, G.; Chen, Z.; Tan, W. *Macromol. Rapid Commun.* **2013**, 34, 1271–1283.

281 Nguyen, K. V.; Holade, Y.; Minteer, S. D. *ACS Catal.* **2016**, 6, 2603–2607.

282 Qu, F.; Zhang, Y.; Rasooly, A.; Yang, M. *Anal. Chem.* **2014**, 86, 973–976.

283 Zhou, J.; Liao, C.; Zhang, L.; Wang, Q.; Tian, Y. *Anal. Chem.* **2014**, 86, 4395–4401.

284 Wu, X.; Li, Z.; Chen, X.-X.; Fossey, J. S.; James, T. D.; Jiang, Y.-B. *Chem. Soc. Rev.* **2013**, 42, 8032–8048.

285 Zhang, C.; Losego, M. D.; Braun, P. V. *Chem. Mater.* **2013**, 25, 3239–3250.

286 Guan, Y.; Zhang, Y. *Chem. Soc. Rev.* **2013**, 42, 8106–8121.

287 Gabai, R.; Sallacan, N.; Chegel, V.; Bourenko, T.; Katz, E.; Willner, I. *J. Phys. Chem. B* **2001**, 105, 8196–8202.

288 Mesch, M.; Zhang, C.; Braun, P. V.; Giessen, H. *ACS Photonics* **2015**, 2, 475–480.

289 Wang, R.; Li, Y. *Biosens. Bioelectron.* **2013**, 42, 148–155.

290 Xu, L.; Wang, R.; Kelso, L. C.; Ying, Y.; Li, Y. *Sens. Actuators B* **2016**, 234, 98–108.

291 Yoon, H. *Sensors* **2013**, 3, 524–549.

292 Xia, L.; Wei, Z.; Wan, M. *J. Colloid Interface Sci.* **2010**, 341, 1–11.

293 Aydemir, N.; Malmstrom, J.; Travas-Sejdic, J. *Phys. Chem. Chem. Phys.* **2016**, 18, 8264–8277.

294 Nguyen, D. N.; Yoon, H. *Polymers* **2016**, 8, 118–157.

295 Martin, J.; Maiz, J.; Sacristan, J.; Mijangos, C. *Polymer* **2012**, 53, 1149–1166.

296 Kros, A.; van Hovell, W. F. M.; Sommerdijk, N. A. J. M.; Nolte, R. J. M. *Adv. Mater.* **2001**, 13, 1555–1557.

297 Barton, A. C.; Collyer, S. D.; Davis, F.; Gornall, D. D.; Law, K. A.; Lawrence, E. C. D.; Mills, D. W.; Myler, S.; Pritchard, J. A.; Thompson, M.; Higson, S. P. J. *Biosens. Bioelectron.* **2004**, 20, 328–337.

298 Barton, A. C.; Davis, F.; Higson, S. P. H. *Anal. Chem.* **2008**, 80, 6198–6205.

299 Barton, A. C.; Davis, F.; Higson, S. P. H. *Anal. Chem.* **2008**, 80, 9411–9416.

300 Barton, A. C.; Collyer, S. D.; Davis, F.; Garifallou, G.-Z.; Tsekenis, G.; Tully, E.; O'Kennedy, R.; Gibson, T. D.; Millner, P. A.; Higson, S. P. J. *Biosens. Bioelectron.* **2009**, 24, 1090–1095.

301 Kwon, O. S.; Park, S. J.; Park, H. W.; Kim, T.; Kang, M.; Jang, J.; Yoon, H. *Chem. Mater.* **2012**, 24, 4088–4092.

302 Jang, J.; Yoon, H. *Small* **2005**, 1, 1195–1199.

303 Hong, J. Y.; Yoon, H.; Jang, J. *Small* **2010**, 6, 679–686.

304 Jang, J.; Yoon, H. *Chem. Commun.* **2003**, 720–721.

305 Long, Y. Z.; Li, M. M.; Gu, C. Z.; Wan, M. X.; Duvail, J. L.; Liu, Z. W.; Fan, Z. Y. *Prog. Polym. Sci.* **2011**, 36, 1415–1442.

306 Zhong, W. B.; Deng, J. Y.; Yang, Y. S.; Yang, W. T. *Macromol. Rapid Commun.* **2005**, 26, 395–400.

307 Liu, Z.; Zhang, X. Y.; Poyraz, S.; Surwade, S. P.; Manohar, S. K. *J. Am. Chem. Soc.* **2010**, 132, 13158–13159.

308 Ding, H. J.; Shen, J. Y.; Wan, M. X.; Chen, Z. J. *Macromol. Chem. Phys.* **2008**, 209, 864–871.

309 Park, H. W.; Kim, T.; Huh, J.; Kang, M.; Lee, J. E.; Yoon, H. *ACS Nano* **2012**, 6, 7624–7633.

310 Zhang, L. J.; Wan, M. X. *Thin Solid Films* **2005**, 477, 24–31.

311 Alam, M. M.; Wang, J.; Guo, Y. Y.; Lee, S. P.; Tseng, H. R. *J. Phys. Chem. B* **2005**, 109, 12777–12784.

312 Yoon, H.; Ko, S.; Jang, J. *J. Phys. Chem. B* **2008**, 112, 9992–9997.

313 Zhao, M.; Wu, X. M.; Cai, C. X. *J. Phys. Chem. C* **2009**, 113, 4987–4996.

314 Forzani, E. S.; Zhang, H. Q.; Nagahara, L. A.; Amlani, I.; Tsui, R.; Tao, N. J. *Nano Lett.* **2004**, 4, 1785–1788.

315 Wu, J. M.; Yin, L. W. *ACS Appl. Mater. Interfaces* **2011**, 3, 4354–4362.

316 Yoon, H.; Lee, S. H.; Kwon, O. S.; Song, H. S.; Oh, E. H.; Park, T. H.; Jang, J. *Angew. Chem. Int. Ed.* **2009**, 48, 2755–2758.

317 Song, H. S.; Kwon, O. S.; Lee, S. H.; Park, S. J.; Kim, U. K.; Jang, J.; Park, T. H. *Nano Lett.* **2013**, 13, 172–178.

318 Yoon, H.; Jang, J. *Mol. Cryst. Liq. Cryst.* **2008**, 491, 21–31.

319 Kwon, O. S.; Ahn, S. R.; Park, S. J.; Song, H. S.; Lee, S. H.; Lee, J. S.; Hong, J. Y.; Lee, J. S.; You, S. A.; Yoon, H.; Park, T. H.; Jang, J. *ACS Nano* **2012**, 6, 5549–5558.

320 Ramanathan, K.; Bangar, M. A.; Yun, M.; Chen, W.; Myung, N. V.; Mulchandani, A. *J. Am. Chem. Soc.* **2005**, 127, 496–497.

321 Bangar, M. A.; Shirale, D. J.; Chen, W.; Myung, N. V.; Mulchandani, A. *Anal. Chem.* **2009**, 81, 2168–2175.

322 Park, S. J.; Song, H. S.; Kwon, O. S.; Chung, J. H.; Lee, S. H.; An, J. H.; Ahn, S. R.; Lee, J. H.; Yoon, H.; Park, T. H.; Jang, J. *Sci. Rep.* **2014**, 4, 1–8.

323 Lin, M.; Cho, M.; Choe, W.; Yoo, J.; Lee, Y. *Biosens. Bioelectron.* **2010**, 26, 940–945.

324 Gangadharan, R.; Anandan, V.; Zhang, A.; Drwiega, J. C.; Zhang, G. *Sens. Actuators B* **2011**, 106, 991–998.

325 Kan, J. Q.; Pan, X. H.; Chen, C. *Biosens. Bioelectron.* **2004**, 19, 1635–1640.

326 Zhu, N. N.; Chang, Z.; He, P. G. Fang, Y. Z. *Electrochim. Acta* **2006**, 51, 3758–3762.

327 Huang, J.; Weib, H.; Chen, J. *Sens. Actuators B* **2008**, 134, 573–578.

328 Morrin, A.; Ngamna, O.; Killard, A. J.; Moulton, S. E.; Smyth, M. R.; Wallace, G. G. *Electroanalysis* **2005**, 17, 423–430.

329 Rahman, M. A.; Park, D. S.; Chang, S. C.; McNeil, C. J.; Shim, Y. B. *Biosens. Bioelectron.* **2006**, 21, 1116–1124.

330 Özcan, L.; Sahin, M.; Sahin, Y. *Sensors* **2008**, 8, 5792–5805.

331 Deore, B.; Freund, M. S. *Analyst* **2003**, 128, 803–806.

332 Bandodkar, A. J.; Wang, J. *Trends Biotechnol.* **2014**, 32, 363–371.

333 Bandodkar, A. J.; Jeerapan, I.; Wang, J. *ACS Sens.* **2016**, 1, 464–482.

12

Applications of Magnetic Nanomaterials in Biosensors and Diagnostics

Zeynep Altintas

Technical University of Berlin, Berlin, Germany

12.1 Introduction

Nanoparticles are the materials that possess dimension between 1 and 100 nm at least in one dimension and comprise from a couple of hundreds to 10^5 atoms. Magnetic nanoparticles are among these materials that display a response toward an applied magnetic field. These nanostructures are classified to five main types: ferromagnetic, ferrimagnetic, diamagnetic, paramagnetic, and antiferromagnetic. Ferromagnetic materials constitute the most commonly utilized ones and are derived from iron, cobalt, or nickel [1–4]. The unpaired electrons lead to a net magnetic moment in these magnetic nanomaterials (MNPs), and they are comprised from domains each involving a great number of atoms that have parallel magnetic moments, resulting in a net magnetic moment for the domain that indicates in some direction [5]. The magnetic moments of these domains are normally distributed randomly, and in case the ferromagnetic particle is located in a magnetic field, the domains' magnetic moments align along the applied magnetic field direction that creates a large net magnetic moment [6]. For paramagnetic materials, which are composed of gadolinium, lithium, tantalum, and magnesium, the atom possesses a net magnetic moment owing to the unpaired electrons; however, magnetic domains do not exist. In case of placing paramagnetic materials in a magnetic field, the atoms' magnetic moments align along the magnetic field that forms a poor net magnetic moment. Being another class of MNPs, diamagnetic materials, which contain copper, gold, silver, and some other elements, do not possess unpaired electrons in the atoms forming zero net magnetic moment; therefore, these MNPs demonstrate a very weak response toward the applied magnetic field since the electron orbits realign under the applied magnetic field. They do not

Biosensors and Nanotechnology: Applications in Health Care Diagnostics, First Edition.
Edited by Zeynep Altintas.
© 2018 John Wiley & Sons, Inc. Published 2018 by John Wiley & Sons, Inc.

display magnetic moment in case the magnetic field is removed. Antiferromagnetic materials, containing MnO, NiO, $CuCl_2$, and CoO, have occupied the different lattice positions and the two atoms possess magnetic moments. The magnitudes of the moments are equal and they have opposite directions that lead to zero net magnetic moment. Involving different atoms (Fe_3O_4 and γ-Fe_2O_3), ferrimagnetic materials also have different lattice positions in their atoms with antiparallel magnetic moments. Nevertheless, the magnitudes of the moments are not equal, leading to a net spontaneous magnetic moment. In case ferromagnetic and antiferromagnetic compounds are located in a magnetic field, they demonstrate a similar behavior to ferromagnetic materials. The size and shape of MNPs also influence their magnetic behavior. For instance, the superparamagnetism of MNPs is stated by the material type, the crystallinity of the structures, and the number of spins; therefore, a general rule cannot be assigned to predict the magnetic characteristics of an MNP. Magnetism is often measured by employing a magnetometer that tracks magnetization as a function of the applied magnetic field [2, 3, 7].

In recent years, various types of MNPs were manufactured: ferrites of magnesium, manganese, nickel, and cobalt; iron oxides (Fe_3O_4 and Fe_2O_3); multifunctional composites of MNPs such as FePt–Ag, CdS–FePt, Fe_3O_4–Ag, and Fe_3O_4–Au [4, 8, 9]. These nanomaterials are synthesized by physical or chemical techniques. Electron-beam lithography and gas-phase deposition are among the most commonly used physical methods. However, majority of the synthesis techniques rely on chemical approaches since it is not easy to control particle size down to nanometer scale using physical methods [1]. Coprecipitation, sol–gel synthesis, aerosol-/vapor-phase technique, high temperature thermal decomposition and/or reduction, flow-injection synthesis, supercritical fluid method, oxidation method, electrochemical technique, and nanoreactor-based synthesis constitute the chemical approaches that can be used for MNP production. Besides these widely employed techniques, microbial methods are also used for MNP synthesis, owing to offering remarkable features such as low cost, high yield, desirable stability, and good reproducibility.

To get rid of irreversible agglomeration and give opportunity for dissociation, MNPs are required to be stabilized. The surface coating of the nanoparticles by utilizing convenient surfactants and polymers can be applied for stabilization purpose [10, 11]. Dextran and polyethylene glycol (PEG) are among the widely used stabilization agents due to their biocompatibility. The agents form polymeric shells that prevent from cluster development after nucleation and keep the material domains against striking forces. The characterization of MNPs after their synthesis has played a pivotal role, and several analytical techniques are employed for this aim. Scanning electron microscopy (SEM), transmission electron microscopy (TEM), atomic force microscopy (AFM), near-field scanning optical microscopy (NSOM), scanning transmission microscopy (STEM), and environmental scanning electron microscopy (ESEM) are appropriate analytical tools to determine the shape and size of the synthesized MNPs. On the other hand, X-ray diffractometry (XRD), electron energy loss spectrometry (EELS), X-ray fluorescence (XRF), and energy-dispersive X-ray transmission-electron microscopy (EDX-EM) are the methods that are utilized to quantify the elemental constitutions of single MNPs. In case of applying MNPs in biosensing systems, these tools are also the most widely employed ones for characterization purposes [12, 13].

Sensing approaches based on MNPs have offered great advantages by providing enhanced sensitivity and specificity, lower detection limit, shorter analysis time, and high signal-to-noise ratio [14–18]. The implementations of MNPs in biosensors can be classified into two groups based on their functions: (i) The transducers, which can be optical, piezoelectric, electrochemical, and colorimetric, are modified with these nanomaterials for biological assays. (ii) Biomolecules are conjugated with MNPs to act as labels for biosensing. Table 12.1 displays the various applications of MNPs in biosensors for the detection of target biological compounds in different samples. Although the table includes different transduction principles, a detailed discussion of these transducers, beyond the scope of this chapter, can be found under Section 2 of this book. MNP-based biosensors have a broad range of applications in many fields such as medical applications, food industry, and environmental investigations. However, there is a need for special considerations in case of biomedical applications: (i) their size should be in the range of 10–50 nm to save their colloidal stability and prevent from aggregation, providing a larger surface area for a certain volume of the materials, which is a desirable feature. In this size range, MNPs are stable in water at pH of 7.0 and it is also possible to prevent from precipitation owing to gravitation forces. (ii) MNPs should be nontoxic and biocompatible. (iii) MNPs should possess a high saturation magnetization, allowing control of the movement of the MNPs in the blood with a moderate external magnetic field and also giving opportunity to the particles to move close to the targeted tissue. The following sections will cover the application areas of MNPs in biosensors and medical diagnostics [2, 3, 8, 37].

12.2 MNP-Based Biosensors for Disease Detection

Magnetic nanoparticles have found important applications in biosensors. They are used either during the development of a transducer or as signal amplification agents to decrease the detection limit and increase the efficiency of a sensor. MNPs can be applied for various biosensor types and among them electrochemical transducers have dominated in the field. An MNP-modified capacitive sensor was recently been constructed for the detection of cancer markers. The nanoparticles were employed as signal amplification agent and they were

Table 12.1 Various applications of MNPs in biosensors for the detection of target biological compounds in different samples.

Transducer type	Biosensor type	MNP type	Target analyte	Investigation range	Detection limit	Reference
Optical	SPR	Fe_3O_4@Au MNPs	α-Fetoprotein	$1–200\,ng\,mL^{-1}$	$0.65\,ng\,mL^{-1}$	[19]
	SPR	MNPs (fluidMAG-ARA) with iron oxide core	β-Human chronic gonadotropin	NA	$0.45\,pM$	[20]
	SPR	Fe_3O_4 MNPs	Thrombin	$0.27–27\,nM$	$0.017\,nM$	[21]
	SPR	Iron oxide carboxyl-modified MNPs	Ochratoxin A	$1–50\,ng\,mL^{-1}$	$0.94\,ng\,mL^{-1}$	[22]
Piezoelectric	QCM	Core–shell Fe_3O_4@ Au-MWCNT composites	Myoglobin	$0.001–5\,ng\,mL^{-1}$	$0.3\,pg\,mL^{-1}$	[23]
	QCM	Iron oxide magnetic nanobeads	Avian influenza virus H5N1	$0.128–12.8$ HA unit	0.0128 HA unit	[24]
	QCM	Fe_3O_4@SiO_2	C-reactive protein	$0.001–100\,ng\,mL^{-1}$	$0.3\,pg\,mL^{-1}$	[25]
Electrochemical	Amperometric	Fe_3O_4@SiO_2/MWCNT	Glucose	$1\,\mu M$ to $30\,mM$	$800\,nM$	[26]
	Amperometric	Core–shell Au–Fe_3O_4@SiO_2	Glucose	$0.05–1\,mM/1–8\,mM$	$0.01\,mM$	[27]
	Voltammetric	Core–shell Au–Fe_3O_4	Carcinoembryonic antigen	$0.005–50\,ng\,mL^{-1}$	$0.01\,ng\,mL^{-1}$	[28]
	Voltammetric	Core–shell Fe_3O_4@SiO_2/ MWCNT	Uric acid	$0.60–100\,\mu M$	$0.13\,\mu M$	[29]
	Electrochemiluminescent	Core–shell Fe_3O_4 Au NPs	α-Fetoprotein	$0.0005–5\,ng\,mL^{-1}$	$0.2\,pg\,mL^{-1}$	[30]

Potentiometric	Core–shell Fe_3O_4	Glucose	$0.5\,\mu M$ to $34\,mM$	$0.5\,\mu M$	[31]
Impedance	Fe@Au NPs-2-aminoethanethiol Functionalized graphene NPs	DNA	1×10^{-4} to 1×10^{-8} M	2.0×10^{-15} M	[32]
Magnetic field					
Giant magnetoresistive	Cubic FeCo NPs	Interleukin-6	125 fM to 41.5 pM	NA	[33]
Giant magnetoresistive	Cubic FeCo NPs	Endoglin	NA	83 fM	[34]
Superconducting quantum	Carboxyl-functionalized iron oxide NPs	MCF7/Her2-18 breast cancer cells	NA	1.3×10^6 cells	[35]
Hall sensor	Manganese-doped ferrite ($MnFe_2O_4$)	Rare cells: MDA-MB-468 cancer cells	10^1–10^5 cells	NA	[36]

coated with a secondary antibody [18]. The biosensor was initially investigated using C-reactive protein that is a cardiac marker and extensively researched by scientific community in the field [38]. The methodology was then transferred into a multiple marker detection platform to provide a precise diagnosis of cancer. The levels of certain molecules called as biomarker have shown difference in the case of a particular disorder [39]. Many critical health problems such as cancer, cardiovascular disorders, infectious diseases, and neurodegenerative conditions can be diagnosed using their particular markers [18, 38–43]. However, the amount of some biomarkers increases in several disease cases. This has a pivotal role especially to discriminate cancer types. For example, the concentration of carcinoembryonic antigen (CEA) has increased both in lung and breast cancers. Therefore, there is a need to investigate a couple of markers of a certain disease together to ensure the type of the cancer [39, 43, 44]. In the research on capacitive sensor, this was successfully achieved, and CEA, epidermal growth factor receptor (hEGFR), and cancer antigen 15-3 (CA15-3) were detected with high sensitivity using MNP-conjugated secondary antibodies. The limit of detection in human serum for each biomarker (CEA: $20 \, \text{pg mL}^{-1}$; hEGFR: $20 \, \text{pg mL}^{-1}$; CA15-3: $10 \, \text{U mL}^{-1}$) was found to be much lower than the threshold levels (CEA: $5 \, \text{ng mL}^{-1}$; hEGFR: $64 \, \text{ng mL}^{-1}$; CA15-3: $50 \, \text{U mL}^{-1}$) in lung cancer cases [18].

Fe_3O_4 at TiO_2 MNPs were synthesized and developed for a biomarker immunoassay to electrochemically quantify organophosphorylated butyrylcholinesterase (BChE), a specific marker of exposure to organophosphorus agents, in plasma. The Fe_3O_4 at TiO_2 particles were produced through hydrolysis of tetrabutyl titanate on the surface of Fe_3O_4 and then characterized by different tools (X-ray diffraction, TEM, and attenuated total reflection Fourier-transform infrared spectra). The functionalized Fe_3O_4 at TiO_2 particles were utilized as capture antibody to selectively increase phosphorylated moiety rather than phosphoserine antibody in the conventional sandwich immunoassays. The secondary recognition was carried out with quantum dot (QD)-tagged antibody (QDs-anti-BChE). By using a magnet, the sandwich complex (Fe_3O_4 at TiO_2/OP-BChE/ODs-anti-BChE) was simply separated from sample solutions, and the freed cadmium ions were quantified on a disposable screen-printed electrode. The developed method gets rid of the limitation arising from the unavailability of the commercial organophosphorus-specific antibody and also enables to amplify the signal during the detection with the aid of QD tags and easy sample separation due to magnetic forces. The sensor is capable to detect OP-BChE in the range of 0.02–$10 \, \text{nM}$, and it provided a detection limit of $0.01 \, \text{nM}$. The successful validation of the biosensor with human plasma samples also indicated that the method can be used for screening exposure to not only OP pesticides but also nerve agents [45].

A very recent study reported graphene/Fe_3O_4@Au nanocomposites for electrochemiluminescence (ECL) biosensing of HeLa cells. The multifunctional nanostructure provided very good electron transfer and high stability and emission intensity. Moreover, an ultrasensitive magnetically controlled solid-state ECL sensor was constructed for label-free determination of HeLa cells by employing the multifunctional nanocomposite. Magnetically controlled ECL biosensing system accomplished a high sensitivity for HeLa cells and resulted in a linear detection range between 20 and 1×10^4 cells mL^{-1} with good reproducibility and high stability [46]. Another nanocomposite biosensor, which is composed of tyrosinase-Fe$_3$O$_4$ nanoparticles-chitosan, was developed for the amperometric detection of dopamine. Dopamine is the most crucial neurotransmitter and has a pivotal role in the function of central nervous, hormonal, renal, and cardiovascular systems. The nanoparticles attached to the glassy carbon electrode surface displayed substantial electrochemical features and also simultaneously acted as a mediator to transfer the electrons between the electrode and the enzyme. The sensor demonstrated broad linear response from 2.0×10^{-8} to 7.5×10^{-5} mol L^{-1}, with a detection limit of 6.0×10^{-9} mol L^{-1} for the quantification of dopamine in the presence of ascorbic acid. The MNP-modified tyrosinase biosensor indicates a promising future for fast, easy, and cost-effective detection of dopamine in the samples. The nanocomposite approach significantly enhanced the stability of the electron transfer mediator and can be used for the construction of bioelectronic devices for medical diagnostic [47].

Sensitive dopamine detection is also possible with MNP-based colorimetric sensor. A visual biosensor was developed using magnetic Fe$_3$O$_4$ particles and dithiobis(sulfosuccinimidylpropionate)-modified gold nanoparticles (DTSSP-AuNPs) as the recognition elements. Dopamine was specifically attached onto the surface of DTSSP-AuNPs through amine coupling reaction between the activated carboxyl group of DTSSP and amino group of the dopamine molecule. Target-anchored DTSSP-AuNPs were captured by Fe$_3$O$_4$ via the interaction of iron and catechol. The formed Fe$_3$O$_4$–dopamine–DTSSP-AuNPs conjugates were easily separated from the solution in a magnetic field, resulting in a decrease of the AuNPs suspension and fading of the UV-Vis signal (Figure 12.1). The sensor achieved a detection limit of 10 nM for dopamine target. This sandwich-type strategy relying on Fe$_3$O$_4$ and AuNPs can be used in various colorimetric detection applications in clinical research by optimizing the surface chemistry of AuNPs and Fe$_3$O$_4$ [48].

CRP is an important human blood biomarker for cardiac diseases and inflammatory conditions. A magnetic biosensor was reported for its quantification in human serum, urine, and saliva. The detection principle is based on a sandwich assay using two different anti-CRP antibodies; one is for CRP capture and the other for the labeling with MNPs. The sensor achieved a linear investigation range from 25 ng mL^{-1} to 2.5 µg mL^{-1} and offered a more sensitive method than a typical CRP-ELISA assay [49]. An interesting work combining a QCM sensor with Fe$_3$O$_4$@SiO$_2$ magnetic capture nanoprobes for the

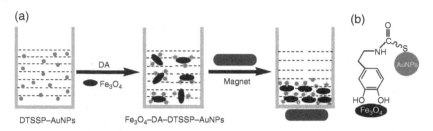

Figure 12.1 Schematic illustration of the strategy for dopamine (DA) detection using dithiobis(sulfosuccinimidylpropionate)-modified gold nanoparticles (DTSSP-AuNPs) and Fe_3O_4 magnetic particles (MPs) (a) and the interaction between DA, Fe_3O_4, and DTSSP-AuNPs (b). *Source:* Wang et al. [48]. http://www.mdpi.com/1996-1944/6/12/5690/htm. Used under CC BY 3.0 http://creativecommons.org/licenses/by/3.0/.

ultrasensitive detection of CRP was recently reported. This piezoelectric immunosensor based on the magnetic capture nanoprobes and HRP-antibody co-immobilized nanogold as signal tags allowed to detect $0.3\,pg\,mL^{-1}$ CRP in human serum with an investigation range of $0.001-100\,ng\,mL^{-1}$. The optimum assay conditions were investigated, and the developed method provided a rapid, simple, and sensitive detection approach for the clinical testing of ultra-trace CRP in cardiac diseases [25]. Many other examples of magnetic biosensors for biomedical applications were listed in Table 12.1.

12.3 MNPs in Cancer Diagnosis and Therapy

Being responsible for 1 out of every 4 deaths, cancer is a very challenging disease to treat and requires a huge cost every year. The best chance for survival is coming from the early diagnosis. Currently, the clinical diagnosis techniques and treatments are able to supply neither timely detection nor curative therapy under many conditions. Chemotherapy and radiotherapy constitute the leading treatments for advanced cancer cases; however, these methods are unable to strictly target tumor cells, causing significant damage to healthy organs and tissues and limiting the required therapeutic dose. Hence, nanotechnology and its applications can provide strong alternatives to the currently available techniques. MNPs serve as well-suited materials for biomedical applications being biocompatible and biodegradable. Several iron oxide-based compounds have currently been found on the market to be used as iron supplement or magnetic resonance imaging (MRI) contrast agents. Coating of the iron core has played a pivotal role for the performance of the magnetic material for cancer diagnosis and therapy. The coating is generally composed of a polymer that protects the core from oxidation and agglomeration. The polymer coat is also very important since it allows further modification of the nanoparticles for target-specific bio-compartments in human body such as the lymph nodes, lung, blood pool, liver, and so on. Moreover, the coating around the nanoparticles functions as a template for imaging tags, targeting macromolecules and distributing therapeutic payloads. There are commonly utilized surface coating materials to stabilize iron oxide particles, and these materials are appropriate for medical applications. PEG, fatty acids, dextran, chitosan, polyvinylpyrrolidone (PVP), gelatin, and polypeptides constitute these coating materials, some of which are synthetic and the others are natural polymers [11, 13]. The selection of the convenient coating is critical to be able to use MNPs as clinical compounds and also to control their size. One of the most important points during the synthesis and preparation of MNPs is to avoid oxidation and protect their magnetic characteristics. This can be achieved by performing the complete process in the presence of an inert gas such as argon or nitrogen and in an oxygen free platform. The coating materials are usually added throughout the coprecipitation period to get rid of the agglomeration and the surface coating may have an influence on superparamagnetic behavior of MNPs; therefore, these nanostructures need to be curvaceously coated to diminish magnetic dipole–dipole interactions. Different biomolecules and coating agents utilized in the manufacturing of MNPs were summarized in Table 12.2 with their advantages and applications.

Using the functionalized MNPs noninvasive diagnosis of cancer is achievable with high sensitivity. Furthermore, the MNPs can be slightly modified to involve a therapeutic component that gives opportunity to combine diagnosis with drug delivery. The mostly investigated area in biomedicine with the aid of MNPs is the image-guided therapy, which brings diagnosis and *in vivo* imaging of drug's bioavailability, function, and potential together. Conjugation of genetic material with MNPs, such as RNA inference to be used in mRNA level for RNAi-based drug research, has offered an excellent technology for the diagnosis of the diseased tissues by possessing the specific targeting groups on the MNPs. A proper conjugation of RNA molecules with MNP can even protect the genetic material from the degradation of nucleases and, therefore, provide robust delivery carrier. A pioneering work reported the modification of MNPs with siRNA to knockdown the expression of a targeted protein (green fluorescent protein (GFP)) in mice that was implanted with GFP expressing cancer cells. The modification was performed via the functionalization of amine terminals of the MNPs with bifunctional linker molecules and then coupled to short siRNA molecules whose 5′ ends were modified with thiol. This thiol-modified siRNA specifically targeted the gene that is responsible for GFP expression. Moreover, the MNPs were also modified with a membrane translocation peptide for intracellular delivery. It was shown that *in vivo* tracking of these probes for tumor uptake by MRI could be achieved, and optical imaging in two different tumor types was also feasible by employing the developed magnetic nanoparticles. This research displays a great headway of siRNA

Table 12.2 Different biomolecules and coating agents utilized in the manufacturing of magnetic nanoparticles.

Coating agents/ biomolecule	Applications	Advantages	References
Antibodies	Specific and strong recognition, immunomagnetic separation, capturing and tracking of cells, determination of *in vivo* enzyme activity, advanced drug discovery	A number of antibodies that bind to specific proteins exist in particular cellular structures and allowing to a wide range of applications. They also gather in tumors	[50–55]
Proteins or enzymes	Detection, separation, and purification of proteins	Biomarkers of many important diseases such as cancer, CVD, and neurodegenerative disorders. Gathering in tumors	[12, 56, 57]
Folic acid	Tumor targeting compound	Favored tumor targeting	[58, 59]
Aptamers	Artificial design of them for medical diagnosis using their recognition capacity	Smaller than antibody receptors. Providing efficient recognition Can be designed for specific targets	[60–62]
Oligonucleotides	Utilized in DNA/RNA detection or separation as probes	Acting as nanoswitches	[63, 64]
Tat peptide	Cellular uptake of MNPs	Enhancing cellular uptake of MNPs	[65]
Adenoviral agents	Gene delivery and MR imaging	Aiding cytosolic release	[66, 67]
PEG	Protein conjugation	Long-term circulation ability in blood vessels	[68–70]
Doxorubicin	Anticancer drug for cancer treatment	Efficient in case of targeted therapy	[71, 72]
Paclitaxel	Anticancer drug for cancer treatment	Highly efficient in case of targeted therapy	[73, 74]

delivery and imaging approaches and proves a significant promise for therapy development for cancer cases [75]. Kumar et al. proved that in case the MNPs were functionalized with a higher affinity target-specific peptide, the tumoral uptake and silencing effect increased significantly. This study also employed *in vivo* noninvasive MRI and optical microscopy [76]. Research that targets different cancer biomarkers and genes in the field was also reported [77].

The use of MNPs is generally restricted to iron-based magnetic oxides since they display the best compromise, owing to desirable magnetic properties with saturation magnetization, stability against oxidation conditions, and more importantly the low toxicity. European Medicines Agency (EMA) has approved the use of magnetic iron oxide-based MNPs (magnetite and maghemite) [78]. Different general features of iron oxide-based MNPs synthesized using different chemical techniques were summarized in Table 12.3. Nevertheless, magnetic ferrites have started to be intensively investigated in recent years for

Table 12.3 Different general features of iron oxide-based MNPs synthesized using different chemical techniques.

	Coprecipitation	Reversed micelles coprecipitation	Hydrothermal coprecipitation	Non-hydrolytic methods
Type of MNPs	Fe_2O_3, Fe_3O_4, MFe_2O_4	Fe_2O_3, Fe_3O_4, MFe_2O_4	Fe_3O_4, Fe_2O_3	Fe_2O_3, Fe_3O_4, MFe_2O_4, Cr_2O_3, Co_3O_4, MnO, NiO
Diameter range	10–50 nm, 50–100 nm	2–30 nm, 20–80 nm	1–0.25 mm, 15–31 nm	3–500 nm
Advantages	Synthesis in water environment; easy surface modification, ferrites formation and maghemite transformation; inexpensive chemicals are needed; scaling-up is easy; reaction conditions are user friendly	Size can be easily adjusted; advanced size control; narrow size distribution; providing uniform magnetic characteristics	Synthesis in water; magnetic features are adjustable; advanced size control is possible; size distribution is narrow	Very good size control and crystallinity; scaling-up can be achieved; magnetic properties can be regulated; size distribution is narrow
Limitations	Wide size distribution; oxidation cannot be controlled; reproducibility is low	Removal of the surfactants is difficult; low yield and crystallinity	Inadequate safety of reactants; high temperature	Requiring phase transfer; use of toxic organic solvents; high temperature
References	[79]	[80]	[81]	[82–85]

biomedical applications [86]. Functionalization of MNPs with affinity ligands offers a wide range of interesting research in the field. Monoclonal antibodies are used by the immune system for the determination of foreign compounds, and they possess specific antigens to bind and react. The most commonly applied strategy is the covalent attachment of these ligands via their most active amine groups. This approach may result in random orientation of the antibody, which leads to a significant reduction in recognition ability [50]. The random orientation can be avoided with unspecific reversible interactions between the antibody and MNP in order to steer this ligand on the magnetic surface prior to its irreversible covalent attachment [51]. Such an MNP-functionalized antibody ligand can be extensively used in cancer diagnosis by improving the stability and performance of the ligands against the target biomarkers. Similar strategies are developed in the area by employing and modifying another class of receptors called aptamers. These nucleic acid-based ligands are designed and synthesized in the laboratory, and they consist of RNA, DNA, or short peptides. Their specific interaction with proteins makes them good candidates to be used in specific uptake of aptamer-coated MNPs by target cells. The aptamers are attached to the MNP surface by several methods: (i) Biotinylated aptamers are conjugated to streptavidin-coated MNPs [61]. (ii) Ethyl(dimethylaminopropyl) carbodiimide/N-hydroxysuccinimide (EDC/NHS) chemistry is used in case the magnetic surface consists carboxyl groups [87]. (iii) Thiolated aptamers are directly mixed with gold nanoparticle-coated MNPs [62].

Having the excellent nanovehicle feature MNPs can be used in cancer therapy by enhancing the drugs bioactivities. This is achieved by delivering the drugs directly to the targeted location in the human body where they are supposed to act. The drugs used in cancer treatment sometimes lose their efficiency due to several drawbacks including low solubility in water environment, low bioavailability in selective targeting of cancer cells, and absence of a proficient method for their detection and following. These limitations can be overcome by using the MNPs as the appropriate carriers for the drugs. For instance, paclitaxel (PTX) is an anticancer, which serves as a mitotic inhibitor, and used for the treatment of many cancer types such as ovarian, lung, melanoma, breast and prostate cancers and solid tumors. This drug is irritant that leads to inflammation in the veins and also tissue damage. Therefore, the appropriate labeling of MNPs with PTX enables to prevent the mentioned limitations as well as the side effects on human health [73, 74]. Doxorubicin (DOX) is subjected to the same limitations as another widely used anticancer drug, and the similar solutions can be considered for its applications. It serves as anthracycline antibiotic that works by intercalating DNA and demonstrates side effects. In case of escaping from the vein, it leads to blistering and massive tissue damage. Majority of approaches to capture DOX rely on the formation of a coating polymer layer in amphiphilic nature and the loading of this polymer layer with the hydrophobic drug through covalent bonds or hydrophobic interactions. During these processes, pH-triggered release of DOX, thermic effects, and enzymatic degradation need to be taken into consideration for the best output [71, 72]. In all these bio-applications with MNPs the activity, stability, and orientation of the biological molecules within the magnetic nanocomposite play the central role in the formation of highly successful systems.

12.4 Cellular Applications of MNPs in Biosensing, Imaging, and Therapy

Cell diagnosis and imaging by employing MRI technique has a great demand for the aim of cell visualization and tracking in the living organisms. *In vivo* applications of cells transplanted into the targeted location in the human body have been already achieved using MNPs [88–90]. These nanostructures are iron oxide-based nanoparticles due to their biocompatibility and low toxicity as this issue was also referred in the previous sections. The contrast agents that are already on the market can be utilized as markers for short- or long-term MR tracking of the transplanted cells noninvasively. There are some important issues that need to be taken into account for labeling of the cells with MNPs. This includes ensuring the cell viability and integrity after labeling. The MNPs uptake by cells is usually proportional to the nanoparticle size [91]. The efficiency of labeling has been tried to increase using some strategies including the coating of the MNPs with translocating agents [92] or cell internalizing antibodies [93] and the usage of transfection agents and albumin-mediated transport into the cells. Zhao et al. reported the differential conjugation of tat peptide as a translocated agent to MNPs and its effect on cellular uptake. Surface modification of MNPs was performed with HIV-1 tat peptide that led to successful intercellular labeling as well as noninvasive tracking of various cell types by employing MRI. The number of attached tat peptides on the MNP surface was studied to determine the efficiency of MNPs on cellular uptake. A modified P2T method to quantify the peptide numbers per particle surface was developed using disulfide linkage, and it demonstrated an easy-to-use approach with good reproducibility. Cells labeled with this method could be detected by MRI with high sensitivity and also allowed *in vivo* tracking of 100-times lower cell concentration than the other reported works [94]. The combination of transfection agents with MNPs was also investigated for the aim of enhanced cell labeling in several applications, and a significant improvement was achieved. Lipofectamine, poly(L-ornithine), poly(L-lysine), heat-activated dendrimer, protamine sulfate, and some other agents were used in these works [95, 96].

Magnetoelectroporation (MEP) is another alternative method for cell labeling and diagnosis. It was proved that the method is capable to realize instant (<1 s) endosomal labeling with the US Food and Drug Administration

(FDA)-approved preparation Feridex, and it does not require the initiating cell cultures or auxiliary agents. Since MEP is harmful at higher voltages or pulse durations, the developed protocol was calibrated using a pulse of 130 V and 17 ms. Stem cells of rats, mice, and humans were successfully labeled with this method, and the iron uptake was found in the picogram levels. The results indicate that MEP method is successful as much as the method in which transfection agents are used. Furthermore, MEP-labeled stem cells displayed a constant viability, mitochondrial metabolic level, and proliferation. The labeled neural stem cells (NSCs) and mesenchymal stem cells (MSCs) differentiated into neural, adipogenic, and osteogenic lineages in the same manner as unlabeled cells, in case of containing Feridex particles [97]. MEP labeling may be a strong alternative to the other methods since it does not rely on the secondary agents, dispensing with the need for clinical approval.

To date imaging of transplanted stem cells, apoptotic cells, dentritic cells, and transplanted pancreatic islet cells were intensively investigated and reported with the aid of magnetic materials. The transplantation of stem cells is very important particularly in the field of regenerative medicine, and it has been researched for the treatment of several key diseases such as ventricular dysfunction, heart failure, neuronal damage, vascular injury, and so on. These cells can be transplanted either by direct injection or by vascular insertion and consequent MR imaging confirmation of their direction to the targeted organ. Tallheden et al. investigated *in vivo* MRI of magnetically labeled human embryonic stem (HES) cells by using dextran-coated MNPs. The direct injection of HES cells into the anterior left ventricular wall of mice myocardium was carried out for transplantation and then imaged by MRI. The clear hypointense regions were observed at the injection site where the labeled HES cells existed [98]. The method has shown a great potential to be widely used in regenerative medicine, and the similar applications have been launched afterwards [99–102]. *In vivo* imaging of islet transplantation was also studied by labeling pancreatic islet cells with MNPs modified with Cy5.5 dye. MRI techniques were combined with near-infrared imaging in this work, and the transplanted cells could be monitored in 6 months. The method achieved the *in vivo* diagnosis of transplanted human pancreatic islets by employing MRI that provided a real-time noninvasive monitoring of islet grafts in diabetic mouse. This preclinical research has indicated a promising future for the individuals who suffer from type 1 diabetes without leading to an increase in hypoglycemic conditions in the body [103].

12.5 Conclusions

Having unique and smart features, MNPs have found a vast range of applications in biomedical arena from the research to the preclinical and clinical applications. Owing to fascinating opportunities provided by surface modification agents, they can be manipulated and fabricated to be specifically used in biosensor-based disease detection, cancer imaging, therapy development, and also cellular applications. Their use in biosensors has offered highly sensitive, reliable, and reproducible detection for many important diseases including cancer, neuro-degenerative disorders, and cardiac problems. They can be fabricated in the form of nanocomposites by combining different nanomaterials together such as gold nanoparticles, QDs, and carbon nanotubes; and this strategy generally increases their performance in diagnostic. Research in cancer imaging, treatment, and cellular applications is also widely conducted using MNPs. However, there is a considerable challenge in these areas due to the unacceptable toxicity levels of many MNPs. Today, iron oxide-based MNPs have dominated in the field owing to nontoxic nature, and their use in clinical applications has been broadly demonstrated with successful implementations. The attempts to obtain the other nontoxic MNPs with desirable features have been carried out. The improvements at this juncture are critical since the tumor cell-targeted therapy using the anticancer drugs displays a pivotal role in effective therapy, and this is possible to achieve with the aid of MNPs by preventing from uncontrolled release of the drugs into the healthy tissues and organs. Being able to develop nontoxic nanocomposite-based MNPs and use these fascinating materials for cellular uptake will also increase the successful applications of cell diagnosis and imaging.

Acknowledgment

Z.A. gratefully acknowledges support from the European Commission, Marie Curie Actions and IPODI as the principle investigator.

References

1 Akbarzadeh, A.; Samiei, M.; Davaran, S. *Nanoscale Research Letters* **2012**, 7, 1–13.
2 Kang, J. H.; Hahn, Y. K.; Kim, K. S.; Park, J. K. *Magnetophoretic Biosensing and Separation Using Magnetic Nanomaterials*, Wiley-VCH Verlag, Weinheim, **2007**.
3 Pankhurst, Q. A.; Connolly, J.; Jones, S. K.; Dobson, J. *Journal of Physics D: Applied Physics* **2003**, 36, 167–181.
4 Lu, A.-H.; Salabas, E. L.; Schueth, F. *Angewandte Chemie International Edition* **2007**, 46, 1222–1244.
5 Kodama, R. H. *Journal of Magnetism and Magnetic Materials* **1999**, 200, 359–372.
6 Townsend, J.; Burtovyy, R.; Galabura, Y.; Luzinov, I. *ACS Nano* **2014**, 8, 6970–6978.

7 Singh, A.; Sahoo, S. K. *Drug Discovery Today* **2014**, 19, 474–481.
8 Gallo, J.; Long, N. J.; Aboagye, E. O. *Chemical Society Reviews* **2013**, 42, 7816–7833.
9 Lee, J.-H.; Chen, K.-J.; Noh, S.-H.; Garcia, M. A.; Wang, H.; Lin, W.-Y.; Jeong, H.; Kong, B. J.; Stout, D. B.; Cheon, J.; Tseng, H.-R. *Angewandte Chemie International Edition* **2013**, 52, 4384–4388.
10 Kainz, Q. M.; Reiser, O. *Accounts of Chemical Research* **2014**, 47, 667–677.
11 Colombo, M.; Carregal-Romero, S.; Casula, M. F.; Gutierrez, L.; Morales, M. P.; Boehm, I. B.; Heverhagen, J. T.; Prosperi, D.; Parak, W. J. *Chemical Society Reviews* **2012**, 41, 4306–4334.
12 Mazzucchelli, S.; Colombo, M.; De Palma, C.; Salvade, A.; Verderio, P.; Coghi, M. D.; Clementi, E.; Tortora, P.; Corsi, F.; Prosperi, D. *ACS Nano* **2010**, 4, 5693–5702.
13 Reddy, L. H.; Arias, J. L.; Nicolas, J.; Couvreur, P. *Chemical Reviews* **2012**, 112, 5818–5878.
14 Holzinger, M.; Le Goff, A.; Cosnier, S. *Frontiers in Chemistry* **2014**, 2, 1–10.
15 Martin, M.; Salazar, P.; Villalonga, R.; Campuzano, S.; Manuel Pingarron, J.; Luis Gonzalez-Mora, J. *Journal of Materials Chemistry B* **2014**, 2, 739–746.
16 Urbanova, V.; Magro, M.; Gedanken, A.; Baratella, D.; Vianello, F.; Zboril, R. *Chemistry of Materials* **2014**, 26, 6653–6673.
17 Altintas, Z.; Tothill, I. E. Nanomaterial applications in health care diagnostics, In: *Concise Encyclopaedia of Nanotechnology*, Eds. Kharisov, B. I; Kharissova, O. V.; Ortiz-Mendez, U. CRC Press, Boca Raton, **2016**, 317–319.
18 Altintas, Z.; Kallempudi, S. S.; Sezerman, U.; Gurbuz, Y. *Sensors and Actuators B: Chemical* **2012**, 174, 187–194.
19 Liang, R. P.; Yao, G. H.; Fan, L. X.; Qiu, J. D. *Analytica Chimica Acta* **2012**, 737, 22–28.
20 Wang, Y.; Dostalek, J.; Knoll, W. *Analytical Chemistry* **2011**, 83, 6202–6207.
21 Wang, J. L.; Zhu, Z. Z.; Munir, A.; Zhou, H. S. *Talanta* **2011**, 84, 783–788.
22 Zamfir, L. G.; Geana, I.; Bourigua, S.; Rotariu, L.; Bala, C.; Errachid, A.; Jaffrezic-Renault, N. *Sensors and Actuators B: Chemical* **2011**, 159, 178–184.
23 Gan, N.; Wang, L. H.; Li, T. H.; Sang, W. G.; Hu, F. T.; Cao, Y. T. *Integrated Ferroelectrics* **2013**, 144, 29–40.
24 Li, D. J.; Wang, J. P.; Wang, R. H.; Li, Y. B.; Abi-Ghanem, D.; Berghman, L.; Hargis, B.; Lu, H. G. *Biosensors & Bioelectronics* **2011**, 26, 4146–4154.
25 Zhou, J.; Gan, N.; Li, T. H.; Zhou, H. K.; Li, X.; Cao, Y. T.; Wang, L. H.; Sang, W. G.; Hu, F. T. *Sensors and Actuators B: Chemical* **2013**, 178, 494–500.
26 Baby, T. T.; Ramaprabhu, S. *Talanta* **2010**, 80, 2016–2022.
27 Chen, X. J.; Zhu, J. W.; Chen, Z. X.; Xu, C. B.; Wang, Y.; Yao, C. *Sensors and Actuators B: Chemical* **2011**, 159, 220–228.
28 Li, J. P.; Gao, H. L.; Chen, Z. Q.; Wei, X. P.; Yang, C. F. *Analytica Chimica Acta* **2010**, 665, 98–104.

29 Arvand, M.; Hassannezhad, M. *Materials Science & Engineering C: Materials for Biological Applications* **2014**, 36, 160–167.
30 Zhou, H. K.; Gan, N.; Li, T. H.; Cao, Y. T.; Zeng, S. L.; Zheng, L.; Guo, Z. Y. *Analytica Chimica Acta* **2012**, 746, 107–113.
31 Yang, Z. P.; Zhang, C. J.; Zhang, J. X.; Bai, W. B. *Biosensors & Bioelectronics* **2014**, 51, 268–273.
32 Yola, M. L.; Eren, T.; Atar, N. *Electrochimica Acta* **2014**, 125, 38–47.
33 Li, Y. P.; Srinivasan, B.; Jing, Y.; Yao, X. F.; Hugger, M. A.; Wang, J. P.; Xing, C. G. *Journal of the American Chemical Society* **2010**, 132, 4388–4392.
34 Srinivasan, B.; Li, Y. P.; Jing, Y.; Xing, C. G.; Slaton, J.; Wang, J. P. *Analytical Chemistry* **2011**, 83, 2996–3002.
35 Hathaway, H. J.; Butler, K. S.; Adolphi, N. L.; Lovato, D. M.; Belfon, R.; Fegan, D.; Monson, T. C.; Trujillo, J. E.; Tessier, T. E.; Bryant, H. C.; Huber, D. L.; Larson, R. S.; Flynn, E. R. *Breast Cancer Research* **2011**, 13, R108.
36 Issadore, D.; Chung, J.; Shao, H. L.; Liong, M.; Ghazani, A. A.; Castro, C. M.; Weissleder, R.; Lee, H. *Science Translational Medicine* **2012**, 4, 141ra92.
37 Issa, B.; Obaidat, I. M.; Albiss, B. A.; Haik, Y. *International Journal of Molecular Sciences* **2013**, 14, 21266–21305.
38 Altintas, Z.; Fakanya, W. M.; Tothill, I. E. *Talanta* **2014**, 128, 177–186.
39 Altintas, Z.; Tothill, I. *Sensors and Actuators B: Chemical* **2013**, 188, 988–998.
40 Altintas, Z.; Uludag, Y.; Gurbuz, Y.; Tothill, I. E. *Talanta* **2011**, 86, 377–383.
41 Altintas, Z.; Tothill, I. E. *Sensors and Actuators B: Chemical* **2012**, 169, 188–194.
42 Kallempudi, S. S.; Altintas, Z.; Niazi, J. H.; Gurbuz, Y. *Sensors and Actuators B: Chemical* **2012**, 163, 194–201.
43 Altintas, Z.; Kallempudi, S. S.; Gurbuz, Y. *Talanta* **2014**, 118, 270–276.
44 Altintas, Z.; Tothill, I. E. *Nanobiosensors in Disease Diagnosis* **2015**, 4, 1–10.
45 Zhang, X.; Wang, H. B.; Yang, C. M.; Du, D.; Lin, Y. H. *Biosensors & Bioelectronics* **2013**, 41, 669–674.
46 Gu, W. L.; Deng, X.; Gu, X. X.; Jia, X. F.; Lou, B. H.; Zhang, X. W.; Li, J.; Wang, E. K. *Analytical Chemistry* **2015**, 87, 1876–1881.
47 Wang, Y.; Zhang, X.; Chen, Y.; Xu, H.; Tan, Y.; Wang, S. *American Journal of Biomedical Sciences* **2010**, 2, 209–216.
48 Wang, Z. Y.; Bai, Y. Y.; Wei, W. C.; Xia, N.; Du, Y. H. *Materials* **2013**, 6, 5690–5699.
49 Meyer, M. H. F.; Hartmann, M.; Krause, H. J.; Blankenstein, G.; Mueller-Chorus, B.; Oster, J.; Miethe, P.; Keusgen, M. *Biosensors & Bioelectronics* **2007**, 22, 973–979.
50 Chou, S. W.; Shau, Y. H.; Wu, P. C.; Yang, Y. S.; Shieh, D. B.; Chen, C. C. *Journal of the American Chemical Society* **2010**, 132, 13270–13278.
51 Puertas, S.; Batalla, P.; Moros, M.; Polo, E.; del Pino, P.; Guisan, J. M.; Grazu, V.; de la Fuente, J. M. *ACS Nano* **2011**, 5, 4521–4528.
52 Ryan, S.; Kell, A. J.; van Faassen, H.; Tay, L. L.; Simard, B.; MacKenzie, R.; Gilbert, M.; Tanha, J. *Bioconjugate Chemistry* **2009**, 20, 1966–1974.

53 Yu, M. K.; Kim, D.; Lee, I. H.; So, J. S.; Jeong, Y. Y.; Jon, S. *Small* **2011**, 7, 2241–2249.

54 Lee, J.; Yang, J.; Ko, H.; Oh, S. J.; Kang, J.; Son, J. H.; Lee, K.; Lee, S. W.; Yoon, H. G.; Suh, J. S.; Huh, Y. M.; Haam, S. *Advanced Functional Materials* **2008**, 18, 258–264.

55 Yang, J.; Lee, C. H.; Ko, H. J.; Suh, J. S.; Yoon, H. G.; Lee, K.; Huh, Y. M.; Haam, S. *Angewandte Chemie International Edition* **2007**, 46, 8836–8839.

56 Razgulin, A.; Ma, N.; Rao, J. H. *Chemical Society Reviews* **2011**, 40, 4186–4216.

57 Garcia, I.; Gallo, J.; Genicio, N.; Padro, D.; Penades, S. *Bioconjugate Chemistry* **2011**, 22, 264–273.

58 Kohler, N.; Fryxell, G. E.; Zhang, M. Q. *Journal of the American Chemical Society* **2004**, 126, 7206–7211.

59 Chen, F. H.; Zhang, L. M.; Chen, Q. T.; Zhang, Y.; Zhang, Z. J. *Chemical Communications* **2010**, 46, 8633–8635.

60 Chen, T.; Shukoor, M. I.; Wang, R. W.; Zhao, Z. L.; Yuan, Q.; Bamrungsap, S.; Xiong, X. L.; Tan, W. H. *ACS Nano* **2011**, 5, 7866–7873.

61 Bamrungsap, S.; Shukoor, M. I.; Chen, T.; Sefah, K.; Tan, W. H. *Analytical Chemistry* **2011**, 83, 7795–7799.

62 Liang, G. H.; Cai, S. Y.; Zhang, P.; Peng, Y. Y.; Chen, H.; Zhang, S.; Kong, J. L. *Analytica Chimica Acta* **2011**, 689, 243–249.

63 Uhlen, M. *Nature* **1989**, 340, 733–734.

64 Tanaka, T.; Sakai, R.; Kobayashi, R.; Hatakeyama, K.; Matsunaga, T. *Langmuir* **2009**, 25, 2956–2961.

65 Wunderbaldinger, P.; Josephson, L.; Weissleder, R. *Bioconjugate Chemistry* **2002**, 13, 264–268.

66 Tresilwised, N.; Pithayanukul, P.; Mykhaylyk, O.; Holm, P. S.; Holzmuller, R.; Anton, M.; Thalhammer, S.; Adiguzel, D.; Doblinger, M.; Plank, C. *Molecular Pharmaceutics* **2010**, 7, 1069–1089.

67 Huh, Y. M.; Lee, E. S.; Lee, J. H.; Jun, Y. W.; Kim, P. H.; Yun, C. O.; Kim, J. H.; Suh, J. S.; Cheon, J. *Advanced Materials* **2007**, 19, 3109–3112.

68 Tromsdorf, U. I.; Bruns, O. T.; Salmen, S. C.; Beisiegel, U.; Weller, H. *Nano Letters* **2009**, 9, 4434–4440.

69 Xie, J.; Xu, C.; Kohler, N.; Hou, Y.; Sun, S. *Advanced Materials* **2007**, 19, 3163–3166.

70 Karakoti, A. S.; Das, S.; Thevuthasan, S.; Seal, S. *Angewandte Chemie International Edition* **2011**, 50, 1980–1994.

71 Lim, E. K.; Huh, Y. M.; Yang, J.; Lee, K.; Suh, J. S.; Haam, S. *Advanced Materials* **2011**, 23, 2436–2442.

72 Bakandritsos, A.; Mattheolabakis, G.; Chatzikyriakos, G.; Szabo, T.; Tzitzios, V.; Kouzoudis, D.; Couris, S.; Avgoustakis, K. *Advanced Functional Materials* **2011**, 21, 1465–1475.

73 Santra, S.; Kaittanis, C.; Grimm, J.; Perez, J. M. *Small* **2009**, 5, 1862–1868.

74 Hwu, J. R.; Lin, Y. S.; Josephrajan, T.; Hsu, M. H.; Cheng, F. Y.; Yeh, C. S.; Su, W. C.; Shieh, D. B. *Journal of the American Chemical Society* **2009**, 131, 66–68.

75 Medarova, Z.; Pham, W.; Farrar, C.; Petkova, V.; Moore, A. *Nature Medicine* **2007**, 13, 372–377.

76 Kumar, M.; Yigit, M.; Dai, G. P.; Moore, A.; Medarova, Z. *Cancer Research* **2010**, 70, 7553–7561.

77 Bartlett, D. W.; Su, H.; Hildebrandt, I. J.; Weber, W. A.; Davis, M. E. *Proceedings of the National Academy of Sciences of the United States of America* **2007**, 104, 15549–15554.

78 Cabuil, V. *Encyclopedia of Nanoscience and Nanotechnology*, American Scientific Publishers, Valencia, **2004**, 1715–1730.

79 Gupta, A. K.; Gupta, M. *Biomaterials* **2005**, 26, 3995–4021.

80 Lee, Y. J.; Lee, J. W.; Bae, C. J.; Park, J. G.; Noh, H. J.; Park, J. H.; Hyeon, T. *Advanced Functional Materials* **2005**, 15, 2036–2036.

81 Niederberger, M.; Krumeich, F.; Hegetschweiler, K.; Nesper, R. *Chemistry of Materials* **2002**, 14, 78–82.

82 Puntes, V. F.; Krishnan, K. M.; Alivisatos, A. P. *Science* **2001**, 291, 2115–2117.

83 Hyeon, T.; Lee, S. S.; Park, J.; Chung, Y.; Bin Na, H. *Journal of the American Chemical Society* **2001**, 123, 12798–12801.

84 Redl, F. X.; Black, C. T.; Papaefthymiou, G. C.; Sandstrom, R. L.; Yin, M.; Zeng, H.; Murray, C. B.; O'Brien, S. P. *Journal of the American Chemical Society* **2004**, 126, 14583–14599.

85 Carta, D.; Casula, M. F.; Floris, P.; Falqui, A.; Mountjoy, G.; Boni, A.; Sangregorio, C.; Corrias, A. *Physical Chemistry Chemical Physics* **2010**, 12, 5074–5083.

86 Sun, S. H.; Zeng, H.; Robinson, D. B.; Raoux, S.; Rice, P. M.; Wang, S. X.; Li, G. X. *Journal of the American Chemical Society* **2004**, 126, 273–279.

87 Min, K.; Jo, H.; Song, K.; Cho, M.; Chun, Y. S.; Jon, S.; Kim, W. J.; Ban, C. *Biomaterials* **2011**, 32, 2124–2132.

88 Zhang, C.; Liu, T.; Gao, J. N.; Su, Y. P.; Shi, C. M. *Mini-Reviews in Medicinal Chemistry* **2010**, 10, 194–203.

89 Bulte, J. W. M. *Journal of Magnetism and Magnetic Materials* **2005**, 289, 423–427.

90 Bulte, J. W. M.; Zhang, S. C.; van Gelderen, P.; Herynek, V.; Jordan, E. K.; Duncan, I. D.; Frank, J. A. *Proceedings of the National Academy of Sciences of the United States of America* **1999**, 96, 15256–15261.

91 Shapiro, E. M.; Skrtic, S.; Koretsky, A. P. *Magnetic Resonance in Medicine* **2005**, 53, 329–338.

92 Lewin, M.; Carlesso, N.; Tung, C. H.; Tang, X. W.; Cory, D.; Scadden, D. T.; Weissleder, R. *Nature Biotechnology* **2000**, 18, 410–414.

93 Bulte, J. W. M.; de Cuyper, M.; Despres, D.; Frank, J. A. *Journal of Magnetism and Magnetic Materials* **1999**, 194, 204–209.

94 Zhao, M.; Kircher, M. F.; Josephson, L.; Weissleder, R. *Bioconjugate Chemistry* **2002**, 13, 840–844.

95 Arbab, A. S.; Yocum, G. T.; Kalish, H.; Jordan, E. K.; Anderson, S. A.; Khakoo, A. Y.; Read, E. J.; Frank, J. A. *Blood* **2004**, 104, 1217–1223.

96 Janic, B.; Rad, A. M.; Jordan, E. K.; Iskander, A. S. M.; Ali, M. M.; Varma, N. R. S.; Frank, J. A.; Arbab, A. S. *PLoS ONE* **2009**, 4, e5873.

97 Walczak, P.; Kedziorek, D. A.; Gilad, A. A.; Lin, S.; Bulte, J. W. M. *Magnetic Resonance in Medicine* **2005**, 54, 769–774.

98 Tallheden, T.; Nannmark, U.; Lorentzon, M.; Rakotonirainy, O.; Soussi, B.; Waagstein, F.; Jeppsson, A.; Sjogren-Jansson, E.; Lindahl, A.; Omerovic, E. *Life Sciences* **2006**, 79, 999–1006.

99 Bulte, J. W. M.; Douglas, T.; Witwer, B.; Zhang, S. C.; Strable, E.; Lewis, B. K.; Zywicke, H.; Miller, B.; van Gelderen, P.; Moskowitz, B. M.; Duncan, I. D.; Frank, J. A. *Nature Biotechnology* **2001**, 19, 1141–1147.

100 Li, Z. J.; Suzuki, Y.; Huang, M.; Cao, F.; Xie, X. Y.; Connolly, A. J.; Yang, P. C.; Wu, J. C. *Stem Cells* **2008**, 26, 864–873.

101 Neri, M.; Maderna, C.; Cavazzin, C.; Deidda-Vigoriti, V.; Politi, L. S.; Scotti, G.; Marzola, P.; Sbarbati, A.; Vescovi, A. L.; Gritti, A. *Stem Cells* **2008**, 26, 505–516.

102 Frank, J. A.; Miller, B. R.; Arbab, A. S.; Zywicke, H. A.; Jordan, E. K.; Lewis, B. K.; Bryant, L. H.; Bulte, J. W. M. *Radiology* **2003**, 228, 480–487.

103 Evgenov, N. V.; Medarova, Z.; Dai, G. P.; Bonner-Weir, S.; Moore, A. *Nature Medicine* **2006**, 12, 144–148.

13

Graphene Applications in Biosensors and Diagnostics

Adina Arvinte[1] and Adama Marie Sesay[2]

[1] *"Petru Poni" Institute of Macromolecular Chemistry, Centre of Advanced Research in Bionanoconjugates and Biopolymers, Iasi, Romania*
[2] *Unit of Measurement Technology, Kajaani University Consortium, University of Oulu, Oulu, Finland*

13.1 Introduction

Certain biomarkers' concentration can be used to define the disease type or the body response to a treatment. Usually, an efficient and precise diagnosis demands the determination of more than one biomarker, therefore increasing the number of clinical tests. As a consequence of these challenges, the elaboration of innovative biosensor-based strategies that could allow reliable analysis of biomarkers relevant in medical diagnostics has found an increasing interest, since biosensor devices could replace the time-consuming, large, and automated analyzers from centralized dedicated laboratories and could enable a massive number of clinical tests.

Biosensor research is a fast-growing field, gaining enormous attention in the last years both in nanoscience and nanotechnology, since biosensors have been developed as smart analytical devices capable of detecting and quantifying numerous analytes. In particular, the medical applications and demands have been the main driving force for a blooming development of biosensors. Advances in the fabrication and engineering of nanomaterials have yielded nanostructured materials with distinctive properties, which can be applied in biosensor design, in order to meet specific requirements needed by particular applications.

Biosensors are generally defined as analytical devices that convert molecular recognition of a target analyte into a measurable signal via a transducer. Micro- and nanofabrication technologies combined with diverse sensing strategies including optical, electrical, and mechanical transducers are exploited in

Biosensors and Nanotechnology: Applications in Health Care Diagnostics, First Edition.
Edited by Zeynep Altintas.

modern biosensor development. Despite clinical need, translation of biosensors from research laboratories to clinical applications has remained limited to a few notable examples [1].

At the confluence of nanoscience and nanotechnology is the field of nanomedicine, which aims to use the properties and physical characteristics of nanomaterials for the diagnosis and treatment of diseases at molecular level [2]. In order to increase sensitivities and to lower limits of detection (LOD) to even individual molecules, a wide range of novel nanomaterials have been developed as promising candidates due to their capability to immobilize more bioreceptor units at reduced volumes and even to act itself as a transduction element. Among such nanomaterials, metal nanoparticles (NPs), oxide particles, polymers, quantum dots, carbon nanotubes (CNTs), graphene, and their nanocomposites are intensively studied.

13.2 Graphene and Biosensors

In recent years, graphene has drawn much attention both from fundamental and applied researches. Due to its unique physical and chemical properties, graphene is very promising for the fabrication of novel and ultrasensitive sensors. Application of graphene in sensor construction is a predictable combination, since their excellent electrical conductivity, large surface-to-volume ratio, optical properties, and high thermal conductivity [3] could bring new functions and properties beneficial for sensor performance. The large surface area of graphene ($>2500 \, m^2 g^{-1}$) [4] can enhance the surface loading of desired biomolecules, while the excellent conductivity and small bandgap can be beneficial for conducting electrons between biomolecules and the electrode surface.

13.2.1 Structure

Graphene is a one-atom-thick layer of sp2-bonded carbon organized in a two-dimensional (2D) honeycomb lattice. It is a new carbon material, distinctly different from 1D CNTs, 0D fullerenes, and three-dimensional (3D) bulk graphite [5], but it is considered as the basic structural element of those carbon allotropes. Since its discovery in 2004 by A. Geim and K. Novoselov, the research on graphene has exploded, and the structure of graphene at the atomic level was studied in many scientific institutions [3, 6–8]. On the basis of their structure (Figure 13.1), the distance between carbon atoms was calculated to be 0.142 nm, which is the average of the single (C—C) and double (C=C) covalent σ bonds, and the density of carbon atoms is $3.9 \times 10^{15} \, cm^{-2}$ [6]. It is a kind of ideal 2D atomic crystal, which was successfully prepared and identified in recent years.

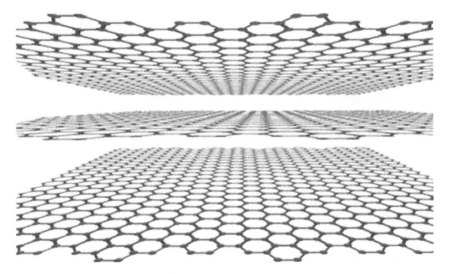

Figure 13.1 A visual depiction of the structure of a layered microscopic segment of graphene. *Source:* From http://www.graphenomenon.com/

13.2.2 Preparation

Since its material's discovery, the mass production of graphene has been advanced in the last decade, but with a rather small number of established methods available (Figure 13.2). Among the existing procedures devoted to the preparation of graphene, the oxidation of graphite to graphite oxide with a subsequent reduction is considered to be a scalable and low-cost method for bulk production of graphene [9–12].

A recently described method for the isolation of graphene or reduced graphene oxide (rGO) is the micromechanical exfoliation. There are several carbon sources available for this, but highly ordered pyrolytic graphite (HOPG) is often chosen because of its high atomic purity and smooth surface, which enables the facile delamination of carbon layers as a result of the weak van der Waals forces that hold the layers together [13, 14]. The micromechanical exfoliation is an important method for producing "pristine" graphene, free of defects and free from significant functional groups, and appropriate for electronic studies or other fundamental measurements [14].

A major inconvenience related to the micromechanical preparation of graphene is represented by its lack of scalability. For the preparation of these materials in relatively large quantities (grams or higher scale), analogous methods have been developed, where graphite can be exfoliated in liquid media into few-layer or even monolayer graphene [15–17]. Alternative approaches have been explored for the preparation of pristine graphene, such as chemical vapor deposition (CVD) and epitaxial growth on SiC substrates [18]. A brief outline

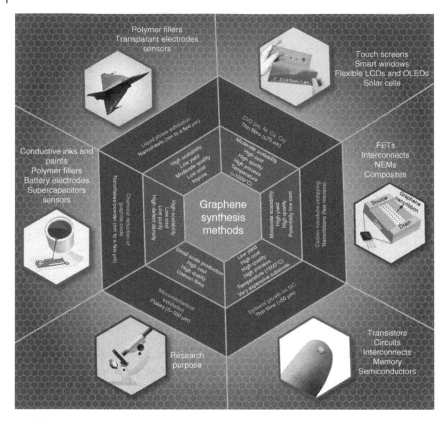

Figure 13.2 A schematic showing the conventional methods commonly used for the synthesis of graphene along with their key features and the possible applications. *Source:* Image CKMNT, http://www.nanowerk.com/spotlight/spotid=25744.php (*See insert for color representation of the figure.*)

of the mentioned methods with the good and the weak points, along with their specific application, is suggestively overviewed in Figure 13.1. Each method can have its own advantages as well as limitations depending on its target applications.

According to Nanowerk's detailed report of global manufacturing patents [19], the most patented graphene production methods include CVD and exfoliation, followed by a trail of other methods including epitaxial growth and chemical synthesis. But the concern is related to the involvement of some toxic chemicals in the synthesis of graphene by conventional methods and the concomitant generation of hazardous waste and poisonous gases. Consequently, there is still a need to develop green methods to produce high-quality graphene by environmentally friendly approaches at a large-scale production.

A recently published paper [20] reports the synthesis of 4–5-layer graphene sheets by a novel lamp ablation method that simultaneously fulfills the virtues of being a one-step, high-yield process that is catalyst-free, devoid of toxic reagents, with short reaction times, and potentially scalable. The possible mechanism for graphene formation relies on the ultrahot reactor conditions produced by the continuous irradiation, which are weakening the van der Waals interactions between graphite layers, followed by the sublimation of graphite. As the products move out of the irradiated zone toward the ampoule's colder distal end, they reconfigure and quench, forming few-layer graphene.

13.2.3 Properties

As carbon is a nonmetal, it is expected that graphene will have the same property. In fact, it behaves much more like a metal, although it conducts electricity in a very different way (with both holes and electrons as charge carriers). For this reason it was described as a semimetal or a semiconductor (a material midway between a conductor and an insulator).

Graphene exhibits unique electronic, optical, magnetic, thermal, and mechanical properties arising from its strictly 2D structure and, thus, has many important technical applications. It has extremely high specific surface area and high porosity. Some theoretical calculations predict a large surface area of single-layer graphene close to $2600 \, m^2 g^{-1}$ [21]. Compared with CNTs, graphene has a larger surface area, fewer impurities, and superior mechanical and electrical properties [22, 23]. Its high porosity makes it ideal for adsorption of different gases such as hydrogen, methane, and carbon dioxide, and there are several groups who studied the adsorption of H_2 by graphene and other carbon nanomaterials [21, 24, 25].

An important reason for the interest in graphene is the particular unique nature of its charge carriers. The electronic mobility of graphene at room temperature is as high as $15\,000 \, cm^2 V^{-1} s^{-1}$, and the charge carriers can be tuned continuously between electrons and holes [23]. The electronic properties of graphene change with the number of layers and relative position of atoms in adjacent layers determined by the stacking order [26]. As the number of layers increases, the band structure becomes more complicated, leading to the appearance of several charge carriers and the overlapping of the conduction and valence bands.

The surface of graphene is relatively smooth and highly optically transparent. Graphene has the ability to absorb a rather large 2.3% of white light, and the optical transmission through the graphene surface has been experimentally observed in the visible range and is more than 95% [27].

Owing to the strong C—C bonds, monolayer of graphene has superior mechanical properties, such as high Young's modulus and strong fracture strength [28]. The elastic modulus of graphene sheets prepared by the chemical exfoliation method was determined to be about 0.25 TPa [29]. Also, it has been demonstrated that the yielding strength of graphene (about 6.4 TPa) is several times higher than that of carbon steel [30].

The surface of pure graphene usually interacts with other molecules via physical adsorption. The ability to tailor the physicochemical properties of graphene for specific functions by immobilizing a variety of molecules on them has led to successful applications in sensors, smart surfaces, and molecular electronics [31, 32]. Functionalized graphene particularly appears to be exceptionally promising for chemical and biological sensor applications.

13.2.4 Commercialization in the Field of Graphene Sensors

The research of the field is expanding rapidly, generating more interest commercially than any other material since silicon. A literature survey reveals thousands of new patents and numerous companies already entering the graphene domain [33]. From publicly available online information, one can see the existence of numerous groups of small- and medium-sized enterprises involved with graphene and its orientation toward science.

According to a website (www.graphene-info.com) [34], in November 2013, Nokia's Cambridge research center developed a humidity sensor based on GO that proved to be incredibly fast, thin, optically transparent, and flexible and have great response and recovery times. They also found that the response rate of sensors depends on the thickness of GO films—an increase in film thickness resulted in a reduction in the speed of the sensor. Nokia also filed a patent in August 2012 for a graphene-based photodetector that is transparent and thin and should ultimately be cheaper than traditional photodetectors. On the same application trend, in September 2014, the German AMO (GmbH) company developed a graphene-based photodetector in collaboration with Alcatel Lucent Bell Labs, which is said to be the world's fastest photodetector. The graphene is used as the active element in this design, and the detector also uses nickel/aluminum, hydrogen silsesquioxane, and buried oxide.

In 2014, Graphene Frontier from the University of Pennsylvania announced raising $1.6m to expand the development and manufacturing of their highly sensitive graphene field-effect transistor (GFET)-based biological and chemical sensors [34]. Graphene Frontiers also launched the "six sensors" brand for highly sensitive chemical and biological GFET-based sensors. These are especially promising to diagnose diseases with multiple markers such as cancer and other illnesses currently diagnosed using enzyme-linked immunosorbent assay (ELISA) technologies. Their "six sensors" brand for highly sensitive chemical and biological sensors can be used to diagnose diseases, with sensitivity and efficiency unequaled by traditional sensors.

In June 2015, a collaboration between another German engineering giant involved in sensor technology, Bosch, and scientists from the Max Planck Institute for Solid State Research yielded a graphene-based magnetic sensor 100 times more sensitive than an equivalent device based on silicon. The sensor performance is defined by two parameters: (i) sensitivity, which depends on the number of charge carriers, and (ii) power consumption, which varies inversely with charge carrier mobility. The high carrier mobility of graphene makes them useful in such applications, and the results achieved by the Bosch team confirmed this. In short, graphene provides for a high-performance magnetic sensor with low power and footprint requirements.

13.2.5 Latest Developments in Graphene-based Diagnosis

Researchers at the Korea Institute of Science and Technology (KIST) have developed a highly sensitive biosensor manufacturing technology that can diagnose various diseases by designing a large panel of a 4-inch wafer board with graphene, which is called the "dream material." This graphene-based biosensor can reportedly check the blood concentration of amyloid-beta peptide, which is known as the most remarkable biomarker of Alzheimer's disease. With the sensor, the research team confirmed the diagnostic capability for dementia through the blood samples of transgenic mice and normal mice models. The sensor also detected the amyloid-beta peptide level from human plasma [34, 35]. The research group led by Dr. Hwang Kyo-sun, which developed and transferred the technology utilizing a blood test for early dementia diagnosis to a company earlier this year, is putting more effort into the follow-up studies in order to assess the early diagnostic capability for numerous diseases, such as cancer, diabetes, and depression as well as dementia, and commercialize the technology.

Despite the impressive number of publications on biosensors in the diagnostics field, the commercialization of this technology is feasible only minimally in the near future, besides the biosensors for blood glucose and lactate and a few others. In particular, the biosensor market is dominated by only a few products. For medical diagnostics, approximately 90% of biosensors are glucose monitors, blood gas monitors, and electrolyte or metabolite analyzers.

13.3 Medical Applications of Graphene

The clinical utility of a biochemical test is determined by its sensitivity, ability to detect the disease with no false negatives, and its specificity (i.e., the ability to avoid false positives). In this context, biosensors combining the exceptional electrical/optical properties of graphene, and the selectivity of a biocomponent with the processing power of nanoelectronics and diversity of fabrication, offer new powerful diagnostic tools with much greater precision for medical science. The nano-dimensions of graphene and its electronic properties make it an ideal candidate substrate for anchoring the biologic element involved in biochemical sensing and a promising electrode material for intensifying the electron transfer. Special attention has been paid on the potential use of graphene in electrochemical and optical sensors for sensitive detection of clinical relevant analytes. There is enormous research progress as regards graphene in sensors and biosensors published in recent years and rationally reviewed by many researchers [36, 37]. In this chapter we outline the latest development of biosensors based on graphene focusing on the detection of molecular biomarkers in real samples, with promising applications in diagnostics.

13.3.1 Electrochemical Graphene Biosensors for Medical Diagnostics

Electrochemical biosensors provide unlimited opportunities for monitoring biomarkers of medical interest. Electrochemical detection is of particular significance in the development of biosensors since it allows for high sensitivity and selectivity, simple instrumentation, tailored architecture of electrode surfaces, fast response, and possibility for miniaturization. Electrochemistry studies the chemical response of a system to an electrical stimulation, involving the loss of electrons (oxidation) or gain of electrons (reduction) that a material undergoes during the electrical stimulation. Graphene has significant potential for enabling the development of electrochemical (bio)sensors, based on direct or mediated electron transfer between the biologic component and the electrode surface. An outline of the recent advances in electrochemical sensors and biosensors based on graphene has been made, depending on the type of biomarker detected.

13.3.1.1 Glucose Detection

Monitoring the glucose levels is critical for the evaluation of the body condition and health to confirm that a treatment is working effectively and to avoid a diabetic emergency, such as hypoglycemia (low blood sugar: <3 mM) or hyperglycemia (high blood sugar: >7 mM). Today's biosensor market is dominated by glucose biosensors [38], and several manufacturers are engaged in strong competition. The electrochemical detection of glucose can be accomplished in an enzymatic approach (by using glucose oxidase (GOD) as the mediator and recognition element) and nonenzymatic approach (using an electrocatalytic material). While the enzyme catalyzes the production or depletion of an electroactive species, a voltage is applied to the electrode in amperometric sensors, which induces the redox reaction of the electroactive species, generating a signal [39]. This electrical signal correlates to the concentration of the analyte in the sample.

Although the high-quality graphene has many advantages as presented in previous sections, bare graphene has limited uses in electrochemical devices for the detection of glucose or other biomarkers. Graphene is usually involved in electrochemical sensing, in combination with inorganic NPs, or included in different nanocomposites. Some of the recent reports on glucose sensing based on graphene show that they have the potential to be used for a noninvasive one due to their broad sensing range and low LOD, but most of the reported research works are feasible only under the laboratory conditions and seldom find applications in real-life scenario.

Ultrasensitive electrochemical biosensor for glucose was reported by Zhiguo et al. based on CdTe–CdS core–shell quantum dot ensuring the ultrafast electron transfer between graphene–gold nanocomposite and gold NP [40]. The authors proved the improvement of the conductivity between graphene nanosheets due to the introduction of gold NPs, ultrafast charge transfer from CdTe–CdS core–shell quantum dot to graphene nanosheets and gold NP due to unique electrochemical properties of the CdTe–CdS core–shell quantum dot, and good biocompatibility of gold NP for GOD. The biosensor exhibited high sensitivity toward glucose sensing, ensuring an LOD of 3 pM and a wide linear response range (from 1×10^{-11} to 1×10^{-8} M). Further experiments showed the detection of glucose in human saliva sample diluted with buffer solutions. The obtained results are in good agreement with those obtained by a routine enzymatic method using Beckman SYNCHRON CX7 biochemical analyzer for glucose.

A glucose biosensor based on hybrid conjugates of reduced graphene oxide and nickel NPs (rGO-Ni Np) deposited onto glassy carbon electrode as a nanocomposite film of chitosan and GOD was found to have good linearity and sensitivity up to $129 \, \mu A \, cm^{-2} \, mM^{-1}$ [41]. This approach brings the advantages of the enhanced surface area and balanced electrical/electrochemical properties of the rGO-Ni NPs conjugate, the excellent biocompatibility, and the enhanced density of enzyme molecules accommodated in the chitosan scaffold. The results suggest its possible applications in noninvasive assays using body fluids such as saliva and tears. The application of constructed electrode as a glucose biosensor was explored for the first time in proof-of-concept tests.

Combination of graphene with the glucose-sensing enzyme and chitosan to construct a glucose detector was reported by Kang et al. [42]. The hybrid nanocomposite of graphene–chitosan was prepared and modified on the surface of glassy carbon electrode, and then GOD was absorbed on the nanocomposite film, which provides a favorable microenvironment for GOD to realize the direct electrode transfer. The resulting biosensor not only retained the bioactivity of GOD but also exhibited a wide linear range from 0.08 to 12 mM, with an LOD of 0.02 mM and much higher sensitivity compared with other nanostructured supports.

Integrating graphene-based composites with enzyme provides a potent strategy to enhance biosensor performance due to their unique physicochemical properties. Wang et al. [43] reported on the utilization of graphene–CdS (G–CdS) nanocomposite as a novel immobilization matrix for GOD. Based on the decrease of electrocatalytic response of the reduced form of GOD to dissolved oxygen, a glucose sensor has been developed, which displays satisfactory analytical performance over a linear range from 2.0 to 16 mM along with good stability because of the large surface area and fast electron transfer of GR and CdS nanocrystals. Interference from uric acid and ascorbic acid (AA) that usually coexist with glucose in real samples has been found to be limited. The G–CdS–GOD-modified electrode has been employed in the detection of blood sugar concentration in human plasma utilizing standard addition method without sample pretreatment, and recoveries are between 92.2 and 105.1%, demonstrating good accuracy for the glucose sensing in real samples. The results ascertain the practical application of the proposed glucose biosensor in clinical analysis.

An electrochemical platform based on the one-step synthesis of GO decorated with platinum nanocubes and copper oxide nanoflowers was designed for amperometric sensing of glucose by Dhara et al. [44]. Pt-CuO/rGO nanocomposite dispersed in *N,N*-dimethylformamide was drop casted onto the screen-printed electrode system and allowed the oxidation of glucose at a low potential of +0.35 V. The sensor showed very high sensitivity ($3577 \, \mu A \, mM^{-1} \, cm^{-2}$) and selectivity toward glucose through the explained mechanism, in which copper oxide catalyzes glucose oxidation and platinum NPs act as a cocatalyst to enhance the electron transfer during the oxidation of glucose. The sensor was employed for the testing of glucose in blood serum, and the results obtained were comparable with other standard test methods.

The search for a minimally invasive or noninvasive method is extremely intense, and the detection of glucose in sweat seems to be an alternative route to acquire glucose data that correlate with blood sugar levels. Such a less intrusive device for glucose monitoring that integrates components that capture sweat from the skin and sensors for glucose, pH, humidity, and temperature was reported by Lee et al. [45], with promising application in diabetes monitoring. They show that graphene doped with gold and combined with a gold mesh has improved electrochemical activity over bare graphene, sufficient to form a wearable patch for sweat-based diabetes monitoring and feedback therapy. The stretchable device features a serpentine bilayer of gold mesh and gold-doped graphene that forms an efficient electrochemical interface for the stable transfer of electrical signals. Graphene biochemical sensors with an Ag/AgCl counter electrodes show enhanced electrochemical activity, sensitivity, and selectivity in detecting important biomarkers contained in human sweat. The wireless connectivity further highlights the practical applicability of the current patch system. These advances using nanomaterials and devices provide new opportunities for the treatment of chronic diseases such as diabetes mellitus.

Recently, an interesting study developed a novel and highly sensitive disposable glucose sensor strip incorporating direct laser-engraved graphene decorated with pulse-deposited copper nanotubes (CuNCs) and observed that it had high reproducibility (96.8%), stability (97.4%), and low cost of fabrication [46]. The sensor exhibited high selectivity and sensitivity ($4532.2\,\mu A\,mM^{-1}\,cm^{-2}$). It also had low LOD of 250 nM and a linear range of 25 μM to 4 nM. The study suggested that this electrochemical sensor could aid glucose detection within the physiological range for saliva, sweat, and tears.

13.3.1.2 Cysteine Detection

L-Cysteine belongs to sulfur-containing amino acid molecules that play a crucial role in biological systems. Biological thiols, such as cysteine, homocysteine, and glutathione, play a central role in metabolism and cellular homeostasis. Homocysteine may be catabolized to cysteine, which is a precursor to glutathione. Altered levels of this compound have been implicated in hyperhomocysteinemia [47] and in a number of pathological conditions including Alzheimer's and Parkinson's diseases [48], autoimmune deficiency syndrome [49], depigmentation of hair, edema, liver damage, loss of muscle and fat, and weakness [50].

An electrochemical sensor for L-cysteine that is based on the use of Y_2O_3 nanoparticles (Y_2O_3 NPs) supported on nitrogen-doped reduced graphene oxide (N-rGO) was described by Yang et al. [51]. When deposited on carbon paste electrode, the material displays a strong oxidation peak for L-cysteine at pH 7.0. The voltammetric response is much higher compared with only Y_2O_3 or N-rGO modified electrode, demonstrating the synergistic effect, which includes high catalytic active sites of well-distributed Y_2O_3 NPs directly anchoring on the surface of N-rGO. The current, measured at 0.7 V potential, increases linearly in the concentration range 1.3–720 μM L-cysteine, and the LOD is 0.8 μM. The sensor shows a remarkable selectivity for L-cysteine with no or very limited response to other interfering species, as displayed in Figure 13.3. The sensor was successfully applied to the determination L-cysteine in spiked syrup.

Graphene nanosheets uniformly decorated with Au NPs have been successfully used in electrochemical sensing of L-cysteine [52]. A robust method for the synthesis of Au NP-decorated graphene nanosheets using chloroauric acid ($HAuCl_4$) and poly(vinylpyrrolidone) (PVP) covalently functionalized graphene oxide (PGO) as precursor and template was developed (Figure 13.4).

The aqueous dispersion of graphene–Au hybrid nanosheets was then directly coated onto the surface of glassy carbon electrode to prepare the chemically modified electrode, which exhibited enhanced electrochemical properties, such as low LOD, good stability, resistance to interference, and satisfactory recovery, and these were attributed to the synergistic effects of the conductive graphene and the uniform Au NPs. The linear response for L-cysteine was ranging from 0.1 to 24 μM with a low LOD of 20.5 nM under the optimized conditions. The graphene–Au hybrid-based electrode was then applied to determine L-cysteine in human urine with recovery results.

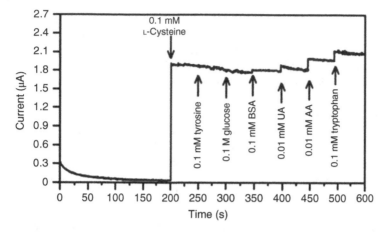

Figure 13.3 Amperometric responses of modified electrode to successive additions of 0.1 mM L-cysteine, 0.1 mM tyrosine, 0.1 M glucose, 0.1 mM BSA, 0.01 mM UA, and 0.01 mM AA in 0.1 M phosphate buffer of pH 7.0 at 0.7 V. *Source:* Yang et al. [51]. Reproduced with permission of Springer.

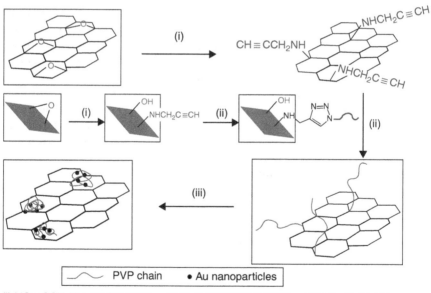

(i) HC≡CCH₂NH₂/NaOH, 80°C;   (ii) CuBr/PMDETA/PVP-N₃;   (iii) HAuCl₄/NaBH₄

Figure 13.4 Schematic synthesis of graphene–Au hybrid nanosheets. *Source:* Xu et al. [52]. Reproduced with permission of Elsevier.

A combination of GO with Au NPs, but in the form of composite, was also used for the oxidation of L-cysteine in human urine samples [53]. The graphene oxide–Au nanocluster (GO–Au NCs) composite was prepared under sonication of GO solution with Au NCs in a ratio of 1 : 1. The as-prepared GO–Au NCs composite electrode demonstrated excellent conductivity, extraordinary electron transport properties, and large specific surface area, which resulted in high electrochemical response toward L-cysteine in a large range from 0.05 to 20.0 μM, with a remarkably low LOD of 0.02 μM and low oxidation potential (+0.387 V). The direct determination of free reduced and total CySH in human urine samples has been successfully carried out without the assistance of any separation techniques.

A very novel nanocomposite of graphene quantum dot–β-cyclodextrin was fabricated on the surface of glassy carbon electrode using one-step and green electrodeposition methods [54]. The modified electrode showed an efficient electrocatalytic activity toward the oxidation of L-cysteine through a surface-mediated electron transfer, but the applied potential in the detection step is rather high (+0.6 V), and the application in real samples was not investigated.

13.3.1.3 Cholesterol Detection

Determination of cholesterol is of high importance in the medical sector since abnormal concentrations of cholesterol are related with hypertension, hyperthyroidism, anemia, and coronary artery diseases. Cholesterol and its esters are the essential components found in the cell membranes of all human and animal cells. The normal cholesterol limit in human serum is in the range of 1.0–2.2 mM, and its excessive accumulation in blood results in fatal diseases.

The determination of free cholesterol in human serum samples was achieved by a bienzymatic cholesterol biosensor based on the co-immobilization of cholesterol oxidase (ChOx) and catalase (CAT) on a graphene/ionic liquid-modified glassy carbon electrode [55]. The H_2O_2 generated during the enzymatic reaction of ChOx with cholesterol could be reduced electrocatalytically by immobilized CAT to obtain a sensitive amperometric response to cholesterol. An excellent sensitivity of 4.16 mA mM^{-1} cm^{-2} and response time less than 6 s enabled the determination of free cholesterol in human serum samples, proving to be useful for preliminary determination of cholesterol in clinical diagnostics.

A new approach for the one-pot synthesis of reduced graphene oxide–dendritic Pd nanoparticle (rGO–nPd) hybrid material and the development of biosensing scaffold for the amperometric determination of H_2O_2 and total cholesterol in human serum are described by R.S. Dey [56]. *In situ* reduction of both GO and $PdCl_4^{2-}$ in acidic solution was achieved for the fabrication of rGO–nPd hybrid material, which was further investigated for sensing H_2O_2 at 0.2 nM level without any redox mediator. The cholesterol biosensing platform was developed by integrating ChOx, which catalyzes the oxidation of

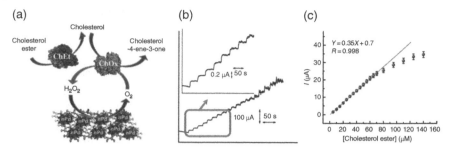

Figure 13.5 (a) The biosensing of cholesterol ester with enzyme-integrated rGO–nPd-based biosensor is illustrated. Amperometric (b) response obtained for the sensing of cholesterol ester. Corresponding calibration plot is shown in (c). *Source:* Dey and Raj [56]. Reproduced with permission of Elsevier.

cholesterol in the presence of O_2, and cholesterol esterase, which hydrolizes the cholesterol ester, with the hybrid material. The enzymatically generated H_2O_2 was detected by the highly sensitive rGO–nPd electrode according to Figure 13.5a. The biosensor is highly sensitive ($5.12\,\mu A\,\mu M^{-1}\,cm^{-2}$) and stable, with a fast response time of 4 s (Figure 13.5b and c) that enables to detect cholesterol ester as low as 0.05 μM. The biosensor is successfully used for the analysis of total cholesterol in human serum and butter.

Inclusion of graphene in nanocomposite together with PVP and polyaniline (PANI) was explored for the modification of paper-based biosensors via electrospraying [57]. Interestingly, the presence of a small amount of PVP ($2\,mg\,mL^{-1}$) in the nanocomposites can substantially improve the dispensability of graphene and increase the electrochemical conductivity of electrodes, leading to enhanced sensitivity of the biosensor. This modified electrode also exhibits excellent electrocatalytic activity toward the oxidation of hydrogen peroxide (H_2O_2). Furthermore, ChOx is attached to G/PVP/PANI-modified electrode for amperometric determination of cholesterol. Under optimum conditions, a linear range of 50 μM to 10 mM is achieved, with an LOD of 1 μM for cholesterol. The proposed system can be applied for the determination of cholesterol in a complex biological fluid like human serum.

13.3.1.4 Hydrogen Peroxide (H_2O_2)

H_2O_2 is a biomarker of oxidative stress [58], which is closely related to the physiological and pathological events such as aging, cancer, ischemia/reperfusion injury, traumatic brain injury, and memory functions [59, 60]. It is also known that H_2O_2 is abnormally produced in the inflammation process by causing oxidative damage [61]. Therefore, new strategies for real-time monitoring of H_2O_2 *in vivo*, which can enable researchers to understand the chemical nature of oxidative stress in physiological and pathological events, have gained extreme attention recently. The key point in developing electrochemical sensors for H_2O_2 is to decrease the oxidation/reduction overpotentials in order to eliminate the potential interference of other electroactive substances existent in human samples. A great number of papers [36, 37] have meticulously revised the graphene-based sensors for electrochemical detection of H_2O_2, but only few examples proved the practical application in human samples, with promising usability for the medical sector.

A highly sensitive and selective hydrogen peroxide sensor was successfully constructed with Pt–Au bimetallic nanoparticles (Pt–Au NPs) electrodeposited onto reduced graphene sheets (rGSs) [62]. Various molar ratios, Au/Pt, and different electrodeposition conditions were evaluated to control the morphology and electrocatalytic activity of the Pt–Au bimetallic NPs. Upon optimal conditions, the constructed sensor can sensitively detect H_2O_2 over wide linear ranges from 1 μM to 1.78 mM and 1.78 to 16.8 mM, with a low LOD (0.31 μM). Due to the synergetic effects of the bimetallic NPs and rGSs, the amperometric H_2O_2 sensor can operate at a low potential of 0 V, limiting the common anodic interferences induced from AA, uric acid (UA), and dopamine (DA) and also the cathodic interference induced from endogenous O_2. Furthermore, the proposed sensor had been successfully used in the detection of H_2O_2 released from the cancer cells, using rat pheochromocytoma cells (PC 12) as model. This method can provide a promising alternative for *in vivo* H_2O_2 monitoring in the fields of physiology, pathology, and diagnosis.

An interesting decoration of rGO with CuS NPs was successfully used as electrochemical sensor for reliable detection of H_2O_2 [63]. The new electrocatalyst, CuS/RGO composites, was prepared by heating a mixture of $CuCl_2$ and Na_2S aqueous solutions in the presence of PVP-protected GO at 180°C. Application of CuS/RGO composites-modified electrode as a sensor to H_2O_2 revealed a linear increase of the current response with the concentration from 5 to 1500 μM, with a fast response less than 2 s. The LOD was calculated as 0.27 μM. The as-made sensor was applied to determine the H_2O_2 levels in human serum and urine samples as well as H_2O_2 released from human cervical cancer cells with satisfactory results.

A flower-like MoS_2 nanostructure was grown on graphene and CNTs (GR-MWCNTs) via *in situ* hydrothermal method, and the resulting composite was employed for determination of H_2O_2 [64]. The MoS_2/GR-MWCNTs composite film-modified electrode showed excellent electrocatalytic ability to the reduction of H_2O_2, exhibiting a high sensitivity of $5.184 \, \mu A \, \mu M^{-1} \, cm^{-2}$, an LOD of 0.83 μM, appreciable stability, repeatability, and reproducibility. Practical applicability of the MoS_2/GR-MWCNTs/GCE was demonstrated in human serum sample.

A novel strategy for real-time monitoring of cellular small biomolecules such as H_2O_2 based on graphene hybrid composite film was just released to the scientific community [65]. The sensor was fabricated by one-step electrodeposited reduced graphene oxide (ERGO) incorporating electroactive methylene blue (MB) onto glassy carbon electrode. Under physiological condition, the ERGO-MB/GCE showed a linear amperometric response from 0.5 μM to 11.68 mM for H_2O_2, with the LOD of 60 nM. The developed method has been successfully applied to the determination of H_2O_2 released from living cells, which could clearly distinguish cancer cells including Hep3B, HepG2, and HeLa cells from normal cells (HEK 293). This assay shows great potential for nonenzymatic determination of H_2O_2 in physiological and pathological investigations.

13.3.1.5 Glycated Hemoglobin

Hemoglobin (Hb) is the most important part of the blood, responsible for transporting O_2 throughout the circulatory system. The majority of Hb is not glycated (~94%), and approximately 6% of Hb can be categorized as glycated hemoglobin (HbA1c) [66], which is a good marker for glycine level in the blood. HbA1c tests in the diagnosis, prevention, treatment, and monitoring of type 2 diabetes mellitus are of great importance [67].

Very recently, a sensitive electrochemical detection of HbA1c by a gold NP-embedded N-doped graphene (NG) nanosheet was reported by Jain and Chauhan [68]. The biosensor is based on fructosyl amino acid oxidase (FAO) immobilized on NG/gold NPs (AuNPs/GNs) deposited on fluorine-doped tin oxide (FTO) glass electrode. This architecture confers to the biosensor a wide linear range from 0.3 to 2000 μM in response to HbA1c, applying a +0.2 V potential. Furthermore, the constructed biosensor exhibits a low LOD of 0.2 μM and good long-term stability (4 months). The results obtained through this work demonstrated that FAO/AuNPs/GNs composite film provides an effective electrochemical response for the selective detection of HbA1c in human blood samples.

13.3.1.6 Neurotransmitters

Brain neuronal communication occurs through the exocytotic release of neurotransmitters into synaptic junctions and the surrounding extracellular fluid. Dopamine (DA) is an important catecholamine neurotransmitter in the mammalian central nervous system that influences several physiological functions. The impact of DA levels within the human body significantly affects the body function. A very comprehensive review makes a survey of graphene (functionalized graphene and NG) and its composite-modified electrodes (metal, metal oxide, polymer, carbonaceous materials, clay, zeolite, and metal–organic-based graphene composites) with their improved sensing performance toward DA along with several interfering species [69].

The challenge in electrochemical detection of DA is given by its low physiological concentration (0.01–1 mM) and interference from much more abundant AA and UA. The highly selective and sensitive detection of DA was ensured by the functionalization of graphene by porphyrin [70]. The aromatic π–π stacking and electrostatic attraction between positively charged DA and negatively charged porphyrin-modified graphene can accelerate the electron transfer while weakening AA and UA oxidation on the porphyrin-functionalized graphene-modified electrode. The detection of DA reached the low limit of 0.01 μM, and detection was possible even in the presence of large excess of AA and UA (Figure 13.6). Satisfactory results were obtained when the sensor

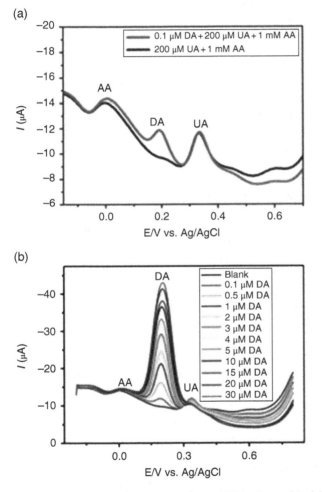

Figure 13.6 (a) DPVs of 1 mM AA and 200 μM UA mixture (black line), 1 mM AA, 200 μM UA, and 0.1 μM DA in a ternary mixture. The line with two peak points represents the data for 200 μM UA + 1 mM AA. The other line with three peak points are for 0.1 μM DA + 200 μM UA + 1 mM AA. (b) DPV responses of different concentrations of DA in PBS solution (pH = 7.0) containing 1 mM AA and 200 μM UA. The bottom line shows the data for blank. The height of the peak increases with the increased DA concentration. *Source:* Wu et al. [70]. Reproduced with permission of Elsevier.

was applied for the determination of DA in real hydrochloride injection sample, human urine, and serum samples.

A 3D interpenetrating electrode of electrochemically reduced graphene was proposed by Shi et al. [71] for the application of DA detection from real samples. This electrode can efficiently lower the oxidation potential of AA, allowing, thus, to selectively detect DA in the presence of AA and UA. The electrochemically rGO-based sensor exhibited a linear response toward DA in the concentration range of 0.1–10 mM with a low LOD of 0.1 mM and was able to detect DA from human serum sample.

Magnetite (Fe_3O_4) nanorods anchored over rGO were synthesized through a one-pot synthesis method [72], where the reduction of GO and *in situ* generation of Fe_3O_4 nanorods occurred concomitantly. The constructed rGO/Fe_3O_4/glassy carbon electrode exhibited excellent electrocatalytic activity toward electrooxidation of DA with a quick response time of 6 s, wide linear range between 0.01 and 100.55 µM, high sensitivity of 3.15 µA µM^{-1} cm^{-2}, and low LOD of 7 nM. Furthermore, the fabricated sensor exhibited a practical applicability in the quantification of DA in urine samples with an excellent recovery rate.

Norepinephrine (NE) or noradrenaline is a catecholamine, which acts as hormone and neurotransmitter. NE promotes the conversion of glycogen to glucose in the liver and helps in converting the fats into fatty acids [73]. The application of graphene-modified palladium (Pd) sensor for the electrochemical analysis of NE was investigated by Rosy et al. [74]. The modified sensor exhibited excellent electrocatalytic activity for the oxidation of NE, leading to a remarkable enhancement in the peak current and lowering the peak potential. The sensor responds linearly in the range of 0.0005–0.5 mM NE in phosphate buffer, at pH 7.2. Also, low detection and quantification limits obtained were 67.44 and 224.8 nM, respectively. The application of the proposed sensor in pharmaceutical dosage forms and human urine samples in the presence of high concentration of UA was achieved with excellent recovery results.

Serotonin (5-hydroxytryptamine (5-HT)) is an important neurotransmitter responsible for the regulation of mood, sleep, and appetite. It is an electroactive compound, but its electrochemical determination is limited by its low concentration in human tissue and in the presence of easily oxidizable AA, UA, and DA. A double-layered membrane of rGO/PANI nanocomposites and molecularly imprinted polymers (MIPs) embedded with gold NPs was employed as an electrochemical sensor for 5-HT [75]. The rGO/PANI nanocomposites were synthesized via electrodeposition process. The prepared sensor displayed remarkable selectivity to 5-HT and a low LOD of 11.7 nM, with a linear range of the response between 0.2 and 10.0 µM. Furthermore, the obtained biomimetic sensor was employed to detect 5-HT in human serum sample with good recoveries results.

13.3.1.7 Amyloid-Beta Peptide

Aβ42 is considered as an important biomarker for the early diagnosis of Alzheimer's disease, which is a progressive neurodegenerative disease affecting a large percentage of the aging population. Cognitive function and synaptic integrity of Alzheimer's disease patients will gradually drop, and abnormal neurotic and core plaques will form in the brains of patients [76]. Aβ 42, a peptide of 42 amino acids, is the major constituent of the abnormal plaques in the brains of Alzheimer's disease patients, thus being considered as a biomarker for Alzheimer's disease diagnosis.

A simple, rapid, reusable, and noninvasive screening strategy for early Alzheimer's disease diagnosis using magnetic N-doped graphene (MNG)-modified Au electrode was developed [77]. Superparamagnetic magnetite (Fe_3O_4) NPs were deposited onto NG to form MNG material. The antibodies of Aβ 1–28 (Aβab) are used as the specific biorecognition element for Aβ42 that were conjugated on the surface of MNG through the cross-linking method to form magnetic immunocarriers (Aβ ab-MNG) (Figure 13.7) that were dropped onto the Au electrode, where they were trapped by placing an external magnet underside of the electrode to carry out electrochemical Aβ detection, which was directly related to the diagnosis of Alzheimer's disease. The reusable biosensor with good reproducibility and stability has a linear response within the range from 5 to $800\,pg\,mL^{-1}$, covering the cutoff level of Aβ42 and a LOD of $5\,pg\,mL^{-1}$. The high sensitivity and selectivity toward Aβ 42 detection of the developed immunomagnetic biosensor brings benefits for early Alzheimer's disease diagnosis and provides a useful platform for bioanalytical and biomedical application.

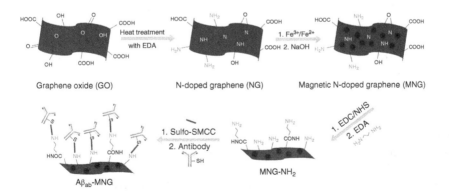

Figure 13.7 Schematic representation of the Aβab-MNG platform construction. *Source:* Li et al. [77]. https://www.ncbi.nlm.nih.gov/pmc/articles/PMC4844990/figure/f1/. Used under CC BY 4.0 http://creativecommons.org/licenses/by/4.0/. (*See insert for color representation of the figure.*)

13.3.2 Electrochemical Graphene Aptasensors

Aptamers are synthetic nucleic acids (DNA or RNA) that can possess specific binding characteristics to their targets, and biosensors that employ aptamers as a recognition element are called aptasensors. In a typical electrochemical aptasensor, the electrode surface serves as the platform to immobilize the biological aptamer, and the analyte-binding event is monitored based on electrochemical current variations. Aptamers can be modified chemically to undergo specific conformational changes depending on the purposes. In graphene-based aptasensor, the graphene is usually used for increasing the surface area of the electrode or serving as the carbon matrix for a tailored decoration with NPs involved in the sensing process.

13.3.2.1 Nucleic Acids

Graphene-based nanoelectronic devices have also been researched for use in DNA sensors (for detecting nucleobases and nucleotides). An impressive number of DNA-based electrochemical sensors with inventive design have been described in the literature and reviewed [78]. These sensors combine nucleic acid layers with electrochemical transducers to produce a biosensor, which provides a simple, sensitive, and accurate tool for diagnosis. DNA analysis is the most promising application of biosensors to clinical chemistry as DNA is well suited for biosensing due to the base pairing interactions between complementary sequences that are both specific and robust. DNA biosensors employ the immobilized relatively short synthetic single-stranded oligodeoxynucleotides that hybridize to a complementary target DNA in the sample [79]. Hybridization can be performed either in solution or on solid supports.

Graphene-based aptasensors present a promising potential for the detection of certain biomarkers, such as thrombin. An advanced sandwich-type electrochemical aptasensor assay coupling the use of CoPt alloy NP-decorated graphene with redox probe thionine, horseradish peroxidase, and secondary aptamer (Apt II) conjugates has been developed for the ultrasensitive detection of thrombin [80]. Good sensitivity and selectivity to thrombin detection with an LOD of 3.4×10^{-13} M was obtained with the proposed method.

A novel and simple synchronous electrochemical synthesis of the graphene/poly(xanthurenic acid) nanocomposite was reported by Jiao et al. [81], where GO and xanthurenic acid (Xa) monomer were adopted as precursors. The $\pi-\pi^*$ interactions between the conjugated GO layers and aromatic ring of Xa enhanced the electropolymerization efficiency accompanied with an increased electrochemical response of PXa. The rich carboxyl groups of PXa–ERGO film were applied to stably immobilize the probe DNA via covalent bonding as shown in Figure 13.8. The captured probe could sensitively and selectively recognize its target DNA. Due to the synergistic effect, this graphene-based electrochemical platform showed an intrinsic advantage in highly sensitive

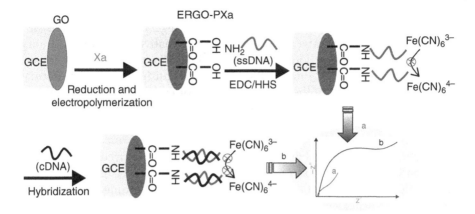

Figure 13.8 Schematic of synchronous electrosynthesis of PXa–ERGO for DNA EIS detection. *Source:* Yang et al. [81]. Reproduced with permission from American Chemical Society.

detection of DNA, in a dynamic detection range from 1.0×10^{-14} to $1.0 \times 10^{-8}\,M$, with an LOD of $4.2 \times 10^{-15}\,M$.

A facile label-free electrochemical strategy for the detection of DNA was reported by Guo et al. [82] based on a newly synthesized water-soluble electro-active dye azophloxine-functionalized graphene nanosheets (AP-GNs). Azophloxine was attached on the graphene surface through non-covalent charge transfer and π–π stacking interactions with graphene. The attached azophloxine on the graphene surface acts both as a stabilizer for graphene and also as the *in situ* probe for the electrochemical DNA detection. Under optimum conditions, this biosensor exhibited high sensitivity for DNA on the wide linear range from 1.0×10^{-15} to $1.0 \times 10^{-11}\,M$ and a low LOD of $4.0 \times 10^{-16}\,M$. Furthermore, this biosensor showed extraordinary capability for single nucleotide polymorphisms assay. The results demonstrate that this AP-GN-based biosensor has potential application in sensitive and selective DNA detection.

A sandwich-type DNA biosensor based on graphene-3D nanostructure gold nanocomposite-modified glassy carbon electrode (G-3D Au/GCE) was fabricated for detection of the surviving gene that was correlated with osteosarcoma [83]. The G-3D Au film was prepared with one-step electrochemical coreduction with graphite oxide and $HAuCl_4$ at cathodic potentials. The response of this sandwich-type DNA biosensor was measured by voltammetry and amperometric current-time techniques, and it showed a good linear relationship between the current signal and the logarithmic function of complementary DNA concentration in a range of 50–5000 fM with an LOD of 3.4 fM. This new biosensor exhibited a fast amperometric response and high sensitivity and selectivity and has been used in a polymerase chain reaction assay of real-life sample with a satisfactory result.

13.3.2.2 Cancer Cell

The nanotechnology-enabled detection of a change in individual cells, for instance, cell surface charge, presents a new alternative and complementary method for disease detection and diagnosis. Since diseased cells, such as cancer cells, frequently carry information that distinguishes them from normal cells, accurate probing of these cells is critical for early detection of a disease. However, up to now not many sensors have been developed for cancer-related tests because only a few of the biomarkers have shown clinical relevance, and the performance of the sensor systems is not always satisfactory.

Feng et al. [84] reported an example for label-free cancer cell detection by using an electrochemical sensor based on aptamer AS1411 used during the first clinical oncology trial II and graphene-modified electrode. The aptamer-perylenetetracarboxylic acid was utilized as nanoscale anchorage substrates to effectively capture cells on electrode surface through the specific binding between cell surface nucleolin and aptamer AS1411. The electrochemical aptasensor can distinguish cancer cells and normal ones and detect as low as one thousand cells. With DNA hybridization technique, this E-DNA sensor can be regenerated and reused for cancer cell detection.

A reusable magnetic graphene oxide (MGO)-modified biosensor for the detection of a cancer biomarker protein in serum was developed by Lin et al. [85]. It utilized Avastin as the specific biorecognition element and MGO as the carrier for Avastin loading. The detected biomarker is the vascular endothelial growth factor (VEGF) in human plasma essential for cancer diagnosis. This biosensor provides the appropriate sensitivity for clinical diagnostics and has a wide range of linear detection, from 31.25 to 2000 nM, compared with ELISA analysis. The proposed biosensor is evaluated by determining the VEGF levels in human serum samples from both healthy people and patients with brain tumors.

A graphene-based electrochemical immunosensor for sensitive detection of the cancer biomarker α-fetoprotein (AFP) was developed by Du et al. [86]. Functionalized graphene sheets have been used as a sensor platform to increase the electrode surface and capture a large amount of Ab_1. Further amplification was achieved by multienzyme-Ab_2-functionalized carbon nanospheres. On the basis of the dual signal amplification strategy of graphene sheets and the multienzyme labeling, the resulting immunosensor showed a sevenfold increase in detection signal compared with the immunosensor without graphene modification and could detect as low as $0.02\,\mathrm{ng\,mL^{-1}}$ AFP in serum samples. This amplification strategy is a promising platform for clinical screening of cancer biomarkers and point-of-care diagnostics.

There are thousands of sensor papers published in the past years, where graphene-based electrochemical sensors represent the most rapidly growing class. A wide variety of strategies are used to improve the efficacy of sensing. However, there is still much effort needed to implement these ultrasensitive sensors and biosensors for real clinical applications. The integration of electro-chemical sensors into microfluidic formats with the incorporation of unique materials for detection needs to be extensively explored in the future.

13.3.3 Optical Graphene Sensors for Medical Diagnostics

Graphene has astonishing optical and electrical properties and therefore has attracted much interest in the optoelectronic field due to its high mobility, optical transparency, flexibility, robustness, and environmental stability [87]. These extraordinary properties are due to its unique crystal lattice structure and can be utilized in innovative electronic and photoelectronic devices.

Presently, graphene has been coupled to optical transduces for optical bio-sensing. Its high thermal conductivity renders it sensitive to ambient tempera-ture variances [88–91]. Therefore, if coated on an optical surface, its refractive index (RI) would change accordingly with temperature. A graphene-modified transparent surface can be used as an efficient optical thermometer based on interferometry or evanescent wave coupling principle [90, 91].

Due to graphene's semimetal delocalization of electrons, it can be considered a plasmonic material [89, 92]. Therefore it is possible to replace more precious metals like gold and silver in plasmonic-based sensors of which surface plas-mon resonance (SPR) is the most well known. Optical detection methods are of the oldest and well-established techniques for sensing biomolecular interactions.

Optical transducers based on SPR are extremely sensitive to minute changes in RI that occur within 100 nm of the sensing metal surface of the transducer. SPR is an optical quantum phenomenon that occurs when an evanescent elec-tromagnetic field is generated at the interface of the metallic sensor surface and a nonconducting dielectric medium (i.e., aqueous sample) when excited by an incident beam of light at an appropriate wavelength and at an angle just beyond the critical angle (θc) of total internal reflection (Figure 13.9).

The evanescent wave produced by the incident beam of light penetrates a short distance into the metallic sensing surface (i.e., usually gold or silver) where the electrons are in turn excited and transformed into what is known as surface-bound plasmons. The coupling of the light "photon" energy into the electrons only occurs at specific wavelengths; at this point the light energy is transferred to the "free" plasmon electrons that have a particular resonance that is different from that of the bulk of the metal film. This coupling leads to a decrease in the amount of reflected light. If the light intensity is plotted against the angle of incidence, a characteristic SPR dip can be seen. The angle of inci-dence that occurs at this point is known as the surface resonance angle (θSPR), which is dependent on the local RI. The RI is directly proportional to any mate-rial (e.g., layer thickness) bound to the metal surface or roughness. The absorp-tion of energy by the surface plasmons (i.e., excited electrons) induces a

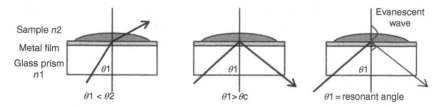

Figure 13.9 Schematic diagram of SPR principle. Light is directed through a prism of high RI into a surface layer with low RI (sample). At a particular angle total internal refection of the impinging light occurs. Although the light does not enter into the sample medium, the intensity at the interfacial boundary is not equal to zero. The photon energy from the light is transferred to the metal electrons, causing them to oscillate and produce surface-bound plasmons. This produces an exponential evanescent wave that penetrates a defined distance (~100 nm) into the low index medium, resulting in a characterized decrease in reflected light intensity.

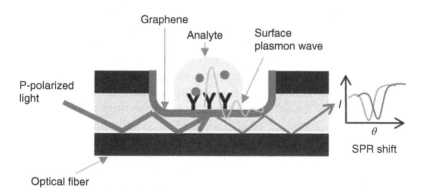

Figure 13.10 Principle of graphene-based fiber optic SPR biosensor and experimental setup. *Source:* Kim et al. [97]. Reproduced with permission from Elsevier.

decrease in energy of the reflected beam and thus creates a reflectance minimum. This causes an alteration in the position of the plasmon angle and therefore a change in the angle at which SPR occurs [93]. It is this angle shift that is used as the basis of SPR detection devices. Because SPR is sensitive to the adsorbed mass, it can be used to detect biomolecular interactions at the sensing surface interface without the use of labels and subsequent reagents. Optical transducers can be used to monitor affinity reactions and have been applied to quantitate antigenic species of interest in clinical chemistry and to study the kinetics and affinity of antigen–antibody and DNA interactions [94–96] (Figure 13.10).

Graphene can be used to enhance the conventional gold film SPR sensors [87–89, 92, 97–103] and has been shown to be able to totally replace the gold surface altogether [97]. Kim et al. were the first to demonstrate that graphene could replace gold and showed it had superior properties as it created a desirable large biocompatible surface sensing area for biological receptor binding [97]. Based on SPR principle, they measured the concentration of biotinylated double crossover DNA (DXB) lattice and protein streptavidin (SA). The graphene layer was used to replace the coating layer of plastic optical fiber, and a small amount of analyte was dropped on the sensing area to measure its concentration. The RI of analyte changed with concentration, so it was found that the resonance absorption could be demodulated by the variation of emergent light wavelength.

Due to graphene stable chemical properties and abundance, it is foreseeable in the future that graphene could not only replace more valuable metals but also enhance the SPR signal and sensor applications. To date graphene has been used and demonstrated in a variety of optical sensors and biosensor [97–103], usually utilizing its plasmonic evanescent wave properties.

A novel graphene–gold metasurface-based biosensing architectures were designed for detecting ssDNA with high sensitivity, which appears to be far beyond relevant values for any state-of-the-art plasmonic sensors available in the markets and bench side prototypes [99]. The proposed metasurface architecture consists of a gold film coated by a single (or multi) sheets of graphene. It is noticeable that more than fourfold enhancement of electric field takes place under the deposition of a single graphene layer on gold. Using graphene-coated film, the system achieves directly label-free ssDNA detection through the π stacking interaction between graphene and ssDNA without the need of any modification steps. The developed system was subjected to continue evaluation for six months for determining the setup stability and showed same reproducibility for all the repeated experiments. The graphene–gold biosensing structure provided the LOD of 1×10^{-18} M for 7.3 kDa 24-mer ssDNA at a signal-to-noise ratio of $3:1$. It is worth noticing that the graphene sensing substrate was fabricated based on pristine graphene layers, which are known to have better optical and electronic properties in comparison with the layers obtained by self-assembly of GOs.

Graphene coupled with SPR transducer can also display enhanced absorption advantages in improving the sensitivity and optimize resonance wavelength and detection range. Xing et al. showed that cancer cells can be distinguished from health cells due to volume differences [101]. They designed a graphene-based optical RI sensor with high resolution of 1.7×10^{-8} and sensitivity of 4.3×10^7 mV RIU^{-1}, as well as an extensive dynamic range. This highly sensitive graphene optical sensor enables label-free, live-cell, and highly accurate detection of a small quantity of cancer cells among normal cells at the single-cell level and the simultaneous detection and distinction of two cell lines without separation. Using this method, the authors demonstrated the ultrasensitive and extensive dynamic range sensing of a single Jurkat cell (type of lymphocyte cancer cell) on the graphene optical sensor platform. This is quite significant as it is relatively difficult to monitor cells and see this phenomenon with a traditional SPR setup. Biological sensors based on the optical adsorption property of the graphene can utilize the light intensity signal, which would greatly simplify the experimental setup.

Optical sensors based on graphene modification are still very much in its infancy, and still much work on evanescent affinity biosensing and optical absorption sensing is needed. However, this area of research shows promising results that may revolutionize molecular interaction detection and cell interrogation, especially in the detection of cancer cells and relevant biomarkers for clinical and medical diagnostics.

13.4 Conclusions

This chapter reviews the applications of graphene in nanotechnology since it came to the field particularly in sensing and biosensing applications. It updates the reader with the scientific progress of the current use of graphene as sensors and biosensors with specific application in human samples. Applications dealing with "real" clinical samples are still rare. Many papers describing the use of biosensors in this field have only exemplary character. Detailed data as well as validation with established methods for particular parameters are missing in most cases. The analytical potential of biosensors in the medical diagnostics field still has to be strengthened by the demonstration of their applicability to real matrices testing.

Research and development of commercial biosensors tend to focus on the creation of new sensors design and the miniaturization of new sensors, which is most suited with electrochemical platforms. The most research takes place at both universities and private business, but the commercialization of biosensors has lagged behind their research and development.

Acknowledgment

Author A. Arvinte is grateful for the financial support from the European Union's Horizon 2020 research and innovation program under grant agreement No 667387 SupraChem Lab.

References

1 M.L.Y. Sin, K.E. Mach, P.K. Wong, J.C. Liao, *Expert Rev Mol Diagn* **2014**, 14, 225–244.
2 B. Kim, J. Rutka, W. Chan, *N Engl J Med* **2010**, 363, 2434–2443.

3 A.K. Geim, K.S. Novoselov, *Nat Mater* **2007**, 6, 183–191.

4 H.K. Chae, D.Y. Siberio-Pérez, J. Kim, Y. Go, M. Eddaoudi, A.J. Matzger, M. O'Keefe, O.M. Yaghi, *Nature* **2004**, 427, 523–527.

5 K.S. Novoselov, A.K. Geim, S.V. Morozov, D. Jiang, Y. Zhang, S.V. Dubonos, I.V. Grigorieva, A.A. Firsov, *Science* **2004**, 306, 666–669.

6 R. Heyrovska, Atomic structures of graphene, benzene and methane with bond lengths as sums of the single, double and resonance bond radii of carbon, **2008**, https://arxiv.org/ftp/arxiv/papers/0804/0804.4086.pdf (accessed on July 26, 2017).

7 S. Stankovich, D.A. Dikin, G.H.B. Dommett, K.M. Kohlhaas, E.J. Zimney, E.A. Stach, R.D. Piner, S.T. Nguyen, R.S. Ruoff. *Nature* **2006**, 442, 282–286.

8 C. Berger, Z. Song, X. Li, X. Wu, N. Brown, C. Naud, D. Mayou, T. Li, J. Hass, A.N. Marchenkov, E.H. Conrad, P.N. First, W.A. de Heer, *Science* **2006**, 312, 1191–1196.

9 S. Stankovich, D.A. Dikin, R.D. Piner, K.A. Kohlhaas, A. Kleinhammes, Y. Jia, Y. Wu, S.T. Nguyen, R.S. Ruoff, *Carbon* **2007**, 45, 1558–1565.

10 S. Stankovich, R.D. Piner, X.Q. Chen, N.Q. Wu, S.T. Nguyen, R.S. Ruoff, *J Mater Chem* **2006**, 16, 155–158.

11 H.C. Schniepp, J.L. Li, M.J. McAllister, H. Sai, M. Herrera-Alonso, D.H. Adamson, R.K. Prud'Homme, R. Car, D.A. Saville, I.A. Aksay, *J Phys Chem B* **2006**, 110, 8535–8539.

12 S. Gilje, S. Han, M. Wang, K.L. Wang, R.B. Kaner, *Nano Lett* **2007**, 7, 3394–3398.

13 J.-C. Charlier, X. Gonze, J.-P. Michenaud, *Europhys Lett* **1994**, 28, 403–408.

14 D.R. Dreyer, R.S. Ruoff, C.W. Bielawski, *Angew Chem Int Ed* **2010**, 49, 9336–9344.

15 J.N. Coleman, *Adv Funct Mater* **2009**, 19, 3680–3695.

16 S. Vadukumpully, J. Paul, S. Valiyaveettil, *Carbon* **2009**, 47, 3288–3294.

17 U. Khan, A. O'Neill, M. Lotya, S. De, J.N. Coleman, *Small* **2010**, 6, 864–871.

18 M.J. Allen, V.C. Tung, R.B. Kaner, *Chem Rev* **2010**, 110, 132–145.

19 Mass production of high quality graphene: an analysis of worldwide patents, http://www.nanowerk.com/spotlight/spotid=25744.php (accessed on June 20, 2017).

20 H. Lu, W.S. Woi, X. Tan, C.T. Gibson, X. Chen, C.L. Raston, J.M. Gordon, H.T. Chua, *Carbon* **2015**, 94, 349–351.

21 A. Ghosh, K.S. Subrahmanyam, K.S. Krishna, S. Datta, A. Govindaraj, S.K. Pati, C.N.R. Rao, *J Phys Chem C* **2008**, 112, 15704–15707.

22 K.S. Novoselov, D. Jiang, F. Schedin, T.J. Booth, V.V. Khotkevich, S.V. Morozov, A.K. Geim, *Proc Natl Acad Sci USA* **2005**, 102, 10451–10453.

23 Y.B. Zhang, Y.W. Tan, H.L. Stormer, P. Kim, *Nature* **2005**, 438, 201–204.

24 G. Gundiah, A. Govindaraj, N. Rajalakshmi, K.S. Dhathathreyan, C.N.R. Rao, *J Mater Chem* **2003**, 13, 209–213.

25 D.J. Collins, H.C. Zhou, *J Mater Chem* **2007**, 17, 3154–3160.

26 M. Pumera, *Energy Environ Sci* **2011**, 4, 668–674.
27 S. Bae, H. Kim, Y. Lee, X. Xu, J.-S. Park, Y. Zheng, J. Balakrishnan, T. Lei, H. Ri Kim, Y.I. Song, Y.-J. Kim, K.S. Kim, B. Ozyilmaz, J.-H. Ahn, B.H. Hong, S. Iijima, *Nat Nanotechnol* **2010**, 5, 574–578.
28 Y.W. Zhu, S. Murali, W.W. Cai, X.S. Li, J. W. Suk, J.R. Potts, R.S. Ruoff, *Adv Mater* **2010**, 22, 3906–3924.
29 C. Gomez-Navarro, M. Burghard, K. Kern, *Nano Lett* **2008**, 8, 2045–2049.
30 A.R. Ranjbartoreh, B. Wang, X.P. Shen, G.X. Wang, *J Appl Phys* **2011**, 109, 014306.
31 H.B. Akkerman, P.W.M. Blom, D.M. de Leeuw, B. de Boer, *Nature* **2006**, 441, 69–72.
32 J.N. Al-Aqtash, I. Vasiliev, *J Phys Chem C* **2011**, 115, 18500–18510.
33 P. Shapira, A. Gok, F. Salehi, *J Nanopart Res* **2016**, 18, 269–293.
34 Graphene-Info: The graphene experts, http://www.graphene-info.com/ (accessed on June 20, 2017).
35 Graphene Frontier's G-FET based chemical sensor, https://ecs.confex.com/ecs/230/webprogram/Paper91827.html (accessed on June 20, 2017).
36 T. Gan, S. Hu, *Microchim Acta* **2011**, 175, 1–19.
37 Y. Liu, X. Dong, P. Chen, *Chem Soc Rev* **2012**, 41, 2283–2307.
38 J.D. Newman, A.P. Turner, *Biosens Bioelectron* **2005**, 20, 2435–2453.
39 N.J. Ronkainen, H.B. Halsall, W.R. Heineman, *Chem Soc Rev* **2010**, 39, 1747–1763.
40 G. Zhiguo, Y. Shuping, L. Zaijun, S. Xiulan, W. Guangli, F. Yinjun, L. Junkang, *Electrochim Acta* **2011**, 56, 9162–9167.
41 R. Krishna, J.M. Campiña, P.M.V. Fernandes, J. Ventura, E. Titus, A.F. Silva, *Analyst* **2016**, 141, 4151–4161.
42 X. Kang, J. Wang, H. Wu, I.A. Aksay, J. Liu, Y. Lin, *Biosens Bioelectron* **2009**, 25, 901–905.
43 K. Wang, Q. Liu, Q.M. Guan, J. Wu, H.N. Li, J.J. Yan, *Biosens Bioelectron* **2011**, 26, 2252–2257.
44 K. Dhara, J. Stanley, T. Ramachandran, B.G. Nair, T.G.S. Babu, *Sens Actuators B* **2014**, 195, 197–205.
45 H. Lee, T.K. Choi, Y.B. Lee, H.R. Cho, R. Ghaffari, L. Wang, H.J. Choi, T.D. Chung, N. Lu, T. Hyeon, S.H. Choi, D.-H. Kim, *Nat Nanotechnol* **2016**, 11, 566–572.
46 F. Tehrani, B. Bavarian, *Sci Rep* **2016**, 6, 27975.
47 A. Anderson, A. Lindgren, B. Huldberg, *Clin Chem* **1995**, 41, 361–366.
48 M.T. Heafield, S. Fearn, G.B. Sterevton, R.H. Waring, A.C. Williams, S.G. Sturman, *Neurosci Lett* **1990**, 110, 216–220.
49 P.M. Ueland, M.A. Mansoor, A.B. Guttormsen, F. Muller, P. Aukrust, H. Refsum, A.M. Svardal, *J Nutr* **1996**, 126, 1281s–1284s.
50 S. Shahrokhian, *Anal Chem* **2001**, 73, 5972–5978.
51 S. Yang, G. Li, Y. Wang, G. Wang, L. Qu, *Microchim Acta* **2016**, 183, 1351–1357.

52 F. Xu, F. Wang, D. Yang, Y. Gao, H. Li, *Mater Sci Eng C* **2014**, 38, 292–298.
53 S. Ge, M. Yan, J. Lu, M. Zhang, F. Yu, J. Yu, X. Song, S. Yu, *Biosen Bioelectron* **2012**, 31, 49–54.
54 N. Shadjou, M. Hasanzadeh, F. Talebi, A.P. Marjani, *Nanocomposites* **2016**, 2, 18–28.
55 M.B. Gholivand, M. Khodadadian, *Biosens Bioelectron* **2014**, 53, 472–478.
56 R.S. Dey, C.R. Raj, *Biosen Bioelectron* **2014**, 62, 357–364.
57 N. Ruecha, R. Rangkupan, N. Rodthongkum, O. Chailapakul, *Biosens Bioelectron* **2014**, 52, 13–19.
58 S. Izawa, K. Kono, K. Mimura, Y. Kawaguchi, M. Watanabe, T. Maruyama, H. Fujii, *Cancer Immunol Immunother* **2011**, 60, 1801–1810.
59 K.A. Youdim, J.A. Joseph, *Free Radic Biol Med* **2001**, 30, 583–594.
60 T.P. Szatrowski, C.F. Nathan, *Cancer Res* **1991**, 51, 794–798.
61 K. Zhang, R.J. Kaufman, *Nature* **2008**, 454, 455–462.
62 G. Yu, W. Wu, X. Pan, Q. Zhao, X. Wei, Q. Lu, *Sensors* **2015**, 15, 2709–2722.
63 J. Bai X. Jiang, *Anal Chem* **2013**, 85, 8095–8101.
64 M. Govindasamy, V. Mani, S.-M. Chen, R. Karthik, K. Manibalan, R. Umamaheswari, *Int J Electrochem Sci* **2016**, 11, 2954–2961.
65 D. Zhang, X. Ouyang, L. Li, B. Dai, Y. Zhang, *J Electroanal Chem* **2016**, 780, 60–67.
66 E. Lenters-Westra R.K. Schindhelm, H.J. Bilo, R.J. Slingerland, *Diabetes Res Clin Pract* **2013**, 99, 75–84.
67 S.H. Ang, M. Thevarajah, Y. Alias, S. Khor, *Clin Chim Acta* **2015**, 439, 202–211.
68 U. Jain, N. Chauhan, *Biosens Bioelectron* **2016**, S0956-5663(16), 30140–30143.
69 A. Pandikumar, G.T. Soon How, T.P. See, F.S. Omar, S. Jayabal, K.Z. Kamali, N. Yusoff, A. Jamil, R. Ramaraj, S. A. John, H. N. Lim, N.M. Huang, *RSC Adv* **2014**, 4, 63296–63323.
70 L. Wu, L. Feng, J. Ren, X. Qu, *Biosens Bioelectron* **2012**, 34, 57–62.
71 X. Yu, K. Sheng, G. Shi, *Analyst* **2014**, 139, 4525–4531.
72 J. Salamon, Y. Sathishkumar, K. Ramachandran, Y.S. Lee, D.J. Yoo, A.R. Kim, G.G. Kumar, *Biosens Bioelectron* **2015**, 64, 269–276.
73 R.N. Goyal, S. Bishnoi, *Talanta* **2011**, 84, 78–83.
74 Rosy, S. Yadav, B. Agrawal, M. Oyama, R.N. Goyal, *Electrochim Acta* **2014**, 125, 622–629.
75 C. Xue, X. Wang, W. Zhu, Q. Han, C. Zhu, J. Hong, X. Zhou, H. Jiang, *Sens Actuators B* **2014**, 196, 57–63.
76 M.P. Murphy, H. LeVine, *J Alzheimers Dis* **2010**, 19, 311–323.
77 S.-S. Li, C.-W. Lin, K.-C. Wei, C.-Y. Huang, P.-H. Hsu, H.-L. Liu, Y.-J. Lu, S.-C. Lin, H.-W. Yang, C.-C.M. Ma, *Sci Rep* **2016**, 6, 25155.
78 T.G. Drummond, M.G. Hill, J.K. Barton, *Nat Biotechnol* **2003**, 21, 1192–1199.
79 E. Paleček, M. Fojta, F. Jelen, *Bioelectrochemistry* **2002**, 56, 85–90.
80 Y. Wang, R. Yuan, Y. Chai, Y. Yuan, L. Bai, Y. Liao, *Biosens Bioelectron* **2011**, 30, 61–66.

81 T. Yang, Q. Li, L. Meng, X. Wang, W. Chen, K. Jiao, *ACS Appl Mater Interfaces* **2013**, 5, 3495–3499.

82 Y. Guo, Y. J. Guo, C. Dong, *Electrochim Acta* **2013**, 113, 69–76.

83 A.L. Liu, G.X. Zhong, J.Y. Chen, S.H. Weng, H.N. Huang, W. Chen, L.Q. Lin, Y. Lei, F.H. Fu, Z.L. Sun, X.H. Lin, J.H. Lin, S.Y. Yang, *Anal Chim Acta* **2013**, 767, 50–58.

84 L. Feng, Y. Chen, J. Ren, X. Qu, *Biomaterials* **2011**, 32, 2930–2937.

85 C.W. Lin, K.C. Wei, S.-S. Liao, C.-Y. Huang, C.-Y. Sun, P.-J. Wu, Y.-J. Lu, H.-W. Yang, C.-C.M. Ma, *Biosens Bioelectron* **2015**, 67, 431–437.

86 D. Du, Z. Zou, Y. Shin, J. Wang, H. Wu, M.H. Engelhard, J. Liu, I.A. Aksay, Y. Lin, *Anal Chem* **2010**, 82, 2989–2995.

87 F. Bonaccorso, F. Sun, T. Hasan, A.C. Ferrari, *Nat Photonics* **2010**, 4, 611–622.

88 H. Fu, S. Zhang, H. Chen, J. Weng, *Sens J IEEE* **2015**, 15, 5478–5482.

89 N. Chiu, T. Huang, *Sens Actuators B* **2014**, 197, 35–42.

90 L. Li, Z. Feng, X. Qiao, H. Yang, R. Wang, D. Su, Y. Wang, W. Bao, J. Li, Z. Shao, M. Hu, *IEEE Sens J* **2015**, 15, 505–509.

91 C. Li, Q. Liu, X. Peng, S. Fan, *Opt Express* **2015**, 23, 27494–27502.

92 Y. Zhao, X. Lia, X. Zhoua, Y. Zhang, *Sens Actuators B* **2016**, 31, 324–340.

93 A.M. Sesay, *Towards a remote portable bio-affinity surface plasmon resonance analyser for environmental steroidal-pollutants*, thesis/dissertation, **2003**, https://dspace.lib.cranfield.ac.uk/handle/1826/854 (accessed on June 20, 2017).

94 Z. Altintas, Y. Uludag, Y. Gurbuz, I.E. Tothill, *Talanta* **2011**, 86, 377–383.

95 Z. Altintas, Y. Uludag, Y. Gurbuz, I. Tothill, *Anal Chim Acta* **2012**, 712, 138–144.

96 Z. Altintas, I.E. Tothill, *Sens Actuators B Chem* **2012**, 169, 188–194.

97 J.A. Kim, T. Hwang, S.R. Dugasani, R. Amin, A. Kulkarni, S.H. Park, T. Kim, *Sens Actuators B* **2013**, 187, 426–433.

98 J.N. Dash, R. Jha, *Plasmonics* **2015**, 10, 1123–1131.

99 S. Zeng, K.V. Sreekanth, J. Shang, T. Yu, C.K. Chen, F. Yin, D. Baillargeat, P. Coquet, H.P. Ho, A.V. Kabashin, K.T. Yong, *Adv Mater* **2015**, 27, 6163–6169.

100 J.N. Dash, R. Jha, *Photonics Technol Lett IEEE* **2014**, 26, 1092–1095.

101 F. Xing, G.X. Meng, Q. Zhang, L.-T. Pan, P. Wang, Z.-B. Liu, W.-S. Jiang, Y. Chen, J.-G. Tian, *Nano Lett* **2014**, 14, 3563–3569.

102 M. Batumalay, S.W. Harun, F. Ahmad, R.M. Nor, N.R. Zulkepely, H. Ahmad, *IEEE Sens J* **2014**, 14, 1704–1709.

103 S.H. Girei, A.A. Shabaneh, N.L. Hong, M.N. Hamidon, M.A. Mahdi, M.H. Yaacob, *Opt Rev* **2015**, 22, 1–8.

Section 4

Organ-Specific Health Care Applications for Disease Cases Using Biosensors

14

Optical Biosensors and Applications to Drug Discovery for Cancer Cases

Zeynep Altintas

Technical University of Berlin, Berlin, Germany

14.1 Introduction

Drug research is a multidisciplinary field enriched by chemistry, pharmacology, clinical sciences, and bioinformatics. The majority of the advancements have come out in the past decades, owing to the great efforts and new developments taken place in molecular biology, genomics, and proteomics [1–7]. The increasing knowledge on hereditary materials as well as signaling pathways of many central proteins and genes has also an immense impact on drug discovery. When the fascinating influence of nanotechnology takes its place in this complex role, the recent years have witnessed an encouraging interdisciplinary endeavor that brings the academia and biotechnology industry together to fill the gaps and establish much more successful and rapid research and development (R&D) platform in drug discovery. By taking into account this objective, the fabrication of new tools and improvement of the previously existing ones have been achieved for biotherapeutic screening and characterization [5, 7, 8].

Natural resources such as marine environments and tropical rain forests take an enormous attention as sources for prospective new drugs since thousands of animals and plants from these habitations have been utilized in the conventional medicine of native community. Obtaining knowledge on traditional medicine has resulted in several important hints related to bioactivity that may lead to novel nature-origin drugs [9–11]. Therefore, there is a special need for screening systems and their development since they have a central role in drug discovery. In the scope of this chapter, optical-based biosensors and their potential use in the development of anticancer drugs will be discussed. Biosensors serve as a cutting-edge technology with a broad range of applications and momentous success [12, 13]. They have been

Biosensors and Nanotechnology: Applications in Health Care Diagnostics, First Edition.
Edited by Zeynep Altintas.
© 2018 John Wiley & Sons, Inc. Published 2018 by John Wiley & Sons, Inc.

intensively employed in medical diagnosis [14–21], environmental analysis [22–25], and food safety [26–32] using various assay strategies. Biosensors are classified into several main categories based on the transducer systems as optical, piezoelectric, electrochemical, and magnetic biosensors [14, 16]. A detailed discussion on the applications of each biosensor type, beyond this chapter, can be found in Section 2 of this book. Due to the intensive R&D on optical sensors, they serve as the most reliable, robust, and widely used platforms with real-time and fast nature. These features also make them broadly applicable in drug research by providing information about binding kinetics of drugs, drug–ligand interaction, association–dissociation rates, cross-reactivity of a target drug with other ligands, and vice versa [33–39]. Considering the fact that the development of a drug takes around 15–20 years and costs more than US$1 million to produce [40] (Figure 14.1), the pharmaceutical industry is constantly seeking for technologies that can lower the development cost as well as the lead time to market; hence, the sensor area has played a pivotal role at this juncture.

Surface plasmon resonance (SPR)-based platforms are the dominated optical sensors in this area as being sensitive, label-free fast systems [42, 43]. They utilize the indirect detection of surface plasmons—electromagnetic waves that oscillate at the interface of two media such as metal and dielectric—from absorption features (for instance, analyte binding) that correspond to the coupling of light to the surface plasmon. SPR is sensitive to changes in the refractive index of the dielectric upon analyte binding to the substrate–metal interface as it passes through a fluidic channel [35, 42]. There are three methods of surface plasmon excitation: prism-based configuration, waveguide-based approach, and optical fiber-based architecture. Using a prism-based configuration, resonance is monitored by collecting the reflected light as a function of the incidence angle at a fixed wavelength or as a function of the incident wavelength at a fixed angle. In waveguide- or optical fiber-based methods, surface plasmons are excited by a broadband light source, and the transmitted light is collected and analyzed, where a dip in the spectrum represents a change in resonance. These sensors are very convenient for screening purposes in order to discover new drugs and provide information about the bioactivity of drugs with known or unknown chemical composition. Their successful integration into drug research mainly depends on the appropriate surface chemistry that needs to be selected by considering the nature of the drug, functional groups of the drug receptor, and assay type. Therefore, immobilization techniques in biosensors, high-throughput SPR systems launched to the market, detailed information and important strategies in drug discovery using optical biosensors, and new approaches such as computational simulation coupled with SPR sensors for drug–receptor monitoring are discussed in the following sections.

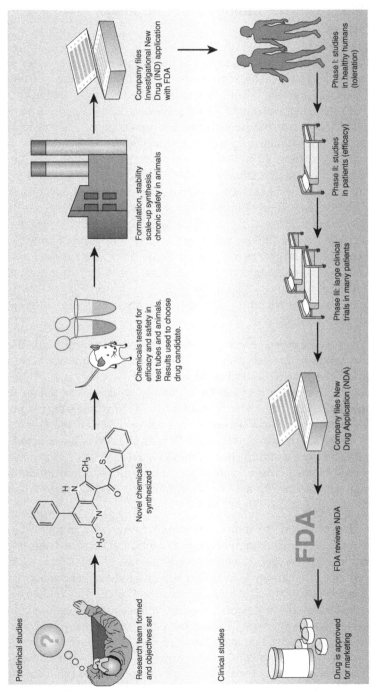

Figure 14.1 Stages of drug discovery process. *Source:* Lombardino and Lowe [41]. Reproduced with permission of Nature Publishing Group. (*See insert for color representation of the figure.*)

14.2 Biosensor Technology and Coupling Chemistries

One of the key components of a sensor is the immobilization of a ligand onto the transducer surface to achieve the stabilization over the course of a binding assay as well as to prevent from biomolecule degradation, owing to direct attachment of the ligands onto a metal sensor surface [19, 44]. A variety of techniques can be used to achieve ligand immobilization; some of which are based on covalent attachment, whereas some others rely on non-covalent binding onto the surface. Commonly applied covalent coupling chemistries to attach a ligand to a surface include (i) *EDC-NHS chemistry*, which is used on self-assembled monolayer surfaces involving an exposed carboxyl group to be activated by EDC-NHS and bound to the receptor via its amino group [15, 19, 45]; (ii) *use of amino-presenting surfaces*, which are suitable to be treated with bifunctional linkers to influence coupling with free sulfhydryl or amino groups on the ligand [46]; and (iii) *sensor surface derivatized with salicylhydroxamic acid*, which is used to create reversible complexes with ligands activated with pyridinyldithioethanamine [47] (Figure 14.2).

Non-covalent immobilization methods also include three different and commonly used strategies:

1) *Streptavidin- or biotin-presenting surfaces* that offer not only high efficiency but also stable complexes and are widely used for immobilization of 5′-biotinylated oligonucleotides [18, 44, 48–50]
2) *Monoclonal antibodies* that can be covalently attached to a sensor surface through EDC-NHS chemistry, and then fusion- or epitope-tagged proteins such as glutathione S-transferase, FLAG epitope, 6×His, and D epitope of herpes simplex virus glycoprotein are directly and reversibly coupled to the surface via antigen-antibody interaction [51, 52]
3) *Metal-coordinating groups* such as nitrilotriacetic acid and iminodiacetic acid that are commonly applied to immobilize 6×His and 6×His-tagged ligands directly [53, 54] (Figure 14.3)

Despite the successful use of these techniques for receptor immobilization in optical biosensors, the detection of a small compound such as drugs is generally difficult in case of applying a direct binding assay. The biosensor assays can be grouped into three methods including direct, sandwich, and nanomaterial-amplified sandwich assays. The limit of detection (LOD) of an analyte of interest is mostly decreased using sandwich assay with respect to the direct method, and the highest signal with lowest LOD is achieved with nanoparticle-functionalized assays that involve a secondary antibody or probe. Although the applications of sandwich assays are very desirable and widely employed with biosensors in many applications from bacteria, virus, and biomarker detection to environmentally hazardous compounds and cells, the preferred assay type for drug

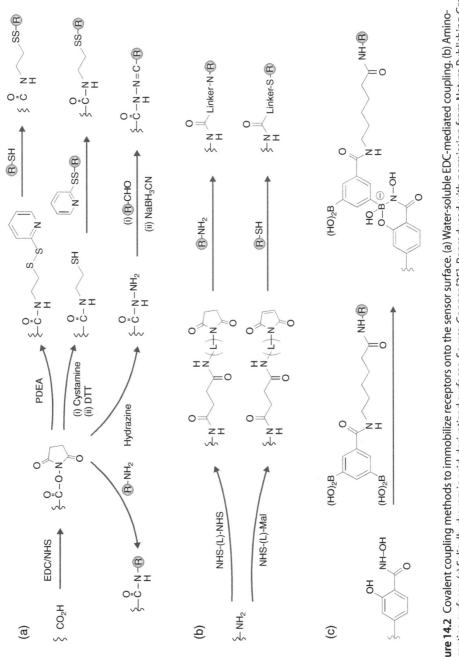

Figure 14.2 Covalent coupling methods to immobilize receptors onto the sensor surface. (a) Water-soluble EDC-mediated coupling. (b) Amino-presenting surfaces. (c) Salicylhydroxamic acid-derivatized surfaces. *Source:* Cooper [35]. Reproduced with permission from Nature Publishing Group.

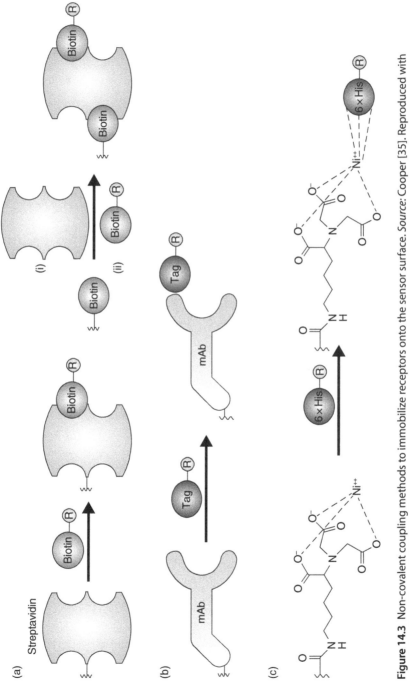

Figure 14.3 Non-covalent coupling methods to immobilize receptors onto the sensor surface. *Source:* Cooper [35]. Reproduced with permission from Nature Publishing Group.

discovery studies is direct assay due to the fact that the complex binding assay may cause conformational changes on the receptor or drug; hence, it may not reflect original interaction between drug and receptor pair in cell or cell membrane. Moreover, it can also obtain lower or higher binding kinetics than the expected, therefore resulting in an unsuccessful screening of drug–receptor interaction for further drug discovery. Two alternative approaches without labeling can be considered for avoiding this problem: (i) *standard direct assay* that requires immobilization of drug receptor onto the sensor surface and then monitoring of drug binding in a certain concentration range and (ii) *indirect assay* that includes the attachment of the drug on the surface and then injection of varying concentration of receptor. In both cases, it is possible to determine binding kinetics, dissociation–association patterns, and mathematical affinity as well as specificity in the case of screening a drug in human serum sample. By employing first models of Biacore sensors (such as Biacore 3000), it is possible to detect the binding response of small molecules with a molecular mass higher than 200 Da. Direct screening of drugs was applied in the selection of DNA-gyrase inhibitors [55], thrombin inhibitors [56], HIV protease inhibitors [57, 58], and many others. Today, the efficacy of optical sensors in terms of small-molecule determination has been greatly improved with the recent years' developments, and the mass limit has been substantially decreased to less than 100 Da. Therefore, the applications of optical-based sensors in drug discovery have intensively increased. Novel SPR-based optical sensors were introduced to the market in the past years by taking into account the high-throughput drug research that will be discussed in the next section; examples of optical-based drug discovery applications will also be given.

14.3 Optical Biosensors for Drug Discovery

The last decade has witnessed momentous advancements of SPR sensors with the release of new throughput systems to the market. Being the owner of the first commercial SPR instrument in 1990, GE Healthcare (Uppsala, Sweden) has launched many other more advanced SPR sensors including Biacore C, Biacore X100, Biacore T200, Biacore S200, Biacore 8K SPR system, and Biacore 4000. They were designed and manufactured as appropriate tools for protein interaction analysis, drug discovery and development, and manufacturing and quality control. All of them serve as automated systems with real-time assays and are allowed to perform kinetic studies. The latest versions such as Biacore 8K SPR system and Biacore 4000 have demonstrated great opportunities to be used in drug discovery, owing to the capability of screening very small molecules (<50 Da) with a high number of parallel analyses (2300 small molecules per day). These systems were particularly developed by taking into consideration the need in drug discovery research. Therefore, some important characteristics of these two SPR biosensors are listed in Table 14.1, and all other SPR sensors with their company names and websites are given in Table 14.2 for further details of each system.

Another company established in 2006 with the name Sierra Sensors GmbH (Hamburg, Germany) has released QCM [18, 26] and SPR sensors [25, 36, 37, 59]. As a result of their main focus on SPR systems, two different SPR biosensors (SPR-2 and SPR-2/4) [36, 59, 60], which are similar to Biacore 3000, were initially launched to the market. These systems have been successfully employed in several different areas [25, 36, 37, 45, 59, 60], and their potential in drug–receptor interactions has been shown in the recent works [36, 37]. The company has manufactured and released new SPR-based sensors (MASS-1 and MASS-2) with great improvements including parallel sample analyses of up to 16 samples, screening over 7000 interactions per day, detailed analysis of greater than 2000 samples per day, wide temperature range between 4 and 40°C for analysis, and detection of much smaller compounds. Hence, MASS-1 and MASS-2 can serve as very suitable and reliable tools in drug discovery applications for cancer cases. Notable features of MASS-1 sensor are listed in Table 14.3, and the detailed information can be found in the company website for other sensor types (Table 14.2). The real-time monitoring ability, high-sensitivity, and throughput nature of the newly developed optical sensors will increase their use in drug research and accelerate the drug discovery process.

Altermann et al. used SPR sensor to screen HIV protease inhibitors, and it was demonstrated that HIV-1 protease is a valuable target for AIDS treatment, owing to its central role in the replication of HIV virus. The scientists performed two different assays for screening: (i) HIV protease was immobilized on a sensor chip and the sample was then injected to the surface. A biotinylated substrate was used to block the free binding regions of the enzyme. (ii) A model inhibitor was immobilized onto the sensor surface, and the sample with HIV inhibitor was incubated with enzyme sample. The immobilized model inhibitor was then bound to the free enzyme. The assays were investigated in the concentration ranges of 0.1 nM to 10 μM and 1–100 nM, respectively. The research indicated the compatibility of these two competitive assays for screening of HIV inhibitors [61]. Hence, HIV protease inhibitors serve as one of the best candidate drugs to be used against AIDS. Similar approach with SPR sensor was also utilized for thrombin inhibitors to avoid blood coagulation. Use of SPR sensors is also successful for DNA-binding drugs' characterization. In a study, DNA probes were immobilized on the sensor surface via their biotin tags, and they demonstrated binding regions for many human signaling proteins. Binding interactions between the DNA oligonucleotides and the targeted compounds including distamycin, chromomycin, and mithramycin could be analyzed using direct assay approach [62]. The developed approaches can be extensively applied in a rapid screening of molecular interactions between DNA-binding drugs and the targeted DNA

Table 14.1 Key features of Biacore 8K SPR system and Biacore 4000 as potential high-throughput drug discovery tools.

	Biacore 8K	Biacore 4000
Key features	A single solution for interaction analysis in both screening and characterization	High-sensitivity and high-quality interaction data for better informed decision making
	Screening of 2300 small-molecule fragments in a day	60 h unattended operation with parallel analysis of up to 16 targets
	High-quality kinetic characterization of 64 interactions in 5 h	Large-scale molecular analysis generating up to 4800 interactions in 24 h
	60 h unattended runtime with queueing abilities and rapid multi-run evaluations	Seamless LIMS integration with convenient data import and export functionality
	Confident interaction analysis of small molecules binding to complex targets such as GPCRs	Innovative software solutions for increased productivity and performance
Main applications	Selection of biotherapeutic or small-molecule hits based on affinity and kinetic ranking	Fragment and LMW compound screening
	Characterization and optimization of selected binders based on detailed kinetic and affinity information	Compound hit validation, hit-to-lead characterization, and lead optimization of compounds
		Antibody and antibody fragment screening and characterization
Detection technology	Surface plasmon resonance (SPR) biosensor	Surface plasmon resonance (SPR) biosensor
Information provided	Kinetic and affinity data (k_a, k_d, K_D), specificity, selectivity, and screening data	Binding kinetic, affinity, selectivity, and specificity data. Concentration measurements
Data presentation	Result tables, result plots, and real-time monitoring of sensorgrams	
Analysis time per cycle	Typically 2–15 min	4 targets per flow cell, cycle time 5 min
Sample type	Small-molecule drug candidates to high molecular weight proteins (also DNA, RNA, polysaccharides, lipids, cells, and viruses) in various sample environments (e.g., in DMSO-containing buffers, plasma, and serum)	LMW drug candidates to high molecular weight proteins in various sample environments, for example, in DMSO-containing buffers, cell culture supernatants, or serum

(Continued)

Table 14.1 (Continued)

	Biacore 8K	Biacore 4000
Required sample volume	Injection volume plus 20–50 μL (application dependent)	N/A
Injection volume	1–200 μL	Typically 60 μL per flow cell (range 30–425 μL)
Sample/reagent capacity	4 × 96- or 384-well microplates, normal and deep-well	Antibody screening: 10 × 96-well microplates
		LMW compound screening: 6 × 384-well microplates
		Kinetic characterization: 4 × 384-well microplates
Number of flow cells	16 in 8 channels	Parallel processing of four independent flow cells, each containing five detection spots
Analysis temperature range	4–40 °C	4–40 °C
Molecular weight detection	No lower limit for organic molecules	Analytes with relative molecular mass $(M_r) > 50$ (Da)

sequences in cancer cases. The novel SPR sensors are enormously capable to increase the efficacy of this approach owing to the possibility of a large number of targets' simultaneous screening. Kampranis et al. evaluated the binding kinetics of cyclothialidine and coumarin drugs to mutant and wild-type DNA gyrases. The research demonstrated high affinity drug–protein interactions with dissociation constant of 1–150 nM and the critical role of residues on gyrase for drug binding [63].

G protein-coupled receptors (GPCRs) constitute the major class of drugs. In contrast to traditional approach for the discovery of GPCR ligands, fragment screening is now taking a great attention as a new approach. The principle of fragment-based drug discovery relies on relatively small number of low molecular weight fragments that display large areas of chemical space [39, 64–66]. Nevertheless, this approach has not been routinely utilized for membrane proteins, and it is limited to some currently approved GPCR targets. Screening of a small library is the initial step of fragment-based drug discovery that generally includes only several hundred to a couple of thousand low molecular weight compounds called as fragments. The molecular weight of these fragments is usually in the range of 100–300 Da. SPR technology has currently

Table 14.2 Developers of label-free optical sensor platforms.

Provider	Sensor type	Product	Website
Axela	Diffractive optic sensors	dot°—Avidin sensor, dot°—Covalent sensor, panelPlus™, and Axela custom sensors	http://www.axelabiosensors.com
Affinité Instruments	SPR	P4SPR	http://affiniteinstruments.com
Biacore	SPR	Q, 3000, C, X100, T200, S200, SK SPR system, and 4000	https://www.biacore.com/lifesciences
Bio-Rad	SPR	ProteOn™	http://www.bio-rad.com
BioNavis	SPR	MP-SPR	http://www.bionavis.com
Bioptix	Enhanced SPR	404pi	http://www.bioptix.com
Biosensing Instrument	SPR	BI-2500 and BI-4500	http://biosensingusa.com
IBIS Technologies BV	SPR imager	MX96	http://www.ibis-spr.nl
Plexera	SPR imaging	PlexArray° HT System	http://www.plexera.com
Nicoya Lifesciences	SPR	OpenSPR sensors	https://nicoyalife.com
Reichert	SPR	SR7500DC and Reichert4SPR	http://www.reichertspr.com
Sensia	SPR	Indicator-G	http://www.sensia.es
Sierra Sensors	SPR	SPR-2/4, MASS-1, and MASS-2	http://www.sierrasensors.com

offered the dominant screening methods for fragment-based drug discovery. Aristotelous et al. reported on the utility of SPR-based fragment screening of wild-type GPCRs by the invention of new high affinity antagonists of β2 adrenoreceptor. It is demonstrated that fragment screening using SPR can be achieved on tagged native GPCRs without the necessity for comprehensive protein engineering. The method has offered a great advantage by screening tagged native receptors, without introducing stabilizing mutations; thus, the pharmacology of the wild-type ligand can be protected. SPR-based fragment screening not only supplies an effective novel approach to the discovery of new GPCR receptors but it also demonstrates exceptional opportunities to screen for ligands against receptor signaling complexes and biased signaling conformations of the receptors [67].

Table 14.3 Characteristics of MASS-1 (Sierra Sensors, Hamburg, Germany) surface plasmon resonance biosensor.

Detection technology	SPR imaging
Working principle	Parallel processing of up to eight simultaneous injections over eight dual-sensor flow cells
	Fully individual addressing of all 16 sensor spots in the 8 × 2 array
Type of information obtained	Binding specificity and selectivity, binding kinetics, affinity, concentration, and thermodynamics
Sample types	Large and small molecules in a variety of sample buffers, solvents (e.g., DMSO, MeOH) and matrices (e.g., cell culture supernatants, serum, plasma)
Molecular weight detection limit	>100 Da
Sample concentration	≥50 pM
Kinetic analysis	Association rate, k_a: 10^3–$10^7\,M^{-1}\,s^{-1}$
	Dissociation rate, k_d: 10^{-6} to $10^{-1}\,s^{-1}$
Affinity	0.1 mM to 1 pM
Number of flow cells	8 dual-sensor flow cells
Number of sensors	16, individually addressable
Flow rate	5–50 μL min^{-1}
Injection volume	1–300 μL
Sample consumption	Injected volume + 10–25 μL (injection type dependant)
Analysis temperature	4–40 °C (or 15 °C below ambient)
Sample Temperature	Ambient
Sample capacity	Two 96-well plates + one 96-vial reagent rack + one 96 well
Automated buffer exchange	Yes, up to five different buffer solutions
Sample throughput	2300 samples per 24 h (5 min cycle^{-1}) with dedicated controls
Cycle throughput	Simultaneous processing of up to eight samples per cycle

Some localized SPR (L-SPR) techniques have also been developed for rapid screening of drug candidates against cytochrome P450s that plays a critical role in drug–drug interactions [68]. Moreover, the method is also employed in the screening of oligomerization-blocking drugs against Alzheimer's disease or other diseases that shows a great potential to be used in a wide range of application [69]. Rapid recognition of drug residues in clinical samples is also possible by employing SPR immunosensors that target low levels of amphetamines or steroids [70, 71]. More examples of SPR-based drug discovery are summarized in Table 14.4.

14.4   Computational Simulations and New Approaches for Drug–Receptor Interactions

The use of computational programs to simulate the interactions between drugs and ligands is providing theoretical and useful information prior to the real applications. Recently, artificial drug receptors were developed with the aid of supramolecular chemistry, and they were screened against their target drugs using several SPR-based sensors (SPR-2 from Sierra Sensors, Biacore 3000 from GE Healthcare) to determine the specific interaction, binding affinity, and nonspecific binding [37, 80]. The real-time monitoring of the drugs down to low $ng\,mL^{-1}$ levels could be successfully achieved. The commonly prescribed medicines—diclofenac as a painkiller, metoprolol as a β-blocker, and vancomycin as an antibiotic—were investigated in this research. The targeted drug diclofenac was quantified in a wide linear range from 1.24 to $80\,ng\,mL^{-1}$, and the sensor surface could be regenerated up to 30 sample analyses without a significant loss on the signal. Moreover, a very high binding affinity between diclofenac and its specific receptor was calculated using SPR-2 software analyzer. The dissociation constant (K_D) was found to be $4.27 \times 10^{-10}\,M$. The sensor results were confirmed by LC–MS that was combined with a solid-phase extraction (SPE) system. A diclofenac detection assay was initially developed in LC–MS, and an investigation range of $1–5000\,ng\,mL^{-1}$ with an LOD of $18\,ng\,mL^{-1}$ was achieved. On the other hand, diclofenac in the amount of $200\,ng\,mL^{-1}$ was filtrated through SPE columns that were previously immobilized with drug-specific receptors using covalent attachment techniques involving EDC-NHS or glutaraldehyde chemistry. Control experiments were also conducted by filtering the drug through SPE columns that were processed in the absence of the receptor. All filtrates were then evaluated using the previously developed method in LC–MS. The uptake of diclofenac by the receptor on SPE column was found to be 100%, which means that $200\,ng\,mL^{-1}$ of diclofenac was captured by its receptors, and it could not be found in the filtrate solution. Both surface chemistries resulted in identical output. In case of the control experiment, complete amount of the diclofenac was observed in the filtrate due to the absence of the receptors in the columns [37].

The further improvement of this research was investigated by integrating computational chemistry approach into the drug–receptor interactions. Artificial drug receptors were designed using computational simulation for the manufacturing of high affinity receptors for drug recognition and monitoring

Table 14.4 Examples of SPR-based drug discovery works from the literature.

SPR biosensors in drug analysis	Investigated drug compounds	Objective	References
High-throughput screening (HTS)	Candidate drugs and drug fragments	Target and assay validation	[72]
	Primary screening of large libraries comprising small (drug) molecules		
Pharmacokinetic drug profiling	Alprenolol, desipramine, dibucaine, homochlorocyclizine, hydrochlorothiazide, imipramine, ketoprofen, oxprenolol, metoprolol, naproxen, propranolol, pindolol, verapamil, salmeterol, suprofen, tetracaine, and tolmetin.	Predictive *in vitro* drug permeability	[73–75]
		Passive transport of drugs through membrane	
	Atenolol, creatinine, D-glucose, mannitol, sulfasalazine, and urea.	Characterization of membrane binding affinity	
	Diverse drugs with a wide range of molecular weight (138.1–664.8 Da) and chemical functionalities		
	Drugs with high transcellular absorption		
Early ADME/T studies	Small molecules (130–800 Da)	Rapid *in vitro* assays for defined ADME/ $T_{parameters}$, as protein-binding studies, for potential drug candidates	[76]
	Drugs binding to alpha(1)-acid glycoprotein (AGP)	Prediction/measuring of drug/human serum albumin binding.	
Fragment-based drug design	Small molecules (100–300 Da)	Identification of lead structures	[55, 62, 70, 77–79]
	Active compounds with Mw < 500 Da, LogD < 5, and solubility >10 μM	Affinity ranking and kinetics for small molecules binding to target protein	
	Thrombin inhibitors		
	HIV protease inhibitors		
	DNA-binding drugs and novel inhibitors of the bacterial enzyme DNA gyrase		

Source: Adapted from Olaru et al. [43].

in combination with optical-based biosensors. The monomer library composed of 21 different chemical compounds was screened for the target drug metoprolol based on binding energy between each monomer and the target for the selection of best monomers to develop successful receptors. The produced receptors were characterized in terms of quality, uniformity, and size prior to the affinity analysis through SPR-based optical biosensors (SPR-2 and SPR-2/4 from Sierra Sensors, and Biacore 3000 from GE Healthcare). The metoprolol-specific receptor was immobilized on the sensor surface via covalent coupling, and this β-blocker drug was successfully detected in the linear concentration range from $1.9\,ng\,mL^{-1}$ to $1\,\mu g\,mL^{-1}$ with a good correlation coefficient of 0.97. A regeneration method developed using 0.1 M glycine-HCl allowed the reuse of sensor for subsequent sample analysis, and cross-reactivity studies with control drugs revealed a highly specific biosensor assay for metoprolol–receptor interaction. The data for kinetic analysis was collected and processed through the Biacore 3000 and SPR-2/SPR-2/4 analyzers to determine the affinity between the receptor and metoprolol. The K_D was $1.35 \times 10^{-10}\,M$ using Langmuir binding model. Additionally, a competitive drug binding assay was also conducted that confirms the reliability of the optical sensor as well as the success of the developed artificial receptor. The receptor affinity and capacity toward metoprolol were also confirmed by SPE method coupled with LC–MS. The achieved results highlight the success of computationally modeled receptor with SPR biosensor for pharmaceutical detection and monitoring [36, 80].

All these achievements on drug monitoring, drug–receptor interaction, and optical sensor-based drug detection have demonstrated the immense importance of computational approaches and new methods to increase the performance of screening processes as well as binding affinity between natural receptors and drugs. Bioinformatic tools and the advanced information obtained from genomic and proteomic research can provide better understanding about drug–receptor interactions. When this knowledge is combined with widely employed and highly trustable characterization tools, the precise interaction points between the drug of interest and its receptor can be visualized *in vivo*; therefore, this leads to faster and more efficient screening of drugs during the drug discovery processes. The potential of optical biosensors with all these improvements and also the development of more advanced optical sensors with their high-throughput nature have shown a great promise in developing new drugs for cancer cases.

14.5 Conclusions

The last decade has witnessed enormous advances in optical sensors, and this enables to conduct fast, reliable, and throughput drug discovery research. SPR sensor technology has become a standard tool employed to characterize pharmaceutical products. Achieving detailed information and not relying on labeling make SPR technology a perfect complement to high-throughput monitoring techniques. Moreover, novel SPR-based biosensors have been particularly developed by taking into account the need in this field that led to successful screening of drugs using direct assay methodologies even if they have very low molecular weight. The receptor–drug interactions as well as nonspecific receptor–drug interactions can easily be investigated simultaneously and more than 7000 interactions per day can be successfully analyzed. Owing to the awareness of pharmaceutical industry on the great potential of SPR sensor technology, it will serve as an indispensable tool in developing new drugs for cancer cases.

Acknowledgment

Z.A. gratefully acknowledges support from the European Commission, Marie Curie Actions and IPODI as the principle investigator.

References

1 Lipinski, C. A.; Lombardo, F.; Dominy, B. W.; Feeney, P. J. *Advanced Drug Delivery Reviews* **2012**, 64, 4–17.

2 Gaulton, A.; Bellis, L. J.; Bento, A. P.; Chambers, J.; Davies, M.; Hersey, A.; Light, Y.; McGlinchey, S.; Michalovich, D.; Al-Lazikani, B.; Overington, J. P. *Nucleic Acids Research* **2012**, 40, D1100–D1107.

3 Famm, K. *Nature* **2013**, 496, 300.

4 Csermely, P.; Korcsmaros, T.; Kiss, H. J. M.; London, G.; Nussinov, R. *Pharmacology & Therapeutics* **2013**, 138, 333–408.

5 Slusher, B. S.; Conn, P. J.; Frye, S.; Glicksman, M.; Arkin, M. *Nature Reviews Drug Discovery* **2013**, 12, 811–812.

6 Kenakin, T.; Christopoulos, A. *Nature Reviews Drug Discovery* **2013**, 12, 205–216.

7 Wang, X. J.; Zhang, A. H.; Wang, P.; Sun, H.; Wu, G. L.; Sun, W. J.; Lv, H. T.; Jiao, G. Z.; Xu, H. Y.; Yuan, Y.; Liu, L.; Zou, D. X.; Wu, Z. M.; Han, Y.; Yan, G. L.; Dong, W.; Wu, F. F.; Dong, T. W.; Yu, Y.; Zhang, S. X.; Wu, X. H.; Tong, X.; Meng, X. C. *Molecular & Cellular Proteomics* **2013**, 12, 1226–1238.

8 Schenone, M.; Dancik, V.; Wagner, B. K.; Clemons, P. A. *Nature Chemical Biology* **2013**, 9, 232–240.

9 Zofou, D.; Ntie-Kang, F.; Sippl, W.; Efange, S. M. N. *Natural Product Reports* **2013**, 30, 1098–1120.

10 Skirycz, A.; Kierszniowska, S.; Méret, M.; Willmitzer, L.; Tzotzos, G. A. *Trends in Biotechnology* **2016**, 34, 781–790.

11 Gomez-Verjan, J.; Gonzalez-Sanchez, I.; Estrella-Parra, E.; Reyes-Chilpa, R. *Scientometrics* **2015**, 105, 1019–1030.

12 Turner, A. P. F. *Chemical Society Reviews* **2013**, 42, 3184–3196.

13 Cetin, A. E.; Coskun, A. F.; Galarreta, B. C.; Huang, M.; Herman, D.; Ozcan, A.; Altug, H. *Light-Science & Applications* **2014**, 3, e122.

14 Altintas, Z.; Fakanya, W. M.; Tothill, I. E. *Talanta* **2014**, 128, 177–186.

15 Altintas, Z.; Kallempudi, S. S.; Gurbuz, Y. *Talanta* **2014**, 118, 270–276.

16 Altintas, Z.; Tothill, I. *Sensors and Actuators B: Chemical* **2013**, 188, 988–998.

17 Altintas, Z.; Kallempudi, S. S.; Sezerman, U.; Gurbuz, Y. *Sensors and Actuators B: Chemical* **2012**, 174, 187–194.

18 Altintas, Z.; Tothill, I. E. *Sensors and Actuators B: Chemical* **2012**, 169, 188–194.

19 Altintas, Z.; Uludag, Y.; Gurbuz, Y.; Tothill, I. E. *Talanta* **2011**, 86, 377–383.

20 Ahmed, M. U.; Saaem, I.; Wu, P. C.; Brown, A. S. *Critical Reviews in Biotechnology* **2014**, 34, 180–196.

21 Grodzinski, P.; Silver, M.; Molnar, L. K. *Expert Review of Molecular Diagnostics* **2006**, 6, 307–318.

22 Jokerst, J. C.; Emory, J. M.; Henry, C. S. *Analyst* **2012**, 137, 24–34.

23 Li, M.; Li, Y. T.; Li, D. W.; Long, Y. T. *Analytica Chimica Acta* **2012**, 734, 31–44.

24 Long, F.; Zhu, A. N.; Shi, H. C. *Sensors* **2013**, 13, 13928–13948.

25 Abdin, M. J.; Altintas, Z.; Tothill, I. E. *Biosensors & Bioelectronics* **2015**, 67, 177–183.

26 Masdor, N. A.; Altintas, Z.; Tothill, I. E. *Biosensors & Bioelectronics* **2016**, 78, 328–336.

27 Li, S. Q.; Chai, Y. T.; Chin, B. A. In *Conference on Sensing for Agriculture and Food Quality and Safety VII*, Baltimore, MD, **2015**.

28 Warriner, K.; Reddy, S. M.; Namvar, A.; Neethirajan, S. *Trends in Food Science & Technology* **2014**, 40, 183–199.

29 Narsaiah, K.; Jha, S. N.; Bhardwaj, R.; Sharma, R.; Kumar, R. *Journal of Food Science and Technology-Mysore* **2012**, 49, 383–406.

30 Sankaran, S.; Panigrahi, S.; Mallik, S. *Sensors and Actuators B: Chemical* **2011**, 155, 8–18.

31 Lavecchia, T.; Tibuzzi, A.; Giardi, M. T. In *Bio-Farms for Nutraceuticals: Functional Food and Safety Control by Biosensors*, Giardi, M. T.; Rea, G.; Berra, B., Eds., Springer Science + Business Media, LLC, New York, **2010**, pp 267–281.

32 Alocilja, E. C. *Abstracts of Papers of the American Chemical Society* **2009**, 237, 77.

33 Siderius, M.; Shanmugham, A.; England, P.; van der Meer, T.; Bebelman, J. P.; Blaazer, A. R.; de Esch, I. J. P.; Leurs, R. *Analytical Biochemistry* **2016**, 503, 41–49.

34 Irannejad, R.; Tomshine, J. C.; Tomshine, J. R.; Chevalier, M.; Mahoney, J. P.; Steyaert, J.; Rasmussen, S. G. F.; Sunahara, R. K.; El-Samad, H.; Huang, B.; von Zastrow, M. *Nature* **2013**, 495, 534–538.

35 Cooper, M. A. *Nature Reviews Drug Discovery* **2002**, 1, 515–528.

36 Altintas, Z.; France, B.; Ortiz, J. O.; Tothill, I. E. *Sensors and Actuators B: Chemical* **2016**, 224, 726–737.

37 Altintas, Z.; Guerreiro, A.; Piletsky, S. A.; Tothill, I. E. *Sensors and Actuators B: Chemical* **2015**, 213, 305–313.

38 Ferrie, A. M.; Wang, C. M.; Deng, H. Y.; Fang, Y. *Integrative Biology* **2013**, 5, 1253–1261.

39 Geschwindner, S.; Dekker, N.; Horsefield, R.; Tigerstrom, A.; Johansson, P.; Scott, C. W.; Albert, J. S. *Journal of Medicinal Chemistry* **2013**, 56, 3228–3234.

40 Vuignier, K.; Veuthey, J. L.; Carrupt, P. A.; Schappler, J. *Drug Discovery Today* **2013**, 18, 1030–1034.

41 Lombardino, J. G.; Lowe, J. A. *Nature Reviews Drug Discovery* **2004**, 3, 853–862.

42 Fang, Y. *Assay and Drug Development Technologies* **2006**, 4, 583–595.

43 Olaru, A.; Bala, C.; Jaffrezic-Renault, N.; Aboul-Enein, H. Y. *Critical Reviews in Analytical Chemistry* **2015**, 45, 97–105.

44 Altintas, Z.; Uludag, Y.; Gurbuz, Y.; Tothill, I. *Analytica Chimica Acta* **2012**, 712, 138–144.

45 Pawula, M.; Altintas, Z.; Tothill, I. E. *Talanta* **2015**, 146, 823–830.

46 Nunomura, W.; Takakuwa, Y.; Parra, M.; Conboy, J.; Mohandas, N. *Journal of Biological Chemistry* **2000**, 275, 24540–24546.

47 Stolowitz, M. L.; Ahlem, C.; Hughes, K. A.; Kaiser, R. J.; Kesicki, E. A.; Li, G. S.; Lund, K. P.; Torkelson, S. M.; Wiley, J. P. *Bioconjugate Chemistry* **2001**, 12, 229–239.

48 Jensen, K. K.; Orum, H.; Nielsen, P. E.; Norden, B. *Biochemistry* **1997**, 36, 5072–5077.

49 Hart, D. J.; Speight, R. E.; Cooper, M. A.; Sutherland, J. D.; Blackburn, J. M. *Nucleic Acids Research* **1999**, 27, 1063–1069.

50 Nilsson, P.; Persson, B.; Uhlen, M.; Nygren, P. A. *Analytical Biochemistry* **1995**, 224, 400–408.

51 Kazemier, B.; deHaard, H.; Boender, P.; vanGemen, B.; Hoogenboom, H. *Journal of Immunological Methods* **1996**, 194, 201–209.

52 Nice, E.; Layton, J.; Fabri, L.; Hellman, U.; Engstrom, A.; Persson, B.; Burgess, A. W. *Journal of Chromatography* **1993**, 646, 159–168.

53 Radler, U.; Mack, J.; Persike, N.; Jung, G.; Tampe, R. *Biophysical Journal* **2000**, 79, 3144–3152.

54 Sigal, G. B.; Bamdad, C.; Barberis, A.; Strominger, J.; Whitesides, G. M. *Analytical Chemistry* **1996**, 68, 490–497.

55 Boehm, H. J.; Boehringer, M.; Bur, D.; Gmuender, H.; Huber, W.; Klaus, W.; Kostrewa, D.; Kuehne, H.; Luebbers, T.; Meunier-Keller, N. *Journal of Medicinal Chemistry* **2000**, 43, 2664–2674.

56 Karlsson, R.; Kullman-Magnusson, M.; Hamalainen, M. D.; Remaeus, A.; Andersson, K.; Borg, P.; Gyzander, E.; Deinum, J. *Analytical Biochemistry* **2001**, 291, 306–306.

57 Markgren, P. O.; Hamalainen, M.; Danielson, U. H. *Analytical Biochemistry* **2000**, 279, 71–78.

58 Hamalainen, M. D.; Markgren, P. O.; Schaal, W.; Karlen, A.; Classon, B.; Vrang, L.; Samuelsson, B.; Hallberg, A.; Danielson, U. H. *Journal of Biomolecular Screening* **2000**, 5, 353–359.

59 Altintas, Z.; Gittens, M.; Guerreiro, A.; Thompson, K.-A.; Walker, J.; Piletsky, S.; Tothill, I. E. *Analytical Chemistry* **2015**, 87, 6801–6807.

60 Altintas, Z.; Abdin, M. J.; Tothill, A. M.; Karim, K.; Tothill, I. E. *Analytica Chimica Acta* **2016**, 935, 239–248.

61 Alterman, M.; Sjobom, H.; Safsten, P.; Markgren, P. O.; Danielson, U. H.; Hamalainen, M.; Lofas, S.; Hulten, J.; Classon, B.; Samuelsson, B.; Hallberg, A. *European Journal of Pharmaceutical Sciences* **2001**, 13, 203–212.

62 Gambari, R.; Feriotto, G.; Rutigliano, C.; Bianchi, N.; Mischiati, C. *Journal of Pharmacology and Experimental Therapeutics* **2000**, 294, 370–377.

63 Kampranis, S. C.; Gormley, N. A.; Tranter, R.; Orphanides, G.; Maxwell, A. *Biochemistry* **1999**, 38, 1967–1976.

64 Murray, C. W.; Rees, D. C. *Nature Chemistry* **2009**, 1, 187–192.

65 Hopkins, A. L.; Groom, C. R.; Alex, A. *Drug Discovery Today* **2004**, 9, 430–431.

66 Navratilova, I.; Hopkins, A. L. *ACS Medicinal Chemistry Letters* **2010**, 1, 44–48.

67 Aristotelous, T.; Ahn, S.; Shukla, A. K.; Gawron, S.; Sassano, M. F.; Kahsai, A. W.; Wingler, L. M.; Zhu, X.; Tripathi-Shukla, P.; Huang, X. P.; Riley, J.; Besnard, J.; Read, K. D.; Roth, B. L.; Gilbert, I. H.; Hopkins, A. L.; Lefkowitz, R. J.; Navratilova, I. *ACS Medicinal Chemistry Letters* **2013**, 4, 1005–1010.

68 Das, A.; Zhao, J.; Schatz, G. C.; Sligar, S. G.; Van Duyne, R. P. *Analytical Chemistry* **2009**, 81, 3754–3759.

69 Hong, Y.; Ku, M.; Lee, E.; Suh, J. S.; Huh, Y. M.; Yoon, D. S.; Yang, J. *Journal of Biomedical Optics* **2014**, 19, 051202.

70 Lee, J. H.; Kim, B. C.; Oh, B. K.; Choi, J. W. *Nanomedicine-Nanotechnology Biology and Medicine* **2013**, 9, 1018–1026.

71 Mitchell, J. *Sensors* **2010**, 10, 7323–7346.

72 Giannetti, A. M. *Methods in Enzymology* **2011**, 493, 170–217.

73 Abdiche, Y. N.; Myszka, D. G. *Analytical Biochemistry* **2004**, 328, 233–243.

74 Frostell-Karlsson, A.; Widegren, H.; Green, C. E.; Hamalainen, M. D.; Westerlund, L.; Karlsson, R.; Fenner, K.; Van De Waterbeemd, H. *Journal of Pharmaceutical Sciences* **2005**, 94, 25–37.

75 Baird, C. L.; Courtenay, E. S.; Myszka, D. G. *Analytical Biochemistry* **2002**, 310, 93–99.

76 Frostell-Karlsson, A.; Remaeus, A.; Roos, H.; Andersson, K.; Borg, P.; Hamalainen, M.; Karlsson, R. *Journal of Medicinal Chemistry* **2000**, 43, 1986–1992.

77 Anraku, K.; Fukuda, R.; Takamune, N.; Misumi, S.; Okamoto, Y.; Otsuka, M.; Fujita, M. *Biochemistry* **2010**, 49, 5109–5116.

78 Geschwindner, S.; Carlsson, J. F.; Knecht, W. *Sensors* **2012**, 12, 4311–4323.

79 Mani, R. J.; Dye, R. G.; Snider, T. A.; Wang, S. P.; Clinkenbeard, K. D. *Biosensors & Bioelectronics* **2011**, 26, 4832–4836.

80 Altintas, Z.; Tothill, I. E. Molecularly Imprinted Polymer-based Affinity Nanomaterials for Pharmaceuticals Capture, Filtration and Detection, GB Patent **2014**, GB1413210.4.

15

Biosensors for Detection of Anticancer Drug–DNA Interactions

Arzum Erdem, Ece Eksin and Ece Kesici

Department of Analytical Chemistry, Faculty of Pharmacy, Ege University, Izmir, Turkey

15.1 Introduction

Deoxyribonucleic acid (DNA) plays an important role in the life process since it carries genetic information, and thus it initiates the biological synthesis of proteins and enzymes through replication and transcription of genetic information in living cells. Research on the binding mechanism of some small molecules to DNA has still been in progress as the one of the key topics during the past few decades [1–3]. Moreover, it is of great help to understand the structural properties of DNA, the mutation of genes, the origin of some diseases, and the action mechanism of some anticancer drugs as well as antitumor and antivirus drugs and also to design novel DNA-targeted drugs dealing with genetic diseases [1–3].

Anticancer drugs interact with DNA in many different ways, such as non-covalent groove binding, covalent binding/cross-linking, DNA cleaving, and nucleoside analog incorporation [4]. Electrostatic interaction, which is generally a nonspecific binding mechanism, occurs between small molecules with negatively charged nucleic acid sugar-phosphate structure. Intercalation of planar aromatic ring systems between base pairs containing several aromatic condensed rings often binds them to DNA in an intercalative mode (e.g., daunomycin, epirubicin, and actinomycin D) and minor and major DNA groove-binding interactions. Minor groove binding makes intimate contacts with the walls of the groove, and, as a result of this interaction, numerous hydrogen binding and electrostatic interactions occur between anticancer drugs and DNA (DNA bases and the phosphate backbone), for example, in the case of mithramycin. Major groove binding occurs via the hydrogen bonding to the

Biosensors and Nanotechnology: Applications in Health Care Diagnostics, First Edition.
Edited by Zeynep Altintas.
© 2018 John Wiley & Sons, Inc. Published 2018 by John Wiley & Sons, Inc.

DNA and can form a DNA triple helix, such as norfloxacin [5]. After the interaction between DNA and drug, there may be changes in the functional properties of DNA.

Biosensor is a sensing device that comprises a biorecognition element and a transducer. A biorecognition element specifically identifies and interacts with an analyte, and the changes in its physicochemical properties (optical, thermal, electrical, and thermodynamic properties) are usually converted into an electrical signal by a transducer. The biorecognition element is a sensing material and may include enzymes, antibodies, microorganisms, tissues, organelles, DNAs, and RNAs. The biosensors can be classified based on the type of biorecognition element or the transducing method used. Based on the biorecognition element, the biosensors can be classified as enzyme sensors, immunosensors, nucleic acid probe sensors, or cell-, tissue-, or organelle-based sensors. Based on the transducing method, biosensors can be classified as piezoelectrical, optical, or electrochemical [6]. A representative scheme for biosensing of anticancer drug–DNA interactions was shown in Figure 15.1.

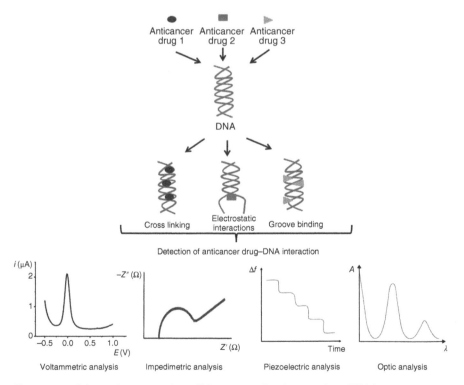

Figure 15.1 Schematic presentation of biosensing of anticancer drug–DNA interactions. (*See insert for color representation of the figure.*)

This chapter is presenting an overview of recent biosensors developed for monitoring the interactions between anticancer drugs and DNA, since there has been a growing interest for the development of different types of biosensors (electrochemical, optical, impedimetrical, piezoelectrical, etc.) on drug–DNA interactions [7–60], and accordingly, some of the representatives were summarized in Table 15.1.

15.2 Voltammetric Techniques

Recently, there has been a great attention for electrochemical investigation of interaction between drugs and DNA [61–65]. One of the practical applications of electrochemistry is the determination of electrode redox process. Due to the existing resemblance between electrochemical and biological reactions, it can be assumed that the oxidation mechanisms taking place at the electrode and in the body share similar principles [66, 67]. By observing the electrochemical signal related to DNA–drug interactions, it is possible to propose a mechanism of interaction and nature of the complex formed and to evaluate the binding constant [68].

Daunomycin (DNR) is an anthracycline antibiotic with antineoplastic activity. The anthracycline moiety of DNR can intercalate between the stacked base pairs of native DNA. Additionally, the sugar-3-amino structure of DNR can bind to the major groove of DNA double helix [69]. Chu et al. [8] investigated the cyclic voltammetric behavior of DNR in aqueous medium in the presence of DNA. It was reported that a decrease at anodic peak currents of DNR was obtained after its intercalating to the DNA, and accordingly, DNA concentrations were determined based on the decrease of the peak current of DNA. In the study of Chu et al. [8], the binding constant and binding site size of DNR–DNA interaction were also examined by a titration curve and nonlinear regression analysis.

Interactions of echinomycin (Echi) with single-stranded DNA (ssDNA) and double-stranded DNA (dsDNA) were studied by Jelen et al. [9] using adsorptive transfer stripping cyclic voltammetry (AdTSCV) with hanging mercury drop electrode (HMDE) and resulted in high Echi signals, suggesting a strong binding of Echi to dsDNA by bis-intercalation at the electrode surface. On the other hand, they reported that under the same conditions, interaction of Echi with ssDNA produced almost no Echi signal. Consequently, Jelen et al. [9] reported that Echi appears to be a good candidate for a redox indicator in electrochemical DNA hybridization sensors. The interaction of cis-DDP and the potential novel chemotherapeutic agent cis-BAFDP was studied with ctdsDNA using DPV technique with pencil graphite electrode (PGE) by Erdem et al. [23]. It was reported that there was a decrease at guanine and adenine signal after the interaction of cis-DDP with dsDNA. Similar results were also

Table 15.1 Summary of the recent biosensor studies of interactions between anticancer drugs and nucleic acids.

Method	Drug	Nucleic acid	Technique	DL	Reference
Voltammetric techniques	Daunomycin	ctdsDNA	CV	$2.4\,\mu g\,mL^{-1}$	[8]
	Echinomycin	ctdsDNA, ctssDNA	AdTSCV	—	[9]
	Wedelolactone	dsDNA	CV, SWV	—	[10]
	Fulvestrant	fsdsDNA	DPV	$0.3\,\mu g\,mL^{-1}$	[11]
	6-Mercaptopurine	dsDNA	DPV	$0.08\,\mu M$	[12]
	Sophoridine	fsdsDNA	CV	—	[13]
	Thiosemicarbazone	fsdsDNA, poly(dA)·poly(dT)	DPV	—	[14]
	6-Thioguanine	dsDNA	AdSV, SWV	$1.1\,\mu M$	[15]
	Leuprolide	dsDNA	DPV	$0.06\,\mu g\,mL^{-1}$	[16]
	Epirubicin	ctdsDNA, ctssDNA	DPV, CV	$0.62\,\mu M$	[17]
	Mitoxantrone	ctdsDNA, ctssDNA	DPV, CV	$56.2\,nM$	[18]
	Mitomycin C	fsdsDNA	DPV	$8.41\,mg\,mL^{-1}$	[19]
	4,4′-Dihydroxy chalcone	ctdsDNA, ctssDNA	DPV	$42\,nM$	[20]
	Lycorine	ctdsDNA, ctssDNA and poly[G]	DPV	$225\,ng\,mL^{-1}$	[21]
	Mitomycin C	fsdsDNA, ctdsDNA	SWV, DPV	$19\,ng\,mL^{-1}$	[22]
	cis-DDP	ctdsDNA	DPV	$122\,nM$	[23]
	cis-BAFDP	ctdsDNA	DPV	$600\,nM$	
	Mitomycin C	dsDNA	DPV	$625\,ng\,mL^{-1}$	[24]
	Mitomycin C	dsDNA	DPV	—	[25]
Optical technique	Mitoxantrone	ctdsDNA ctssDNA	PIET	$10\,nM$	[26]
	Ditercalinium	DNA substrates (nucleotide, base pair, triplet, tetrad)	SPR spectroscopy, ESI MS	—	[27]

Technique	Drug	DNA	Method	Detection limit	Ref.
	The antitumor antibiotic AT2433-B1	The 117bp DNA fragment	SPR spectroscopy	—	[28]
	Oxaliplatin Cisplatin Carboplatin	ctdsDNA	PIET	—	[29]
Voltammetric and spectroscopic techniques	BNN-17	dsDNA	DPV, UV–Vis spectroscopy	0.063 µM	[30]
	Berberine	ssDNA, dsDNA	CV, DPV, UV spectroscopy	14.5 µM	[31]
	Danusertib	ctdsDNA	DPV, UV spectroscopy	—	[32]
	Methotrexate	Salmon sperm DNA	CCIPS, UV–Vis spectroscopy	5 nM	[33]
	Doxorubicin	dsDNA, ssDNA oligomers	SERS, CV, DPV	8 µg mL^{-1}	[34]
	Indirubin	ssDNA, dsDNA	LSV, UV–Vis, IR-spectroscopy	2 µg mL^{-1}	[35]
	Mitoxantrone	dsDNA, oligonucleotides	DPV, SERS, UV–Vis spectroscopy	—	[36]
	Emodin	dsDNA	DPV, CV, UV–Vis spectroscopy	1.3 µg mL^{-1}	[37]
	Formestane	ctssDNA, dsDNA	Fluorescence spectroscopy, UV–Vis spectroscopy, CV, SWV	—	[38]
	Taxol	DNA	CV, SWV, UV–Vis spectroscopy	8.86 µM	[39]
	Lapatinib	ctdsDNA, polyG, polyA,	DPV, CV, UV–Vis spectroscopy, fluorescence spectroscopy	—	[40]
	4-Nitrophenylferrocene	DNA extracted from chicken blood	CV, UV–Vis, fluorescence spectroscopy	—	[41]
	Gemcitabine	dsDNA	DPV, UV–Vis spectroscopy,	0.276 mg L^{-1}	[42]
	Doxorubicin	ctdsDNA	CV, DPV, chronoamperometry, UV–Vis spectroscopy	—	[43]
	Methotrexate, Emodin	ssDNA, dsDNA	DPV, CV, UV–Vis spectroscopy	—	[44]
Impidimetric technique	Bleomycin	dsDNA, ssDNA	EIS	1.63 µg mL^{-1}	[45]
	Topotecan	Healthy DNA, cancer DNA	EIS	—	[46]

(Continued)

Table 15.1 (Continued)

Method	Drug	Nucleic acid	Technique	DL	Reference
Impidimetric and Voltammetric technique	Mitomycin C	dsDNA	DPV, EIS	810 ng mL^{-1}	[47]
	Taxol	Salmon sperm dsDNA	DPV, EIS	80 nM	[48]
	Mithramycin	DNA oligonucleotides	CV, LSV, EIS	10 nM	[49]
	Netropsin			40 nM	
	Nogalamycin			5 nM	
	Daunorubicin	dsDNA	DPV, EIS	125 nM	[50]
	cis-DDP	ctdsDNA	DPV, EIS	—	[51]
	Oxaliplatin			—	
	Mitomycin C	ctdsDNA	DPV, EIS	9.06 µg mL^{-1}	[52]
	Mitomycin C	ctdsDNA	DPV, EIS	4.47 µg mL^{-1}	[53]
	Mitomycin C	ctdsDNA	DPV, EIS	11.01 µg mL^{-1}	[54]
	Topotecan	fsdsDNA	DPV, EIS	0.51 µg mL^{-1}	[55]
	Mitomycin C	fsdsDNA	DPV, EIS	1.12 µg mL^{-1}	[56]
	Temozolamide	fsDNA, ctdsDNA, poly[G], poly[G]-poly[C]	DPV, EIS	6.1 µg mL^{-1}	[57]
	6-Thioguanine	ctdsDNA	DPV, EIS	4.60 µg mL^{-1}	[58]
	Topotecan	fsdsDNA	DPV, EIS	0.37 µg mL^{-1}	[55]
QCM technique	Doxorubicin	DNA oligomers (part of hOGG1 gene)	EQCM, CV	—	[59]
	Cisplatin	Bis-biotinylated dsDNA	QCM, EIS	1 nM	[60]

Abbreviations: *Drugs*: BNN-17, 2-(2-phenyl ethyl)-5-methylbenzimidazole; cis-BAFDP, *cis*-bis(3-aminoflavone)dichloroplatinum(II); cis-DDP, *cis*-diamminedichloroplatinum(II). *Nucleic acids*: ctdsDNA, calf thymus double-stranded DNA; ctssDNA, calf thymus single-stranded DNA; dsDNA, double-stranded DNA; fsdsDNA, fish sperm double-stranded DNA; ssDNA, single-stranded DNA. *Technique*: AdSV, adsorptive stripping voltammetry; AdTSCV, adsorptive transfer stripping cyclic voltammetry; CCPS, constant current potentiometric stripping analysis; CV, cyclic voltammetry; DPV, differential pulse voltammetry; EIS, electrochemical impedance spectroscopy; EQCM, electrochemical quartz crystal microbalance; ESI MS, electrospray ionization mass spectrometry; LSV, linear sweep voltammetry; PIET, photoinduced electron-transfer; QCM, quartz crystal microbalance; SERS, surface enhanced Raman spectroscopy; SPR, surface plasmon resonance; SWV, square wave voltammetry.

found in solution phase after the latter compound interacts with poly[A]. The electrochemical studies were prompted by beneficial biological properties of cis-BAFDP in comparison with cis-DDP, which were proven *in vitro* both in human normal and cancer cells and *in vivo* in mice [23].

Mitomycin C (MC) is a chemotherapeutic agent that is generally effective for upper gastrointestinal, anal, and breast cancers. MC has the ability to covalently bind to DNA, by shielding the oxidizable groups of electroactive DNA bases such as guanine and adenine [24, 25]. The interaction between MC and DNA was investigated using electrochemical biosensors by Erdem's group using different types of electrodes, such as carbon paste electrode (CPE) [22], single-walled carbon nanotube (SWCNT)/poly(vinylferrocenium) (PVF$^+$)-modified PGE [24], graphene oxide (GO)-modified PGE [52], chitosan (CHIT)–ionic liquid-modified PGE [53], CHIT–SWCNT-modified PGE [54], CHIT–MWCNT-modified PGE [56], and sepiolite–SWCNT-modified PGE [47].

One of these studies [19], the interaction between MC and DNA, was investigated by using poly(3,4-ethylenedioxythiophene) (PEDOT)-coated CHIT-modified disposable PGEs that were prepared by using plasma polymerization with RF rotating hydrazine plasma method. Another study [25] detected MC–DNA interaction by gold nanoparticle/polyvinylferrocenium (AuNP/PVF$^+$)-coated electrode using DPV technique, and accordingly, the changes at the magnitude of the oxidation signals of guanine and adenine were monitored as an indicator of the drug–DNA interaction. Fulvestrant (FLV), a new type of estrogen receptor antagonist with no agonist effects in phase II trials, has been shown to be at least as effective as the third-generation aromatase inhibitor anastrozole in the treatment of postmenopausal women with advanced breast cancer progressing on tamoxifen therapy [70–73].

Topal and Ozkan [11] reported the voltammetric detection of interaction between FLV and DNA at dsDNA-modified PGE in combination with DPV technique. It was reported that a decrease was observed at the guanine oxidation signal. This phenomenon was explained in the study [11] by the damage on the oxidizable group of electroactive base guanine because of the adsorption of FLV. Moreover, it was reported that the peak potential of FLV was shifted to a positive direction with the addition of DNA. In another study reported by the same group [16], the interaction between leuprolide (LPR) and dsDNA was investigated based on the changes of guanine signal using DPV technique.

Epirubicin (EPR) is an anticancer chemotherapeutic drug, which exerts its cytotoxic effect by inhibiting DNA synthesis and DNA replication. It is a semisynthetic derivative of doxorubicin, which has been extensively evaluated in patients with breast cancer [74]. The incorporation of drug into double-stranded DNA disturbs its helical structure causing strand breakage. The primary mode of interaction of anthracycline drugs with DNA is intercalation.

Intercalation of anthracycline drugs into the DNA bases leads to inhibition of DNA synthesis and replication [75].

The interaction of EPR with dsDNA and calf thymus ssDNA was studied using DPV and CV at CPE by Erdem and Ozsoz [17]. It was observed that the signal at a bare electrode was higher than that of dsDNA-modified CPE. The signals for EPR in CV were found to decrease in the order of bare CPE and ssDNA-modified and dsDNA-modified CPE. The interaction of EPR with smaller oligonucleotides was also evaluated to use it as possible hybridization indicator.

15.3 Optical Techniques

Mitoxantrone (MTX) is an antitumor drug, active both on proliferative and nonproliferative cells [76]. It has a planar anthraquinone ring intercalating between DNA base pairs and the nitrogen-containing side-chain electrostatic binding that the negatively charged phosphate backbone of DNA [77]. Yuan et al. [26] investigated MTX and DNA interaction by using a photoinduced electron-transfer (PIET) mechanism based on luminescent quantum dots. In their study [26], the kinetics of drug–DNA interaction was investigated relying on the enhancement of QDs PL intensity caused by the binding of quencher with DNA. According to their results, it was concluded that QDs were successfully utilized in the development of a drug–DNA interaction sensing system as an excellent complement to the organic dyes. The advantages of their study are simple, sensitive, and rapid; the proposed sensing protocol was suitable for the investigation of anthracene drug–DNA interaction.

Another study performed by the PIET system was introduced by Zhao et al. [29]. They investigated the mechanism of the different quenching efficiencies among platinum anticancer drugs (cisplatin, oxaliplatin, carboplatin) from the perspectives of molecular structure, hydrolysis rate constant, and stereo-hindrance effect. Accordingly, they reported that cisplatin possesses strongest interaction with DNA with their prior binding sites at guanine, and dsDNA with the longest chain length possesses the strongest combining force with cisplatin.

SPR spectroscopy is a rapidly developing technique for the study of ligand-binding interactions with membrane proteins, which are the major molecular targets for validated drugs and for current and foreseeable drug discovery. SPR is label-free and capable of measuring real-time quantitative binding affinities and kinetics for membrane proteins interacting with ligand molecules using relatively small quantities of materials and has potential to have medium-throughput [78]. Carrasco et al. [27] studied the structural selectivity of the DNA-binding antitumor drug ditercalinium that was investigated by electro-spray mass spectrometry, and a detailed analysis of the binding reaction was performed by surface plasmon resonance (SPR) spectroscopy.

Ditercalinium, a 7*H*-pyridocarbazole dimer, is a synthetic anticancer drug that binds to DNA by bis-intercalation and activates DNA repair processes [79]. It was reported that a strong antitumor activity observed only when 7*H*-pyridocarbazoles are dimerized [80]. In the study of Delbarre et al. [81], 7*H*-pyridocarbazole dimer binds from the major groove side of the double helix to form bis-intercalation complexes with the two pyrido-carbazolium rings inserted between contiguous CpG steps. SPR spectroscopy-binding experiments were accordingly performed to enable an additional and quite different comparison of the interactions between ditercalinium and DNA. They reported that the structure of ditercalinium with that of a macrocyclic bisacridine derivative has been shown to bind strongly to quadruplex DNA and to inhibit telomerase. Also, they investigated the DNA-binding capacity of the glycosyl antibiotic AT2433-B1 by using SPR and DNase I footprinting technique [28].

UV–Vis absorption is one of the spectroscopic techniques generally employed to investigate drug–DNA interactions. Electrochemical investigations of small molecule–DNA interactions can provide useful complement to the results obtained by the spectroscopic methods [68]. The interaction of anticancer herbal drug berberine with dsDNA and ssDNA in solution was investigated using electrochemical techniques (CV, DPV) and UV spectroscopy by Tian et al. [31]. Berberine is known as an important compound in cancer therapy, possessing anticancer activity *in vitro* and *in vivo* [82]. The presence of DNA resulted in a decrease of the currents and a negative shift of the electrode potentials from the DPV curves of berberine, indicating the dominance of electrostatic interactions. The spectroscopy data confirmed that the predominant interaction between berberine and DNA is electrostatic [31]. Another natural anticancer drug, indirubin is an active constituent of a traditional Chinese herbal medicine. Previous studies showed that indirubin or its derivatives had the ability to induce apoptosis in cancer cell and could induce endothelial cell apoptosis and inhibit zebrafish embryo angiogenesis [83]. The interaction of indirubin with DNA was examined by DPV and LSV techniques and UV–Vis or IR spectra. DPV results showed that the peak potential of indirubin was shifted and peak currents were decreased with the addition of DNA. UV–Vis spectra exhibited that the absorption intensity of indirubin was decreased as a result of binding of indirubin to DNA. Additionally, IR spectra of DNA and DNA-indirubin adduct imply that indirubin interacts with the phosphate groups of DNA by hydrogen bond or electrostatic interaction [35].

Formestane is a synthetic steroid acting as a selective aromatase inhibitor *in vitro* and *in vivo*. In postmenopausal patients, estrogens continue to have an important role in breast cancer growth, even if the ovarian function is completely suppressed. Formestane (FMT) reduces the circulating estrogen levels and has shown antitumor activity in postmenopausal women with breast cancer [84, 85]. Temerk et al. [38] investigated the mode of interactions of FMT with dsDNA and ssDNA at different temperatures and two physiological pH values, that is, 7.4 (human blood pH) and 4.7 (stomach pH), using fluorescence spectroscopy, UV–Vis spectroscopy, CV, and SWV. They reported that the FMT intercalates between dsDNA bases, and the strength of interaction is independent on the ionic strength according to the absorption spectra and voltammetric results. Stoichiometric coefficients and thermodynamic parameters of FMT–dsDNA and FMT–ssDNA complexes were also evaluated in the study of Temerk et al. [38]. It was also concluded that the human body temperature provided the most favorable conformation of dsDNA, which binds to the anticancer drug FMT helping this drug to hinder dsDNA replication.

15.4 Electrochemical Impedance Spectroscopy Technique

Electrochemical impedance spectroscopy (EIS) is a method that describes the response of an electrochemical cell to a small amplitude sinusoidal voltage signal as function of frequency. It is used for the analysis of the interfacial properties related to biorecognition events such as reactions catalyzed by enzymes, biomolecular recognition events of specific binding proteins, lectins, receptors, nucleic acids, whole cells, and antibodies or antibody-related substances, occurring at the modified surface [86]. Bleomycin (BLM) is a chemotherapeutic antibiotic, produced by the bacterium *Streptomyces verticillus* [87, 88], and plays an important role in the treatment of lymphoma, squamous cell carcinoma, germ cell tumor, and malignant pleural effusion. It is believed that BLM acts by causing ssDNA and dsDNA breaks in tumor cells and thereby interrupting the cell cycle. It was also reported that it happens by chelation of metal ions and reaction of the formed pseudoenzyme with oxygen, which leads to production of DNA-cleaving superoxide and hydroxide free radicals [89].

Erdem and Congur [45] investigated the surface-confined interaction anticancer drug BLM with single-stranded and double-stranded DNA by using EIS technique in combination with a graphite sensor technology. The most important advantage of this study is that BLM–DNA interaction could be performed without using any extra cofactor agents, such as, Co(II), or Fe (III), that are used for activation of BLM while interacting with nucleic acids. Moreover, some of chemotherapeutic agents, MC and cisplatin, were tested to explore if they affect the characteristic properties of BLM in a mixture sample. Consequently BLM interaction with DNA was performed selectively in the mixture samples containing BLM and MC or BLM and cisplatin under optimum experimental conditions.

One of the recent studies was introduced by Top et al. [46] related to impedimetric biosensor developed for topoisomerase I inhibitor topotecan (TPT)–DNA interaction confirmed by using thin-layer radiochromatography and paper electrophoresis methods. They reported that topotecan, which is an anticancer drug label with 131 I via iodogen method, is labeled with high yield (94.1 ± 5.9%). It was found that intracellular uptake of 131 I-TPT was higher in lung cancer cell line than in healthy cell line. According to the impedimetric results, it was found that 131 I-TPT could interact more effectively with the cancer DNA than healthy DNA. Thus, the authors proposed that the 131 I-TPT is promising in terms of adding a new imaging agent for lung cancer.

The other anticancer drug–DNA interaction study was performed in order to investigate mithramycin, netropsin, and nogalamycin by using LSV, CV, and EIS techniques with gold nanoparticle-modified electrode developed by Li et al. [49]. Their study employs functional electrodes immobilized by synthesized thiol-linked DNA as probes to detect DNA-specific binding drugs differing in binding mode and sequence specificity. Since the electrochemical deposition of gold nanoparticles on flat gold electrode surfaces leads to a significant improvement in detection sensitivity for the assay of DNA-binding drugs, they reported that the changes in the geometry of the electrode surface affected the binding efficiency of the DNA–drug interaction and hence direct the improvement of impedance signal changes. It was concluded that this strategy might be useful for the sensing of other interfacial interaction such as enzyme–DNA, antibody–antigen, and cell antibody.

The biomolecular interactions of some platinum derivatives, *cis*-diamminedichloroplatinum(II) and oxaliplatin (OXP) with dsDNA, were studied by Yapasan et al. [51]. The performance of biomolecular interactions were explored at SWCNT-modified PGEs using EIS and DPV techniques. The R_{ct} values in the absence/presence of interaction between platinum derivatives and dsDNA were recorded. Correspondingly to the surface-confined biomolecular interaction with cis-DDP, they obtained 32.4% as an increase level at the R_{ct} value, which may possibly indicate that there was a specific binding of platinum complexes to the guanine bases in the double helix form of DNA. In their study, the R_{ct} values in the absence/presence of interaction between OXP and dsDNA were recorded, and similar increase at the R_{ct} values was also obtained after interaction of OXP with DNA [51].

The detection of interaction between TPT and dsDNA was examined by using electrochemical techniques in combination with disposable electrodes [55], and consequently, the oxidation signals of TPT and guanine were measured in the same voltammetric scale by using PGEs and SWCNT-modified PGEs. Impedimetric measurements were also performed under the optimum experimental conditions, and the changes at R_{ct} value were evaluated by means of SWCNT modification, immobilization, and drug–DNA interaction process. It was concluded that the intercalation of TPT into dsDNA resulted from the damage of double helix form of DNA, and consequently the decrease was recorded at R_{ct} [55].

A chemotherapy drug, Temozolomide (8-carbamoyl-3-methy-limidazo-[5,1-*d*]1,2,3,5-tetrazin-4(3*H*)-one, TMZ) known as an alkylating agent has been widely used to treat the most common type of adult brain tumor, called glioblastoma, as well as for treating a form of skin cancer, called melanoma in combination with radiotherapy [90, 91]. The surface-confined interaction between TMZ and different types of nucleic acids, such as fish sperm DNA (fsDNA), calf thymus dsDNA, and calf thymus ssDNA, was investigated using different electrochemical techniques in combination with disposable PGEs [57]. The interaction of TMZ with guanine base was also studied in the case of interaction between TMZ and single-stranded poly[G] or double-stranded poly[G]–poly[C]. Moreover, the impedimetric results were also examined in order to understand the interaction mechanisms occurred between TMZ and ssDNA or dsDNA based on the changes in R_{ct} value. It was reported that a higher decrease % was obtained after the interaction of TMZ with ssDNA, and there was a good agreement between EIS results and voltammetric ones for monitoring this interaction [57].

15.5 QCM Technique

The quartz crystal microbalance (QCM) is a well-studied class of sensor based on the piezoelectric properties of quartz crystals, which is ideally suited for the *in situ* study of bimolecular interactions occurring at a solid–liquid interface [92]. Thallium has been known as a toxic element for many years. The crystal structure of the Tl(I) complexes with nucleic bases was already reported in the literature. It appeared that the metal ion is bound mainly at N-3 and N-1 sites to pyrimidine bases, at N-1, N-7, and N-9 positions to purine bases [93, 94]. Nowicka et al. [59] studied the interaction between doxorubicin and DNA in the presence of thallium(I). According to their results, the obtained circular dichroism (CD) spectra and the measurements with an electrochemical quartz crystal microbalance (EQCM) led to a conclusion that in the presence of monovalent thallium cations, the DNA double helix was neither damaged/oxidized nor substantially changed its conformation [59]. Yan and Sadik [60] described the chemistry and methodology for constructing bis-biotinylated dsDNA multilayers on metal substrates after enzyme cleavage, and they demonstrated its use for amplified microgravimetric and impedimetric analyses of the anticancer drug, cisplatin. AC impedance spectroscopy and QCM measurements were implemented, and the results showed that the response to cisplatin increased linearly with target concentration [60].

15.6 Conclusions

The detection of interaction between anticancer drugs and DNA has a key importance among the most important aspects of biological studies for drug discovery and pharmaceutical development processes. Therefore, interaction of small molecules with DNA has been studied extensively, providing insights into the development of effective therapeutic drugs that could control gene expression. Novel and more effective DNA-targeted drugs against to several diseases should be developed urgently. In this direction, there have been many attempts for developing efficient, selective, and sensitive methods for anticancer drug–DNA interactions, including biosensors. A perception of the structural orientations, kinetics, and thermodynamics related to these complexes is fundamental to design of novel "next-generation" chemotherapeutic agents.

Different types of biosensors have been implemented successfully to monitor the anticancer drug–DNA interaction. Process on development of novel biosensor technologies used for monitoring the biointeractions of DNA-targeted drugs will continue to contribute to the growth of these research fields.

Acknowledgments

A.E would like to express her gratitude to the Turkish Academy of Sciences (TUBA) as the Principal member for its partial support.

References

1 Wang, J. *Biosensors and Bioelectronics*, **2006**, 21, 1887–1892.
2 Palecek, E.; Bartosik, M. *Chemical Reviews*, **2012**, 112, 3427–3481.
3 Nararro, J.A.R.; Salas, J.M.; Romera, M.A. *Journal of Medical Chemistry*, **1998**, 41, 332–338.
4 Bleckburn, G.M.; Gait M.N. *Nucleic Acids in Chemistry and Biology*, IRL Press, New York: **1990**, pp. 297–332.
5 Yang, K.; Hu, Y.; Dong, N. *Biosensors and Bioelectronics*, **2016**, 80, 373–377.
6 Srinivasan, B.; Tung, S. *Journal of Laboratory Automation* **2015**, 20, 365–389.
7 Erdem, A.; Ozsoz, M. *Electroanalysis*, **2002**, 14, 965–974.
8 Chu, X.; Shen, G.; Jiang, H.; Kang, T.; Xiong, B.; Yu, R. *Analytica Chimica Acta*, **1998**, 373, 29–38.
9 Jelen, F.; Erdem, A.; Palecek, E. *Bioelectrochemistry*, **2002**, 55, 165–167.

10 Cerven, J.; Havrn, L.; Pecinka, P.; Fojya, M. *Electroanalysis*, **2015**, 27, 2268–2271.

11 Dogan-Topal, B.; Ozkan, S.A. *Electrochimica Acta*, **2011**, 56, 4433–4438.

12 Karimi-aleh, H.; Tahernejad-Javazmi, F.; Atar, N.; Yola, M.L.; Gupta, V.K.; Ensafi, A.A. *Industrial and Engineering Chemistry Research*, **2015**, 54, 3634–3639.

13 Liu, H., Zhao, D.; Zhao, A.; Zou, K.; Li, T.; Wu, Y. *Journal of the Chinese Chemical Society*, **2014**, 61, 897–9002.

14 Bal-Demirci, T.; Congur, G.; Erdem, A.; Erdem-Kuruca, S.; Ozdemir, N.; Akgun-Dar, K.; Varol, B.; Ulkuseven, B. *New Journal of Chemistry*, **2015**, 39, 5643–5653.

15 Mirmomtaz, E.; Zirakbash, A.; Ensafi, A.A. *Russian Journal of Electrochemistry*, **2016**, 52, 320–329.

16 Dogan-Topal, B.; Ozkan, S.A. *Talanta*, **2011**, 83, 780–788.

17 Erdem, A.; Ozsoz, M. *Analytica Chimica Acta*, **2001**, 437, 107–114.

18 Erdem, A.; Ozsoz, M. *Turkish Journal of Chemistry*, **2001**, 25, 469–475.

19 Kuralay, F.; Demirci, S.; Kiristi, M.; Oksuz, L.; Oksuz, A.U. *Colloids and Surfaces B*, **2014**, 123, 825–830.

20 Meric, B.; Kerman, K.; Ozkan, D.; Kara, P.; Erdem, A.; Kucukoglu, O.; Erciyas, E.; Ozsoz, M. *Journal of Pharmaceutical and Biomedical Analysis*, **2002**, 30, 1339–1346.

21 Karadeniz, H.; Gulmez, B.; Sahinci, F., Erdem, A.; Kaya, G.I.; Unver, N.; Kivcak, B.; Ozsoz, M. *Journal of Pharmaceutical and Biomedical Analysis*, **2003**, 33, 295–302.

22 Ozkan, D.; Karadeniz, H.; Erdem, A.; Mascini, M.; Ozsoz, M. *Journal of Pharmaceutical and Biomedical Analysis*, **2004**, 35, 905–912.

23 Erdem A.; Kosmider, B.; Zyner, E.; Osiecka, R.; Ochocki, J.; Ozsoz, M. *Journal of Pharmaceutical and Biomedical Analysis*, **2005**, 38, 645–652.

24 Canavar, P.E.; Kuralay, F.; Erdem A. *Electroanalysis*, **2011**, 23, 2343–2349.

25 Kuralay, F.; Erdem, A. *Analyst*, **2015**, 140, 2876–2880.

26 Yuan, J.; Guo, W.; Yang, X.; Wang, E. *Analytical Chemistry*, **2009**, 81, 362–368.

27 Carrasco, C.; Rosu, F.; Gabelica, V.; Houssier, C.; DePauw, E.; Garbay-Jaureguiberry, C.; Roques, B.; Wilson, W.D.; Chaires, J.B.; Waring, M.J.; Bailly, C. *ChemBioChem*, **2002**, 3, 1235–1241.

28 Carrasco, C.; Facompre, M.; Chisholm, J.D.; Vraken, D.L.; Wilson, W.D.; Bailly, C. *Nucleic Acids Research*, **2002**, 30, 1774–1781.

29 Zhao, D.; Li, J.; Yang, T.; He, Z. *Biosensors and Bioelectronics*, **2014**, 52, 29–35.

30 Tığ, G.A.; Günendi, G.; Bolelli, T.E.; Yalçın, I.; Pekyardımcı, Ş. *Journal of Electroanalytical Chemistry*, **2016**, 776, 9–17.

31 Tian, X., Song, Y.; Dong, H.; Ye, B. *Bioelectrochemistry*, **2008**, 73, 18–22.

32 Diculescu, V.C.; Oliveira-Brett, A.M. *Bioelectrochemistry*, **2016**, 107, 50–57.

33 Rafique, B.; Khalid, A.M., Akhtar, K.; Jabbar, A. *Biosensors and Bioelectronics*, **2013**, 44, 21–26.

34 Ilkhani, H.; Hughes, T.; Li, J.; Zhong, C.J.; Hepel, M. *Biosensors and Bioelectronics*, **2016**, 80, 257–264.

35 Ye, B.; Yuan, L.; Chen, C.; Tao, J. *Electroanalysis*, **2005**, 17, 1523–1528.
36 Meneghello, M.; Papadopoulou, E.; Ugo, P.; Barlett, P.N. *Electrochimica Acta*, **2016**, 187, 684–692.
37 Wang, L.; Lin, L.; Ye, B. *Journal of Pharmaceutical and Biomedical Analysis*, **2006**, 42, 625–629.
38 Temerk, Y.; Ibrahin, M.; Ibrahim, H.; Kotb, M. *Journal of Photochemistry and Photobiology*, **2015**, 149, 27–36.
39 Cheng, W.X.; Peng, D.Y.; Lu, C.H.; Fang, W. *Russian Journal of Electrochemistry*, **2008**, 44, 1052–1057.
40 Dogan-Topal, B.; Bozal-Palabiyik, B.; Ozkan, S.A.; Uslu, B. *Sensors and Actuators B*, **2014**, 194, 185–194.
41 Shah, A.; Zaheer, M.; Qureshi, R.; Akhter, Z.; Nazar, M.F. *Spectrochimica Acta Part A*, **2010**, 75, 1082–1087.
42 Tig, G.A.; Zeybek, B.; Pekyardımcı, Ş. *Talanta*, **2016**, 154, 312–321.
43 Nowicka A.M.; Kowalczyk, A.; Donten, M.; Krysinski, P.; Stojek, Z. *Analytical Chemistry*, **2009**, 81, 7474–7483.
44 Zhou, H.; Shen, Q.; Zhang, S.; Ye, B. *Analytical Letters*, **2009**, 42, 1418–1429.
45 Erdem, A.; Congur, G. *International Journal of Biological Macromolecules*, **2013**, 61, 295–301.
46 Top, M.; Er, O.; Congur, G.; Erdem, A.; Yurt Lambrecht, F. *Talanta*, **2016**, 160, 157–163.
47 Erdem, A.; Muti, M.; Karadeniz, H.; Congur, G.; Canavar, E. *Analyst*, **2012**, 82, 137–142.
48 Tajik, S., Taher, M.A.; Beitollahi, H.; Torkzadeh-Mahani, M. *Talanta*, **2015**, 134, 60–64.
49 Li, C.-Z.; Liu, Y.; Luong, J.T. *Analytical Chemistry*, **2005**, 77, 478–485.
50 Erdem, A.; Karadeniz, H.; Caliskan, A. *Electroanalysis*, **2009**, 21, 464–471.
51 Yapasan, E.; Caliskan, A.; Karadeniz, H.; Erdem, A. *Materials Science and Engineering B*, **2010**, 169, 169–173.
52 Erdem, A.; Muti, M.; Papakonstantinou, P.; Canavar, E.; Karadeniz, H.; Congur, G.; Sharma, S. *Analyst*, **2012**, 137, 2129–2135.
53 Eksin, E.; Muti, M.; Erdem, A. *Electroanalysis*, **2013**, 25, 2321–2329.
54 Canavar, P.E.; Eksin, E.; Erdem, A. *Turkish Journal of Chemistry*, **2015**, 39, 1–12.
55 Congur, G.; Erdem, A.; Mese, M. *Bioelectrochemistry*, **2015**, 102, 21–28.
56 Sengiz, C.; Congur, G.; Eksin, E.; Erdem, A. *Electroanalysis*, **2015**, 27, 1855–1863.
57 Altay, C.; Eksin, E.; Congur, G.; Erdem, A. *Talanta*, **2015**, 144, 809–815.
58 Eksin, E.; Congur, G.; Mese, F.; Erdem, A. *Journal of Electroanalytical Chemistry*, **2014**, 733, 33–38.
59 Nowicka, A.M.; Mackiewicz, M.; Matysiak, E.; Krasnodebska-Ostrega, B.; Stojek, Z. *Talanta*, **2013**, 106, 85–91.
60 Yan, F.; Sadik, O.A. *Journal of the American Chemical Society*, **2001**, 123, 11335–11340.

61 Aleksic, M.M.; Kapetanovic, V. *Acta Chimica Slovenica*, **2014**, 61, 555–573.
62 Guiyun, X.; Jinshi, F.; Kui, J. *Electroanalysis*, **2008**, 20, 1209–1214.
63 Wang, F.; Xu, Y.; Zhao, J.; Hu, S. *Bioelectrochemistry*, **2007**, 70, 356–362.
64 Tian, X.; Li, F.; Zhu, L.; Ye, B. *Journal of Electroanalytical Chemistry*, **2008**, 621, 1–6.
65 Zhang, J.J.; Wang, B.; Li Y.F.; Jia, W.L.; Cui, H.; Wanga, H.S. *Electroanalysis*, **2008**, 20, 1684–1689.
66 Suzen, S.; Dermircigil, B.T.; Buyukbingol, E.; Ozkan, S.A. *New Journal of Chemistry*, **2003**, 6, 1007–1011.
67 Kauffmann, J.M.; Vire, J.C. *Analytica Chimica Acta*, **1993**, 273, 329–334.
68 Kalanur, S.S.; Katrahalli, U.; Seetharamappa, J. *Journal of Electroanalytical Chemistry*, **2009**, 636, 93–100.
69 Wang, Z.Y.; Han, R. *Pharmaceutical Treatment of Tumor* (in Chinese), People's Health Press, Beijing; **1987**, pp. 72–74.
70 Vergote, I.; Abram, P. *Annuals of Oncology*, **2006**, 17, 200–204.
71 Dodwell, D.; Vergote, I. *Cancer Treatment Reviews*, **2005**, 31, 274–282.
72 Thomson, P.R. *Physicians' Desk Reference (PDR)*, PDR.net, Montvale; **2005**, p. 653.
73 Dodwell, D., Coombes, G., Bliss, J.M., Kilburn, L.S., Johnston, S. *Clinical Oncology*, **2008**, 20, 321–324.
74 Ormrod, D.; Holm, K.; Goa, K.; Spencer, C. *Drugs Aging*, **1999**, 15, 389–416.
75 Gianni, L.; Salvatorelli, E.; Minotti, G. *Cardiovascular Toxicology*, **2007**, 7, 67–71.
76 Martinelli, V.; Radaelli, M.; Straffi, L.; Rodegher, M.; Comi, G. *Neurology Science*, **2009**, 30, 167–170.
77 Rosenberg, L.S.; Carblin, M.J.; Krugh, T.R. *Biochemistry*, **1986**, 25, 1002–1008.
78 Patching, S.G. *Biochimica et Biophysica Acta*, **2014**, 1838, 43–55.
79 Gao, Q.; Williams, L.D.; Egli, M.; Rabinovich, D.; Chen, S.L.; Quigley, G.J.; Rich, A. *Proceedings of the National Academy of Sciences*, **1991**, 88, 2422–2426.
80 Lambert, B.; Jones, B.K.; Roques, B.P.; Pecq, J.B.; Yeung, A.T. *Proceedings of the National Academy of Sciences*, **1989**, 86, 6557–6561.
81 Delbarre A.; Delepierre, M.; Garbay, C.; Igolen, J.; Pecq, J.B.; Roques, B.P. *Proceedings of the National Academy of Sciences*, **1987**, 84, 2155–2159.
82 Islam, M.M., Sinha, R., Kumar, G.S. *Biophysical Chemistry*, **2007**, 125, 508–520.
83 Zhang, X.; Song, Y.; Wu, Y.; Dong, Y.; Lai, L.; Zhang, J.; Lu, B.; Dai, F.; He, L.; Liu, M.; Yi, Z. *International Journal of Cancer*, **2011**, 29, 2502–2511.
84 Zilembo, N.; Bajetta, E.; Bichisao, E.; Martinetti, A.; Torre, I.L.; Bidoli, P.; Longarini, R.; Portale, T.; Seregni, E.; Bombardieri, E. *Biomedical Pharmacotherapy*, **2004**, 58, 255–259.

85 Lønning, P.E.; Geisler, J.; Johannessen, D.C.; Gschwind, H.P.; Waldmeier, F.;
    Schneider, W.; Galli, B.; Winkler, T.; Blum, W.; Kriemler, H.P.; Miller, W.R.; Fai,
    J.W. *Journal of Steroid Biochemistry and Molecular Biology*, **2001**, 77, 39–47.
86 Prodromidis, M.I. *Electrochimica Acta*, **2010**, 55, 4227–4233.
87 Adamson, I.Y.R. *Environmental Health Perspective*, **1976**, 16, 119–126.
88 Umezawa, H.; Ishizuka, M.; Maeda, K.; Takeuchi, T. *Cancer*, **1967**, 20,
    891–895.
89 Claussen, C.A.; Long, E.C. *Chemical Reviews*, **1999**, 99, 2797–2816.
90 Cui, B.; Johnson, S.P.; Bullock, N.; Ali-Osman, F.; Bigner, D.D.; Friedman, H.S.
    Journal of Biomedical Research, **2010**, 24, 424–435.
91 Marchesi, F.; Turriziani, M.; Tortorelli, G.; Avvisati, G.; Torino, F.; DeVecchis, L.
    Pharmacological Research, **2007**, 56, 275–287.
92 Wang, J.; Liu, L.; Ma, H. *Sensors and Actuators B: Chemical*, **2017**, 239,
    943–950.
93 Ketabi, S.; Gharib, F. *Russian Journal of Inorganic Chemistry*, **2006**, 51,
    1009–1013.
94 Howerton, S.S.B.; Sines, C.C.; Williams, L.D. *Biochemistry*, **2001**, 40,
    10023–10031.

Index

Biosensors and Nanotechnology: Applications in Health Care Diagnostics, First Edition.
Edited by Zeynep Altintas.
© 2018 John Wiley & Sons, Inc. Published 2018 by John Wiley & Sons, Inc.